How do students learn astronomy? How can the World Wide Web be used to teach? And how do planetariums help with educating the public? These are just some of the timely questions addressed in this stimulating review of new trends in the teaching of astronomy.

Based on an international meeting hosted by the University of London and the Open University (IAU Colloquium 162), this volume presents articles by experts from around the world. Teachers experienced in teaching astronomy at all levels, in Europe, North America, South America, South Africa, the former Soviet Union, India, Japan, Australia and New Zealand provide a global perspective on: university education, distance learning and electronic media, how students learn, the planetarium in education, public education in astronomy and astronomy in schools.

The proceedings of the first IAU Colloquium (105), *The Teaching of Astronomy*, edited by Percy and Pasachoff, were first published in 1990 and soon became established as the definitive resource for astronomy teachers. Astronomy education has advanced enormously in the intervening seven years, and this sequel will inspire and encourage teachers of astronomy at all levels and provide them with a wealth of ideas and experience.

New Trends in Astronomy Teaching

IAU Colloquium 162
University College London and the Open University
July 8–12 1996

Edited by
L. Gouguenheim, D. McNally and J. R. Percy

PUBLISHED BY THE PRESS SYNDICATE OF THE UNIVERSITY OF CAMBRIDGE
The Pitt Building, Trumpington Street, Cambridge CB2 1RP, United Kingdom

CAMBRIDGE UNIVERSITY PRESS
The Edinburgh Building, Cambridge CB2 2RU, United Kingdom
40 West 20th Street, New York, NY 10011–4211, USA
10 Stamford Road, Oakleigh, Melbourne 3166, Australia

First published 1998

Printed in the United Kingdom at the University Press, Cambridge

A catalogue record for this book is available from the British Library

Library of Congress Cataloguing in Publication data

IAU Colloquium (162th: 1996: University College London and the
Open University)
New trends in atronomy teaching : University Collage London and the Open
University, July 8–12 1996 / edited by L. Gouguenheim, D. McNally and J.R. Percy.
p. cm.
ISBN 0 521 62373 1
1. Astronomy–Study and teaching–Congresses. I. Gouguenheim,
L. (Lucienne) II. McNally, Derek, 1934– III. Perct, John R. IV Title.
QB61.I18 1996
520'.7'1–dc21 98–7171 CIP

ISBN 0 521 62373 1 hardback

DEDICATION

To the memory of Edith A. Muller, 1918–1995, Astronomer and Teacher, President of IAU Commission 46 — Teaching of Astronomy, 1967–1973.

Contents

4.   Planetarium Education and Training

5.   Public Education in Astronomy

6.   Teaching Astronomy in the Schools

Posters

Final Address

Preface

Eight years have elapsed since the first IAU Colloquium (No. 105) on astronomical education "The Teaching of Astronomy". In that time there have been substantial changes in the world of education – not just astronomical education. On the one hand, there has been erosion of funding, while on the other there has been an unprecedented opening up of access to information: there has been a change from educational experiment towards more regulation of curricula and determination of standards. But, as a reading of this volume will clearly show, there is still a healthy creativity in astronomy education. There is much important new work being done - there are adventurous schemes in public education, there is new detailed research on how our students and pupils may learn and on the portfolio of misconceptions under which they may be labouring when first confronted with astronomical teaching. One of the new features since 1988 is access to the Internet. An overwhelming variety of information is now readily available from the latest Hubble Space Telescope picture to the Web Page of the local astronomy society. But it is also clear that the sheer richness and variety of the Internet offering creates yet another problem – how to organise that information to maximum teaching and learning benefit. The North American continent is once again in a period of curriculum renewal and it is of great interest to see the interaction between that renewal, electronic media and the Internet. Such enterprises are receiving support in particular from the National Science Foundation in the USA. It is encouraging to see that the Internet is being used to support undergraduate projects. Many scientific databases are freely accessible. This means that actual data can be made available from world-class telescopic facilities for undergraduate work. What is true of the project situation is equally true of the practical class. An exciting new prospect is the use of the robotic telescope for teaching. Clearly a robotic telescope on a top class site can overcome the usual constraints of weather and provide hitherto undreamt of opportunities for astronomy teaching. A robotic telescope separated from the user by 12 hours means that the age old problems of practical astronomy needing to be done at night can be brought within the working day – particularly important for schools and public demonstration. A feature of the present Colloquium is the session on Distance Learning and Electronic Media. In many countries in the developed and developing world alike, social factors are forcing the pace in the provision of Distance Learning. The sheer numbers working for degrees by Distance Learning is phenomenal. Electronic media are a necessity, not a luxury, for Distance Learning and through development of Distance Learning courses and learning structures there will have to be consequent rethinking of conventional tertiary level teaching. This is probably one of the most exciting interfaces for the future. Conventional tertiary education will be given a substantial boost by Distance Learning. To some extent tertiary education has suffered because of declining budgets in the past eight years and it is often difficult to justify the necessary investment of money and time in the development of new electronic media based resources for relatively small advanced astronomy classes. However, that investment is fully justified for Distance Learning and as the papers in this Colloquium make clear, conventional tertiary education will be a major beneficiary of the work that has been invested in Distance Learning. One can only but look forward to a new era where more intensive use of professional data bases, Internet sources and the simulation opportunities offered by electronic media will become a normal component of undergraduate provision. Yet even in the best of electronic wonderlands there is still nothing to beat the educational experience of actual observation of real stars – we hope that formative experience, excitement and stimulation will remain an important element of all forms of astronomical education, enhanced by the prospect of robotic telescopes on excellent sites and despite the rising tide of light pollution and other forms of man-made interference with astronomical observation.

The initial plans for this Colloquium were the outcome of discussions at meetings of IAU Commission 46 – Teaching of Astronomy at the XXIInd General Assembly of the IAU in The Hague in 1994. A Scientific Organising Committee under the Chairmanship of Lucienne Gouguenheim (L. Abati, J. Fierro, A Fraknoi, S. Isobe, B.W. Jones, D. Mc-Nally, J. Narlikar, M. Othman, J. Percy, B. Warner, R. West) devised a programme of Six Sessions – University Education, Distance Learning and Electronic Media, How Students Learn, Planetarium Education, Astronomy in the Schools and Public Education. The Local Organising Committee was chaired by B.W. Jones and D. McNally (and included S.J. Boyle, M.M. Dworetsky, S.J. Fossey, C.A. Newport, A.J. Norton, later joined by K.R. Davies) and oversaw the local arrangements at the Open University and University College London. It is worth noting that the Colloquium was held at two venues. This worked very well and very smoothly. The Open University was the venue for the session on Distance Learning and Electronic Media, where a variety of Distance Learning materials could be put on display. The remaining sessions were at University College London.

The Colloquium was attended by some 130 participants from 40 countries. The southern hemisphere was well represented and it was a pleasure to see representatives from Africa. The Old World was represented from Japan in the east to Ireland in the west and it goes without saying that the New World – North and South – was very well represented as befits a Colloquium on Astronomical Education.

We should like to thank the Open University and University College London for the provision of accommodation and facilities and Mr Ewing and his staff at Ramsay Hall. We were assisted by Valerie Peerless (University of London Observatory), Muna Dar (University College) and Margaret Burgess-Ward, Yvonne McKay and the staff of General Facilities (Open University) – their help, particularly in moments of stress, was greatly appreciated. We are particularly in the debt of Sally Harmsworth at the Open University who single-handedly transformed our papers in a variety of media and digital dialects into a publishable manuscript. Her cheerfulness and good humour are much appreciated. It is a pleasure to acknowledge financial support from the International Astronomical Union, Comitè de Liaison Enseignants et Astronomes (CLEA) and the Royal Astronomical Society for Travel Grants, the Institute of Physics, Apple Computer UK, Easi Bind International for generous support in a range of aspects of the Colloquium and to the Open University for a reception. Thanks are also due to Undine Concannon, Director of the London Planetarium, for an evening event for participants at the Planetarium. Last, but by no means least, many participants enjoyed Alan Pickwick's evening guided walks around London – thoughtful impromptu additions to the Colloquium's ambience.

Sadly for all participants, Lucienne Gouguenheim was prevented by serious illness from attending the Colloquium. Gladly she is making a good recovery. Participants were saddened by the deaths of N.C. Rana and W.F. Wargau in the weeks and months following the Colloquium – they will be much missed by us all.

L. Gouguenheim (Observatoire de Paris)
D. McNally (University of London Observatory)
J.R. Percy (University of Toronto)
Editors

Participants

Abati, L.	Italy	abati@astropd.pd.astro.it
Aleshkina, E.Y.	Russia	aek@ipa.rssi.ru
Bacher, C.	Denmark	bacher@meyer.fys.icu.ku.dk
Bagenal, F.	USA	bagenal@dosxx.colorado.edu
Baruch, J.	UK	j.e.f.baruch@brad.ac.uk
Bennett, M.A.	USA	mbennett@stars.sfsu.edu
Berthomieu, F.	France	berthomi@pelat.ac-nice.fr
Billingham, J.	USA	j_billingham@seti-inst.edu
Borchkhadze, T.	Georgia	tenat@dtapha.kheta.ge
Borel, P.	Denmark	
Boyle, S.J.	UK	sbj@star.ucl.ac.uk
Brake, M	UK	
Brecher, K.	USA	brecher@bu.edu
Bretones, A.L.K.	Brazil	paulo.bretones@mpcbbs.com.br
Bretones, P.S.	Brazil	paulo.bretones@mpcbbs.com.br
Broughton, M.P.V.	UK	mbroughton@plym.ac.uk
Buckland, R.A.	UK	R.Buckland@open.ac.uk
Camino, N.	Argentina	nestor@unpate.edu.ar
Cartellier, R.	France	duval@obmara.cnrs.mrs.fr
Chernin, A.	Russia	chernin@sai.msu.su
Christian, C.	USA	carolc@stsci.edu
Clegg, R.E.S.	UK	srec@ib.re.ac.uk
Comins, N.F.	USA	galaxy@maine.maine.edu
Crutchley, A.	UK	a.cruchley@brockcol.demon.co.uk
Daugherty, B.	UK	bd25@tutor.open.ac.uk
Dous, C. la	Germany	cld@stw.tu-ilmenau.de
Dow, K.	USA	kdow@cra.harvard.edu
Duval, M.F.	France	duval@obmara.cnrs_mrs.fr
Dworetsky, M.M.	UK	mmd@star.ucl.ac.uk
Edelson, R.	USA	edelson@spacly.physics.uiowa.edu
Elliott, I.	Ireland	ie@dunsink.dias.ie
Emerson, J.P.	UK	j.p.emerson@qmw.ac.uk
Fabregat, J.	Spain	fabregat@ma1.upc.es
Fierro, J.	Mexico	julieta@astroscu.unam.mx
Fogh-Olsen, H.J.	Denmark	fogh@astro.ku.dk
Fossey, S.J.	UK	sjf@star.ucl.ac.uk
Fouquet, J.L.	France	
Gear, F.	UK	
Gerbaldi, M.	France	gerbaldi@iap.fr
Gill, P.B.J.	UK	
Greenstein, G.	USA	gsgreenstein@amherst.edu
Gudmundsson, E.H.	Iceland	einar@raunvis.hi.is
Gulyaev, S.A.	Russia	sergei.gulyaev@usu.ru

Hawkins, I.	USA	isabelh@cea.berkley.edu
Hill, P.W.	UK	pwb@st-andrews.ac.uk
Hoff, D.	USA	hoffdarr@martin.luther.edu
Hufnagel, B.	USA	hufnagel@pa.msu.edu
Humpries, S.	UK	
Inglis, M.	UK	m.d.inglis@warwick.ac.uk
Isobe, S.	Japan	isobesz@cc.nao.ac.jp
Iwaniszewska, C.	Poland	cecylia@astri.uni.torun.pl
Jakimice, J.	Poland	
James, P.	UK	paj@staru1.livjm.ac.uk
Jones, B.W.	UK	b.w.jones@open.ac.uk
Jones, M.H.	UK	m.h.jones@open.ac.uk
Kibble, B.	UK	
Kononovich, E.V.	Russia	konov@sai.msu.su
Lacy, T.	UK	
Laing, R.A.	UK	
Lanciano, N.	Italy	laniano@gpxrme.sci.uniromal.it
Lee, Y.B.	S. Korea	yblee@ns.seoul-e.ac.kr
Leedjärv, L.	Estonia	leed@aai.ee
Lomb, N.Q.	Australia	nickl@phm.gov.au
Lovell, A.	UK	alexlovell@dial.pipex.com
McCabe, M.	UK	mccabe@sma.port.ac.uk
McFarland, J.	UK	jmf@star.arm.ac.uk
McNally, D.	UK	dmn@star.ucl.ac.uk
Marschall, L.	USA	marschall@gettysburg.edu
Martinez-Sebastian, B.S.	Spain	isbeui@ctu.es
Martin-Smith, M	UK	
Metaxa, M.	Greece	
Monier, R.	France	rmonier@cdxb6.u_strasbg.fr
Moss, C.M.N.	UK	cml@ast.com.ac.uk
Nash, G.	UK	
Norton, A.J.	UK	a.j.norton@open.ac.uk
Nussbaum, J.	Israel	ntnussba@weizmann,weizmann,ac.il
Olsen, B.	Denmark	
Onuora, L.I.	UK	lonuora@astr.maps.susk.ac.uk
Orchiston, W.	New Zealand	wayne.orchiston@vuw.ac.nz
Othman, M.	Malaysia	
Parkinson, J.H.	UK	e.j.h.parkinson@shu.ac.uk
Pasachoff, J.	USA	jay.m.pasachoff@williams.edu
Penston, M.	UK	mjp@ash.com.ac.uk
Percy, J.R.	Canada	Jpercy@erin.utoronto.ca
Pickwick, A.C.	UK	100316.3710@compuserve.com
Pirotte, E.	Belgium	planetar@eunet.be
Pompea, S.M.	USA	spompea@as.arizona.edu

Raghavan, N.	India	
Rana, N.C.	India	
Ray, T.P.	Ireland	tr@cp.dias.ie
Ros, R.M.	Spain	ros@mat.upc.es
Sadler, P.	USA	psadler@cfa.harvard.edu
Sampson, G.E.	USA	gsampson@omnifest.uwm.edu
Sathe, D.V.	India	
Saygac, A.T.	Turkey	saygac@yunus.mam.tubitak.gov.tr
Seymore, P.A.H.	UK	mbroughton@plym.ac.uk
Sigurdsson, T.	Iceland	thorir@unak.is
Singh, H.P.	India	hps@ttdsvc.ernet.in
Snyder, S.Y.	USA	snyders@pa.msu.edu
Solheim, J.-E.	Norway	janerik@phys.uit.no
Staal, D.H.N.	Netherlands	staal@astro.rug.nl
Stefl, V.	Czech Republic	stefl.astro.sci.muni.cz
Sudzius, J.	Lithuania	sudzius@artz.unc.osf.fr
Swinbank, E.	UK	
Szécsényi-Nagy, G.	Hungary	szena@ludens.elte.hu
Szostak, R.	Germany	
Takvorian, R.	France	
Tamsin, F.	Belgium	
Tidey, S.	UK	stidey@direon.co.uk
Troche-Boggino, A.E.	Paraguay	atroche@dtapha.keta.ge
Touma, J.R.	Canada	touma@cita.toronto.edu
Trevisan, R.H.	Brazil	trevisan@npd.uel.br
Tritton, K.P.	UK	kpt@ast.com.ac.uk
Tritton, S.	UK	sbt@roe.ac.uk
Tsubota, Y.	Japan	tsubota@hc.keio.ac.jp
Vilks, I.	Latvia	astro@acad.latnet.lv
Vincent, F.	UK	
Vujnovic, V.	Croatia	vujnovic@bobi.ifs.hr
Wargau, W.F.	S. Africa	
Welsh, B.	USA	bwelsh@ssl.berkley.edu
Wentzel, D.G.	USA	wentzel@astro.umd.edu
West, R.M.	Germany	rwest@eso.orgh
Williams, P.G.	UK	p.g.williams@qmw.ac.uk
Xerri, A.	France	annie.xerri@cned.fr
Yair, Y.	Israel	yoav_y@mail.cet.ac.i
Zasov, A.V.	Russia	zasov@sia.mus.su
Ziznovsky, J.	Slovak Repub.	ziga@ta3.sk

A welcome from the Chairman

By Lucienne Gouguenheim, Scientific Organising Committee

The beginnings of this second international colloquium on Astronomy Teaching, eight years after the famous one in Williamstown, came during a meeting of Commission 46 in August 1994, in the Hague. It was then submitted as an IAU Colloquium by the President of Commission 46, John Percy, with the support of the newly born European Association for Astronomy Education.

When I was asked to chair the Scientific Organising Committee, I considered this proposal to be a great honour, that I acknowledge, and also an exciting way to learn more about the new developments in astronomy education that you are performing, so many of you, all around the world.

Then came a hard work! Step by step the programme was built, thanks to the help and suggestions from the SOC members, and I would like to mention more particularly Julieta Fierro, Andy Fraknoi, Barrie Jones, Derek McNally, John Percy.

It was my great pleasure, each day, to read your mails on my computer, or on the fax machine a pleasure mixed with some increasing anxiety, when their number began to grow rapidly! The Internet gives this beautiful possibility to interact so easily with people spread out all over the world – you have just to take account of the time zones, which could be also considered as a good astronomical exercise.

John Percy likes to say that Commission 46 is different from the other IAU Commissions in the sense that we are all involved in its activity. This is quite obvious, when looking at the list of participants, astronomers and school teachers, coming from all over the world; I would like to mention that many more, who would have liked to attend, were unable to come because of financial worries. The IAU allocated us the maximum grant given to a colloquium, and we are grateful for it; however it proved to be quite insufficient. We have then looked for complementary ones, which was very difficult, and in most cases unsuccessful. I would like to thank warmly, in the United Kingdom, the Royal Astronomical Society, and in France the Comitè National Francais d'Astronomie, the Comitè de Liaison Enseignants et Astronomes and the Observatoire de Paris for their gifts, the total amount of which reaches roughly half the value of the IAU grant.

I would like to thank the invited speakers, all of those who are presenting contributed talks and posters, with a special mention to Cecylia Iwaniszewska, who has accepted the difficult task to review the posters.

Warm thanks are due also to the members of the Local Organising Committee, and the co-chairmen, Derek McNally and Barrie Jones: they were facing hard work, which even increased at the beginning of May, when I had to stop working, rather suddenly, due to my health problems. In this regard, I would like to make a special mention to my French colleague and friend, Michle Gerbaldi who has stood in for me since that time.

I deeply regret that I am not able to be there, and I wish you a rich and successful colloquium.

Lucienne Gouguenheim
4 July 1996

Astronomy Education: an international perspective

By John R. Percy

Erindale Campus, University of Toronto, Mississauga ON, Canada L5L 1C6

As the first speaker at this Colloquium, it is my pleasure to welcome the participants (and the readers of these Proceedings), on behalf of the International Astronomical Union (IAU) and its Commission 46 (Teaching of Astronomy). It is also my pleasure to thank our hosts University College London, and The Open University; the Scientific Organizing Committee, chaired by Lucienne Gouguenheim, and especially the Local Organizing Committee, chaired by Barrie Jones and Derek McNally. They have made this meeting most enjoyable and successful.

Eight years ago, many of us were in Williamstown, USA, for the first IAU Colloquium on astronomy education. Since then, there have been enormous changes - political, economic, and technological - which have affected our work. There have also been about 100 IAU conferences on research topics, but this is only the second on education. We all agree that we must work to correct that imbalance!

We are here to catch up on what has happened in astronomy education in the last eight years. We are here to teach and learn, through lectures, posters, and discussions - both formal and informal. We are here to renew old friendships, and make new ones. These human dimensions of this Colloquium are only hinted at in these Proceedings, but I assure you that they occurred.

1. Why is Astronomy Education Important?

Education is important to astronomers because it affects the recruitment and training of future astronomers, and because it affects the awareness, understanding and appreciation of astronomy by taxpayers and politicians who support us. We have an obligation to share the excitement and the significance of our work with students and the public. Education is often neglected by the scientific and professional community - not by us, of course - and by many research universities. Our task is not only to be better astronomy educators ourselves, but to convince and train our students and colleagues to do likewise.

There are other reasons why astronomy should be part of our education system and our culture. Astronomy is deeply rooted in the history of almost every society, as a result of its practical applications and its philosophical implications. It still has everyday applications to timekeeping, seasons, navigation and climate, as well as to longer-term issues such as climate change and biological evolution. Astronomy not only contributes to the development of physics and the other sciences, but it is an important and exciting science in its own right. It deals with the origin of the stars, planets, and life itself. It shows our place in time and space, and our kinship with other peoples and species on earth. It reveals a universe which is vast, varied and beautiful. It promotes curiosity, imagination, and a sense of shared exploration and discovery. It provides an enjoyable hobby for millions of people, whether they be serious amateur astronomers, armchair astronomers, or casual skygazers. In a school context, it demonstrates an alternative approach to the "scientific method" - the observation vs. theory approach. It can attract young people to study science and engineering, and can increase public interest and understanding of science and technology - both of which are important in all countries, both developed and developing.

2. What is Astronomy Education?

It is important, before we go further, to define astronomy education. I propose to define it broadly, by quoting Andrew Fraknoi (1996), who posed the following question: "Where does astronomy education take place? Those readers who teach will probably say that it takes place in classrooms like theirs - anywhere from first grade through university. But I want to argue that astronomy education happens in many other places besides the formal classroom. It happens in hundreds of planetaria and museums. It happens at meetings of amateur astronomy groups. It happens when someone reads a newspaper, or in front of television and radio sets. It happens when someone is engrossed in a popular book on astronomy, or leafs through a magazine like *Sky and Telescope*. It happens in youth groups taking an overnight hike, and learning about the stars. It happens when someone surfs the astronomy resources on the Internet. When we consider astronomy education, its triumphs and tribulations, we must be sure that we do not focus too narrowly on academia, and omit the many places that it can and does happen outside the classroom."

I believe that our students are influenced by the astronomy which they learn outside the classroom, as much or more than by the astronomy which they learn from us. Anderson (1996) was recently appalled to find that his college astronomy students, writing essays, did not hesitate to quote popular books and TV programs on pseudoscience as authorities! It is unfortunate that we do not have more educators here from planetariums, science centres, publishers, the news media, and amateur astronomy. Perhaps, at the next IAU Colloquium on education, we can devote much more time to the issue of "informal education".

3. The Problems of Astronomy Education

If astronomy is so interesting and important, and available in so many settings, why is it not taught in more schools? Why are there so many misconceptions about astronomical topics? Astronomy educators all over the world have discovered "universal" barriers to the effective teaching and learning of astronomy. Some of them have discovered solutions!

(*a*) Students have misconceptions about astronomical topics (such as the causes of the seasons), which are not overcome by the teaching methods which are commonly used.

(*b*) Teachers have misconceptions about the teaching of astronomy (and perhaps about astronomical content as well) which discourage them from teaching it well, or teaching it at all.

(*c*) Teachers (especially in elementary schools) have very little knowledge about astronomy or astronomy teaching, while astronomy has progressed by leaps and bounds since these teachers were in school themselves.

(*d*) The most effective teaching tools are simple, inexpensive, hands-on activities, and these are not widely used, or not widely available. These must get around the problem that "the stars come out at night, but the students don't". But what is expensive for the North American teacher is beyond the reach of teachers in the developing world, so we will have to emphasize the use of "available materials" for teaching.

(*e*) Teachers are not always aware of the materials available. There is not a single national or international journal which deals specifically with astronomy education, or an effective network for informing teachers of materials available. The development of the European Association for Astronomy Education is a step in the right direction. It will make good use of the Internet for communication, but we must remember that the Internet is still not available to the vast majority of teachers around the world.

(*f*) The best or most fortunate students may receive good education, but the others - girls, minorities, the disabled, inner-city or rural students - and students in the developing countries - may be left out.

(*g*) The best students must be attracted to astronomy and astronomy education; and provided with career opportunities when they graduate.

(*h*) Teaching often has less status and support than research. Ironically, the problem in elementary school is often that administrators ignore astronomy. In university, they ignore education!

Note that these barriers apply, to a greater or lesser extent, at all levels of education, and in all countries of the world. For instance, even in graduate education, there is a need for professional development of instructors and supervisors. There is debate about the relative importance of coursework and practical work. There is a need to attract women and minorities, and a real need to prepare students for jobs in astronomy and elsewhere.

We need to take a wider look at astronomy education, and its problems. There are innovative projects and programs in many countries, in many cases well-tested. By taking an international perspective, we achieve a deeper historical and cultural understanding, which is especially important in our multicultural societies. We must also work in partnership with teachers, educators in planetariums and science centres, amateur astronomers, and all others who contribute to astronomy education in its broadest sense. We take some pleasure in knowing that, as astronomy educators, we have "kindred spirits" around the world and, in the case of the developing countries, we take some satisfaction in knowing that we can help scientists and educators less fortunate than ourselves.

4. Varieties of Astronomy Education

Even within formal or classroom astronomy education, there are many varieties - as you will quickly learn at this meeting. There are the two systems of education - eloquently described by Don Wentzel (1990) in the introductory lecture at the last IAU Colloquium on astronomy education - the European system, and the North American system. There are also two (or more) methods of education: at one end of the spectrum is the method based on the memorization of lecture and textbook material; at the other end is the inquiry-based or hands- on approach. There are many opinions about astronomy education curriculum: one is to teach a course or unit on general astronomy; another is to introduce astronomy as part of other sciences such as physics. If astronomy is taught as a course unit, should it emphasize classical content or current developments, core concepts or interdisciplinary or speculative topics; topics relevant to everyday or abstract topics like black holes; selective or comprehensive coverage; depth or breadth? If there is practical work, should it be sky-based, or computer-based? If sky-based, should it be daytime or nighttime? Or should there be elements of all of the above?

5. The Needs of the Developing Countries

Whichever systems and methods are used, the developing countries face problems not encountered by astronomy educators elsewhere. Their special needs include: an understanding and interest on the part of astronomy educators elsewhere; opportunities to gain experience in more developed countries, and be visited by astronomers and educators from abroad; access to appropriate books, journals, and data; access to equipment.

Many of the programs and projects of the International Astronomical Union, described elsewhere in this meeting, address these needs.

I use the term "developing countries" broadly. Of the almost 200 states which are recognized by the United Nations, about 100 have some form of organized astronomical activity - either professional or amateur. About half those countries adhere to the IAU. Of those, less than half could be considered fully developed, (considering that the countries of Eastern Europe and the former Soviet Union still face many economic obstacles, despite their rich astronomical heritage). I am including Canada and the US among the developed countries, even though astronomers there have many concerns about the quality of their education system!

In many developing countries, there is only one "lone astronomer" (at most a small group) to do all the educational, research and administrative work which is shared among many in the more developed countries. The accomplishments of these individuals are remarkable. They should be an inspiration to all of us. I am glad that a few of them are here at this Colloquium.

6. Astronomy Education Around The World

It is tempting to try to review all aspects of astronomy education, at all levels, in all countries, but I shall resist the temptation. You will learn all you want to know from this meeting, and from the people who are here. For further reading, I can refer you to "The Teaching of Astronomy" Pasachoff and Percy (1990), the Proceedings of the last IAU Colloquium on education, as well as to the several articles which I have written in the course of my work with IAU Commission 46 (Percy 1994, 1995abc, 1996a).

In the Colloquium, at this point, I showed a large number of slides of astronomy education around the world. They illustrated the education projects and programs of the IAU, as well as local initiatives - some of which are described at this meeting.

7. How Can Astronomers and Educators Support and Improve Education?

(*a*) Make education a part of the professional astronomy organization in your country, and part of your department, observatory or local astronomical society:
- appoint an education co-ordinator, and an education committee
- include an education column in your journal, and education sessions at your annual meeting
- sponsor public lectures and teachers' workshops as part of your annual meeting
- work with other organizations and institutions which have an interest in astronomy education in your area, and in your country.

(*b*) Be aware of developments in astronomy and science education:
- be aware of curriculum changes in your country, and make sure that astronomy is not neglected
- adopt existing astronomy education materials when possible, instead of "rediscovering the wheel".
- organize meetings on astronomy education, inviting all those who are involved in that subject.

(*c*) Seek more funds for science education, from governments and private sources.

(*d*) Make a contribution to astronomy education, and support others who do:
- pass on your knowledge and enthusiasm

- publicize the practical and cultural benefits of astronomy
- give or arrange a public lecture or other such event
- write a popular article on astronomy
- meet with students, teachers and the public
- encourage interested students, especially women and other under-represented groups
- support your local elementary and secondary school

(*e*) Get more and better astronomy in:
- museums, planetaria and science centres
- parks and conservation areas
- day schools and night schools
- the news media: newspapers, magazines, radio and TV.

(*f*) Lobby for, and help develop:
- a planetarium
- a science centre
- a public observatory.

(*g*) Support astronomy education in the developing countries:
- learn about the needs of these countries
- support programs to send surplus books, journals and instruments to these countries
- communicate with, visit, and support the work of the "lone astronomers" in the developing countries
- support the work of the IAU.

This seems like a daunting list, especially when read all at once. I do not ask astronomers to devote all their time to education (unless, of course, that is what they are paid to do). If each astronomer were to spend at least 10 per cent of their time on education, having made a conscious effort to learn more about astronomy education, and to co-ordinate their work with that of others, the results would be quite remarkable! In a recent special issue of the Astronomical Society of the Pacific's popular magazine Mercury, on the future of astronomy, I published a rather optimistic view of the future of astronomy education (Percy 1996b). I urge you to read that article, and help make my dream come true!

REFERENCES

ANDERSON, W.R., 1996, *Sceptical Inquirer* **20**, 5, 59-60.

PASACHOFF, J.M. AND PERCY, J.R. (editors) 1990, *"The Teaching of Astronomy"*, Cambridge, Cambridge University Press.

PERCY, J.R., 1994, *Southern Stars*, **35**, 223-230.

PERCY, J.R., 1995a, *The Planetarian* **14**, 3, 6-9.

PERCY, J.R., 1995b, *MNASSA* **54**, 38-40.

PERCY, J.R., 1995c, *Fifth International Conference on Teaching Astronomy* ed. R.M. Ros, Universidad Politecnica de Catalunya, 63-68.

PERCY, J.R., 1996, *Astronomy Education: Current Developments, Future Co-ordination*, ed. J.R. Percy, ASP Conf. Series, Vol. 89, 1-8.

PERCY, J.R., 1996b, *Mercury* **25**, 1, 34-36.

Sundials in London - Linking architecture and astronomy

By Bob Kibble

Coulsdon College, Placehouse Lane, Old Coulsdon, Surrey, England

1. Sundials in my life

Following the inclusion of Astronomy in the revised National Science Curriculum for England and Wales the Association for Astronomy Education, AAE, embarked on a programme of in-service training workshops for teachers to help them to understand the new ideas and deliver the new curriculum. Teacher confidence and knowledge has been the greatest challenge to establishing astronomy in school curricula. As part of the the AAE team I gave presentations on a host of activities including simple cut and paste sundials for pupil projects. We are now seven years on from the revised Science Curriculum and my interest in sundials has stepped up a gear. I have developed an interest in real dials, both studying existing dials and making dials for the homes of friends and families and for schools. This presentation, which has as its focus, the sundial as an architectural feature, uses slides I have taken of some of the dials to be seen in the central London area including some of my own. I am grateful to the British Sundial Society for a list of dial locations in London.

2. Understanding the hour lines - a model helps

To help explain how hour lines are related to the Suns motion I have developed a three dimensional stick and card model. The model, in four pieces, builds up gradually during a workshop presentation. I start with an equatorial dial showing 15 degree angles marked on an equatorial plane. (360 degrees / 24 hours - the only maths you really need to understand dials.)

The addition of two more planes show how the equatorial lines can be projected onto either a horizontal or vertical surface. Horizontal dials are often found in gardens and cheap brass dials can be bought from suppliers of garden furniture. It must be said that the are unlikely to be accurate as they are mass-produced as ornaments! The vertical plane is that of a south facing wall. The hour lines are symmetrical. Many churches in England are built with a east-west alignment and are the most likely locations for such dials. For those readers who wish to draw up their own lines for either a horizontal or a vertical surface and who do not wish to make a model the lines can be calculated as shown in fig 2.

However most of us do not live in churches and our homes are most likely to have walls which 'decline'to either the west or the east. The model can accommodate such a wall and will show how the original equatorial hour lines can be projected onto a wall of any particular declination. It is instructive to note that a wall facing south-east for example is likely to be complimented by another facing north-west. Both walls are potential surfaces for dials. In this case one will be a morning dial with hours from perhaps 5am to 2pm, the other will show afternoon and evening hours up to sunset. Both examples are illustrated in the slides. Needless to say the hour lines on a declining wall are more difficult to calculate. There are computer programmes available to do the calculations but I have a simple domestic programme which will do the trick if you wish to contact me with your location and declination details.

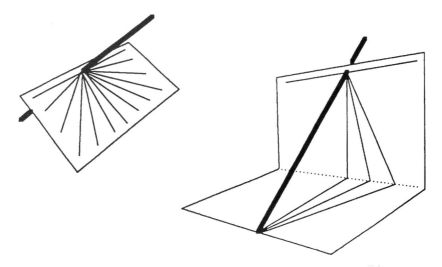

FIGURE 1. Sketches showing the equatorial model and the vertical and horizontal planes. Combining the two together reveals the relationship between hour lines on all surfaces.

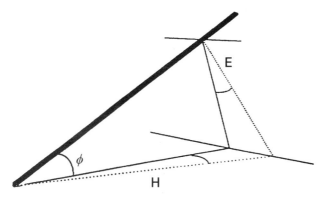

FIGURE 2. To calculate hour angles, H, for a horizontal surface use tan H = sin ϕ tan E where ϕ is the latitude of the location and E the hour angles on the equatorial plane. (For vertical dial angles simply replace by the colatitude, 90 $-\phi$.)

3. Dials in London

A wander about any town will reveal dials to the keen observer. Churches are likely places as are public gardens and museums. However you will find an increasing tendency for dials to appear either as an architectural feature of a building or as a civic commemorative feature, perhaps a centenary dial or a millennium dial. A walk around the City of London will reward the keen observer with a dozen interesting dials. Some of my favorites selected from the slide presentation are listed below. I have presented them in an order such that you can visit them all by foot during a casual Sunday stroll through central London. The walk will take between three and five hours and should be punctuated by refreshment stops in either cafes or pubs. The route will take you to a number of interesting landmarks within London.It is best to have a handy street map with you. I can give only approximate instructions here.

A convenient place to start is at Tower Hill tube station. However I prefer to enjoy a detour at this point to take in the dial high on the wall of the synagogue at the corner of Brick Lane and Fournier Street. To reach this location you must walk northwards towards Spitalfields Market, itself a lively trading attraction on a Sunday morning. The detour will take the best part of an hour but offers an interesting contrast to the visitor, a cosmopolitan part of London steeped in the clothing trade with strong Hugenot connections

Outside Tower Hill tube station, as you face the Tower of London, is an imposing horizontal dial. Use the underpass to cross the road towards the Tower and bear left under Tower Bridge towards St Catherines Dock where you will find an equatorial dial outside the Hotel beside the bridge. Note the dial is calibrated for all hours including midnight! The dock area has its own interest but keep one eye on the clock as you have much walking ahead of you.

Retrace your steps to the Tube station and continue westwards towards St Dunstans and the ruins of St Dunstans church. There behind the hanging overgrowth lies a slate dial - a memorial to the church gardener. Continue westwards and head for St Pauls cathedral. If you are walking on a Sunday these streets will be deserted. Enjoy the variety of architectural styles and imagine the noise and bustle of a trading day. My favourite route would take me a few streets north to Leadenhall market. A really special place. Onwards to Cheapside where above the BT shop you will find two dials on the corner of the wall. One was gnomonless the last time I visited. By all means take in St Pauls but dont linger too long. Head south towards the river and enjoy the views as you walk along the embankment past Blackfriars bridge. A few hundred metres on and the Temple appears on your right. If it is accessible this whole area is worth a visit. There are three horizontal dials in the gardens facing the river and two more high on the walls of the the courts. You are in the nerve centre of the British legal system as you wander through these old courtyards and back streets. Leave at the north end of the Temple which brings you to Fleet Street. Head westwards and take in the dial on the tower of St Clement Danes church located in the centre of the road just before the Aldwych. Note that the wall is not perfectly south facing. Continue westwards along the Strand past Waterloo Bridge. A detour south to the gardens behind the Savoy Hotel will take you to the Savoy centenary dial, a modern armillary sphere.

Back up the hill now to take in the final dial. Head back to the Strand again and cross to Covent garden. If you have not yet stopped for refreshment you will certainly need to. Take in the atmosphere of one of Londons most energetic attractions, once a blooming flower, fruit and vegetable market. Head northwards past Covent Garden tube station to Neal Street and bear left to the Seven Dials. There are only six dials on this renovated obelisk, the seventh apparently being the pillar itself.

This is where my walk ends. If your legs will stand it I can recommend retracing your steps to cross over Waterloo Bridge and head for the South Bank where the National Theatre and Festival Hall complex offers refreshments, evening entertainment and one of the best, no, the best, evening panorama available.

Sadly this walk does not take in my favorite dial. To reach this you will need to take a tube to the Angel, Islington where, high on a wall above the tube station is a morning dial, facing south-east which captures for me the very best in the synthesis of architecture and astronomy.

These are but a few of the many dials in London. I hope that this excursion will encourage you to do something similar for your own town. And of course if your town is rather short of dials then now is the time to make one of your own.

SECTION ONE
University Education

University Education in the Next Century

By Derek McNally

University of London Observatory, Mill Hill Park, London NW7 2QS

1. The Present

There is no doubt that the science of astronomy is now in an exhilarating state. We are in the era of the 10 m optical telescope. Radio astronomy rivals optical astronomy in both positional precision and sensitivity. Observation from space has opened access to a wide range of frequencies in the electromagnetic spectrum. The spectacular achievements of the Hubble Space Telescope underline the success story of space astronomy. At all wavelengths, detector technology has made striking advances in sensitivity and, coupled with cheap, sophisticated and powerful computers, raw data can be transformed into useful scientific data with breathtaking speed. One has only to add up the number of papers published in the three major astronomical journals to realise that one must read 100 journal pages a day (every day) to keep up with the literature in these three journals alone. Astronomy at the close of the 20th century is indeed exhilarating.

But there are indications that all is not well. Not unexpectedly the cost of new astronomical facilities is being called into question. Currently, no one nation can afford to finance a new telescope of the 10 m class and international consortia are now a commonplace to finance such facilities, e.g. the ESO 4 × 8 m telescopes in Chile. The great cost of science more generally, is now being seriously questioned, particularly in those areas of science which are fundamental, e.g. astronomy, particle physics, and which are not regarded as being currently relevant to industrial and commercial activity. International Economic Institutions are raising the issue of whether or not the next stage of fundamental science will be affordable at all. L. Woltjer and F. Praderie at the 1995 European Astronomical Society Meeting (JENAM, Catania) addressed the economic constraints of the future and both concluded that the optimistic outlook would be level funding at present rates. Much greater international co-operation would become the norm - not just confined to major projects. The consequences for the development of the next generation of astronomical facilities are not yet easy to quantify.

A second problem facing astronomy is that of adverse environmental impact. A vibrant and creative technological society has endowed astronomy with outstanding instrumentation both on the ground and in space. Yet that same civilisation, by that same energy that has provided wonderful astronomical tools, could render the science of astronomy impotent. Our civilised life styles are creating impediments to astronomical observation on a wide variety of fronts - ground vibration, increasing atmospheric extinction, electromagnetic pollution at all wavelengths, the increasing amounts of space debris. It is possible to envisage scenarios in the not too distant future when any form of astronomical observation may well be impossible to sustain because the activities of a technological civilisation will overwhelm faint cosmic signals. Astronomers, as much as any other terrestrial citizens, enjoy the fruits of technology - many of us came to this Colloquium by air, exploiting radio technology for safe navigation and good communications, enhancing cloudiness through aeroengine exhaust products. Astronomy has yet to reach a modus vivendi with a society which is not always well disposed to the interest of astronomy - particularly where profit is involved. The future for astronomy education could be impoverished were adverse environmental impact on astronomy to be allowed to worsen significantly.

Student attitudes to education have also changed in the last 20 years. Changes in school education, an unfavourable employment environment and the introduction of the concept of the student as a "customer" have all had a significant impact on university education. It could be argued that increasing undergraduate numbers would have brought about similar changes. In astronomy - a minority interest - one might have been expected to escape much of the adverse consequences of these changes. However, that has not happened. Flexibility, informality, close contact with students has been lost and prescription and formality are now the norm and obtrude between teacher and taught. This has not led to "quality" in educational experience despite emphasis on demonstrating achievement of prescribed quality standards. This is something which I can only regret as a loss. Students on degree studies should have the privilege of being treated as individuals, to explore and develop their potential - particularly so as astronomy is a small subject in terms of numbers but which depends greatly on the creativity and initiative of individuals. As an example of what I mean by "quality" in educational experience let me only cite ability of students to articulate individual courses. There is a strong tendency for students to perceive a "course" as an individual entity which must be learned in order to achieve a satisfactory performance in the examinations. While students cannot be faulted for wishing to perform well, they can be faulted for not being aware of intercourse relationships. We seem to organise degree structures in such a way that intercourse relationships are not made manifest. As customers, students may decide a particular course is not "relevant" only to find that it was essential, but not a pre-requisite, for a subsequent course perhaps in another year of study. To cope with student numbers on the basis of declining resources, even practical courses - an excellent way to provide articulation between courses - are being so streamlined and formalised that the articulation process is being diminished. But it is encouraging that articulation can be quickly resuscitated by placing students in a teaching situation where they must draw widely on their accumulated knowledge. For the future we must try to ensure that articulation within degree courses is given adequate attention.

2. Influence of the Present on the Future

The major uncertainty for the future is funding for major new facilities whether ground based or in space. If level funding is the best that can be expected, what impact will this have on astronomical education? It is unlikely to have significant impact on the provision of astronomy degree courses since they are such a small component of the university sector. But if the pace of astronomical research slows, there will be impact on undergraduate recruitment. Young people find astrophysical and cosmological research, like particle physics, a great incentive to study physical science at university level. If these areas are not producing excitement and stimulation, undergraduate recruitment may remain at current levels. The numbers of school leavers taking A-level physics (A-level is the pre- university, school leaving qualification at age 18 in the UK) is declining year by year. Departments of Physics and Astronomy are struggling hard to maintain current intake rates. But there may be a glimmer of hope. Curtailment of expenditure on large international projects may lead to freeing of support for less capital intensive projects - such as telescopes optimised for particular purposes and, perhaps naively but very hopefully, for stipends for more astronomical personnel. It may be that career prospects and improved access to quality (if not large) facilities may encourage a modest increase in the astronomically motivated in physical science degree courses.

If adverse environmental impact on astronomy continues to grow, and there is no immediate sign of any amelioration, then there will be a very depressing decline in undergraduate numbers with astronomical interest - whether we talk of degree students or students taking an elective astronomy course. If the conditions for astronomical observation become compromised, no undergraduate of ability will waste time on a dying enterprise. Adverse environmental impact is not just a threat to sustained access to high level astronomical observation but could be a real menace to undergraduate recruitment in the future. It is therefore vital, both for astronomy as an observational science and as a worthwhile educational experience, that adverse environmental impact is opposed with vigour at all levels.

We must also endeavour to provide an educational experience of excellence for our students. That, in my view, means providing them with degree courses that reflect the science of astronomy as currently perceived by research scientists and teachers. We do our students, or our science, no favours if we offer anything less. Pressures to make university education more relevant, more geared to wealth production, more geared to the needs of industry, commerce and the community need to be strongly resisted and rebutted. As astronomers we have a duty to maintain the quality of our science and the education of our successors is primary to maintaining that quality. It is an area over which we equivocate at our peril.

But we must also accept that such a policy may well, in the next century, lead to reduced numbers of students taking specialist degrees in the physical sciences. We may have to accept that as an initial condition. What I have found encouraging in recent years is the explosive growth in numbers of students not taking specialist degrees but who still wish to study a diet of physical science. This may well be where the future lies. While there will remain a residual interest in specialist degrees, we may have to accept that many young people are not prepared to make the commitment necessary to sustain them through a demanding specialist degree course. But young people are interested in acquiring scientific literacy. They want to know about cosmology, particle physics, the interior of the Earth.... As a society we need graduates who have a wide range of scientific knowledge and a good grasp of how that knowledge is obtained in the laboratory *and* in the field. The latter is important for the future - physical science if it is to play a real role in environmental matters (as it assuredly must), must educate its students in real life, non-laboratory situations. Astronomy, Geophysics, Meteorology and Oceanography, to name but four, are excellent physical sciences which are not only laboratory based. Students need to appreciate, that even in the physical sciences, there may not be a knowable correct answer. Students need to see that environmental data suffers from all manner of deficiencies and that they must make judgements on the basis of such imperfect data. They must be aware that in some circumstances, the wrong scientific questions are being asked because we do not yet know enough to formulate the correct questions. It seems to me that such training will be of vital importance for the future if there is to be a real increase in the levels of scientific literacy in the next century. Astronomy, in the company of like-minded sciences, must ensure that it has a major input into any new physical science curricula which responds to the widespread need for scientific literacy - a literacy as important as communication and mathematical skills. Astronomy, in particular, should play a central, major part in such curricula.

3. Conclusion

In looking forward I have noted that the outlook for Astronomy in the next century is not universally bright. There are many factors which could operate to the detriment of astronomical research. However, there is also cause for some restrained optimism that there could be improvement for astronomical education even though curtailment of research opportunity could reflect in a reduction of undergraduate numbers taking specialist degrees in the physical sciences.

Where optimism is engendered is in opportunities to develop education for scientific literacy. Given the lack of real public understanding (as opposed to mere "awareness") of science, despite great effort by scientists, teachers and the media, there is a major need to address widely based education for scientific literacy. Indeed this may be a growth area in the next century. Astronomy must lie in the vanguard of such curricula, playing a major and significant role. As a science astronomy is well equipped to do so - the objects of study are available to all, it uses a variety of sciences to deduce astronomically significant information, it deals with observations under circumstances which are frequently not ideal in the laboratory sense and it encourages judgement in analysing situations where the evidence may have to be weighted for significance. In extreme situations, a study of astronomy demonstrates that there are scientific problems so challenging that they may be beyond investigation with the tools available or lie outside current conceptual experience or even both.

We must not be shy about the educational value of astronomy. Our challenge is to ensure that astronomy plays a major role in future physical science education and that the efforts of a small science are not overwhelmed, or sidelined, by necessary partnerships with a range of other, larger, sciences claiming sole rights to setting the framework whereby scientific literacy can become the prerogative of a large number of people. The evidence of public understanding of science now underlines the long road that remains to be travelled and the major challenge which our science must face, and conquer, if our species is to survive through the next millennium. Our task is no less than that.

Who are our Students – And why does it Matter?

By George Greenstein

Amherst College, Amherst, MA 01002 USA

1. Introduction

When designing courses in astronomy – or any other science – there is a tendency to assume that the students whom we are addressing are younger versions of ourselves. As undergraduates we studied astronomy and now we are practicing it: it is natural to assume that the students we teach are destined to go on to become scientists themselves. But while this was a perfectly valid assumption in the past, it is valid no longer; and if we do not adjust our teaching methods accordingly, we do our students a grave disservice.

The sad truth is that most of them cannot possibly go on to become practicing scientists – because there are not enough jobs to accommodate them. We are all familiar with the terrible employment market nowadays: there is no need to belabor the point except to make the obvious observation that the situation is not going to get better in the foreseeable future. It is for Malthusian reasons that the job market for scientists is bad, and is going to stay bad on the average except for temporary fluctuations. If each astronomer guided, say, ten students on to PhDs in the course of his or her entire career, the population of astronomers would have multiplied tenfold over that time span – obviously an impossible situation over the long run.

Among the students who have majored in astronomy at my home institution during the past decade, most have not gone on to graduate school. One is now a computer scientist. Another edits a newspaper. We might assume that these young men and women were not particularly bright, and that they wisely got out of the field because they found it too difficult. But in reality this is not the case: these two were among the best students I have ever had the pleasure to teach.

I am concerned that we – and I include myself here – might have a tendency to ignore such students. We might relegate them to the outskirts of our attention, and regard them as "less serious" in their study of astronomy. And more than this: we might even feel an unconscious kinship with those who share our love of astronomy and plan to make it their life work – and, conversely, to feel an absence of kinship with the others. I do not mention these unpleasant issues to accuse us of misconduct. I mention them to emphasize that, unless we recognize them, they might subtly guide us away from fulfilling our responsibilities to all our students.

How, then, might we design methods of instruction which will serve those of our students who do not plan to become scientists – while at the same time, serving well those who do?

2. Project-oriented courses

Let me briefly describe a course explicitly intended to address these needs, developed by members of the Five-College Astronomy Department in Amherst. The prerequisites of the course are one semester each of astronomy, physics and calculus. It is normally taken in the sophomore year.

The course begins with four images of the same star cluster, in the IR, V, B and UV bands. The students are asked to describe what they see in these images. Their reaction to this question is instructive: they often react with consternation and uncertainty. "We can't figure out what you are trying to get us to say" is a common response. Evidently we are presenting them with a task they have not encountered very many times before. But surely, the ability to recognize patterns in an unfamiliar situation is the very hallmark of an educated person.

After much hemming and hawing, the recognition emerges that the brighter stars are bright in all four images, while the fainter ones show up only in the long-wavelength ones. The course now proceeds to quantify this intuitive result: students are presented with numerical data and, over several weeks, taught how to analyze it on a computer. An essential component of this segment of the course is that the students are learning methods of image analysis, not in a conceptual vacuum, but as a response to their felt need to progress further in chasing down an intriguing regularity which they themselves have spotted. The final product of their efforts is a graph, in which apparent brightness at one wavelength is plotted against the ratio of brightnesses at two wavelengths – with error bars, no less.

The course then proceeds to a demonstration of blackbody radiation from a lamp, and to a series of measurements of these same two quantities for it. Qualitatively similar results are obtained, but a direct comparison with the astronomical data is impossible because the lamp's temperature does not, of course, approach that of stars. So the course now proceeds to a brief series of lectures in which the physics of black body radiation is presented. The students then pass on to a segment in which they learn to write scientific programs for themselves. They integrate the Planck function over the relevant wavelength bands to assemble the theoretical analogue of the graph they have drawn. Here too, they are learning programming skills in response to needs dictated by their ongoing research.

A critical juncture occurs when the theoretical and observational results are compared. By this point the students have invested great time and effort in working out the theory of the regularity they have found in the data, and they are understandably committed to this theory – and they seem to have completely forgotten those error bars in their observational data. The agreement between theory and observation is not perfect, but the students declare the agreement to be "good enough." Only after much prodding do they admit that the theoretical curve falls significantly outside the error bars.

This failure is a crucial component of the course. The goal here is to give the students practice in being honest, and in recovering from mistakes. In this case the mistake was the unstated assumption that all our black bodies were of the same size. The course concludes with a determination of the relative diameters of stars based on the theory.

A further essential element is a heavy emphasis on writing. Students are asked to maintain an intellectual diary: they write about the logical progression the course has been following, they make note of their own questions and unresolved issues and of any important insights which have been reached. The goal here is partially to give students instruction in writing clearly, but more significantly it is to allow them to clarify their own thoughts by writing them down. Expressing something clearly is a good way to reach an understanding of it.

3. Comments

A revealing moment in the course occurs when a student suddenly says of the graph of observational data that has been assembled with so much effort "Wait a minute – I know this graph. Isn't it the HertzsprungRussell diagram?" It is indeed, and it is significant that this recognition does not occur right away. Indeed, the recognition might have occurred the very first day of class, as the students were examining the four images of the star cluster. But it did not – and the crucial question is: why? All students taking this course have taken a previous one in astronomy, and all have encountered the HR diagram before. But somehow, it seems not to have lodged firmly in their minds.

Lightman and Sadler (1993: see also Flick 1987, Lightman and Sadler 1988 and Schnepps 1986) have presented data indicating that this suspicion, unhappily, is precisely the case. Students retain far less of the contents of a course than their instructors believe. Even those who do well on homework and tests seem not to fully internalize what they have learned: an element of rote learning seems to be present. But students in this course, given their semester-long immersion, are sure to remember the HR diagram. They have made it their own, by discovering it for themselves .

But this, while certainly the case, is not a sufficient reason to ask our students to spend so much time on a single facet of astronomy, no matter how important it may be. A further reason to teach in this manner lies in the nature of the understanding of the HR diagram which is imparted in a normal lecture course. Traditional accounts of the HR diagram present it as a plot of luminosity versus temperature, or of apparent magnitude versus color. Surely there is no logical gap between these accounts and the visual appearance of stars in a cluster in the four images with which our course begins. But there is indeed an experiential gap, and it is one which students evidently find difficult to bridge. So a second reason for teaching the HR diagram in this manner is to give our students a greater understanding of the meaning of the somewhat abstract and theoretical accounts which are traditionally given.

But this, too, while valid is not in itself sufficient reason to spend so much time on this one area of astronomy. This brings me to the central point I wish to make about a project-oriented course such as this: the course is only partially about the HR diagram. Indeed. it is only partially about astronomy. Far more. it is about the honing of skills.

The real curricular goals of this course include exposure to some modern astronomy, but they go significantly further. Additional goals are (1) hands-on experience with data, (2) the acquisition of computer skills, (3) the development of independence in both formulating and solving problems, and (4) the development of critical thinking. I want to emphasize that these are essential skills for all of our students, not just those preparing for careers as professional scientists.

One final comment. Were this the only course of this nature a student encountered, it would do little good. Our department has developed two: a discussion of the other, together with further pedagogical reflections, can be found in Greenstein, 1994.

4. Drawbacks

It is obvious that I am personally excited by this method of instruction. But I would never argue that all our teaching should be done in this manner. The method suffers from many drawbacks.

One stems from the fact that this method of instruction is so very slow. An entire semester is spent on a set of topics which would be covered in a mere week or so of an ordinary lecture course. Scanning over the list of purely astronomical topics which our students have learned in this course, as opposed to those they would encounter in a more traditional lecture course, is enough to engender a profound sense of unease. How can we justify spending so much time on these few topics at the expense of all the others?

A second set of drawbacks stems from the fact that most of us are simply not used to this style of teaching. That the course operates in the seminar format is essential to its nature, but few of us have much experience in teaching in this way. It takes time to learn how to do it. Some students speak up hardly at all, and need to be encouraged: others speak far too much, and need to be discouraged without crushing their spirits. Delicate interpersonal skills come to the fore. It is also important that our students are equally unfamiliar with such a loose, wandering method of instruction. They often react initially with incomprehension and, indeed, a certain level of fear. This, too, needs to be handled with delicacy.

So I would never advocate relying entirely on such project- oriented courses. But at the same time, their advantages are many, and it makes sense to leaven our traditional methods of instruction with them from time to time.

REFERENCES

FLICK, L., 1987, Preservice teachers conduct structured interviews with children to improve instructional methods in *Misconceptions and Educational Strategies in Science and Mathematics*, ed J. D. Novak (Cornell University Press, Ithaca), Vol 1, 133.

GREENSTEIN, G.,1994 Teaching science by seminar, *Physics Today*, May, 69.

LIGHTMAN, A. & SADLER, P. 1988, The Earth is round? Who are you kidding? *Sci. Child.* 25 (5), 24.

LIGHTMAN, A. AND SADLER, P., 1993, Teacher predictions versus actual student gains *The Physics Teacher 31* 162.

SCHNEPPS, M., 1986, A private universe, *Pyramid Films*.

The Use of Photographs in Astronomical Instruction

By [1]M.T.Brück AND [2]S.P.Tritton

[1] Craigower, Penicuik, EH26 9LA

[2] Royal Observatory Edinburgh EH9 3HJ

1. The Edinburgh Teaching Packages.

For many years, copies on film of photographs, both direct and through objective prisms, taken with the 1.2 m United Kingdom Schmidt Telescope, have provided teaching material suitable for universities and colleges (Brück and Tritton, 1988). Table 1 outlines the various types of application to which the photographs may be put. With additional data, some real physics can be injected into the exercises, allowing students to perform quite elaborate projects.

1.1. *Uses for UK Schmidt Telescope Film Copies*

Direct photographs
1. Recognition of objects:
    galaxies
    minor planets
    HII regions, SNRs (in external galaxies)
    globular clusters (in the Magellanic Clouds)
2. Statistics
    star-counts, for various purposes
    number-magnitude counts
    star-galaxy counts
    galaxies in clusters
3. Changes in position (from more than one photograph)
    precession
    comet

1.2. *Objective prism photographs*

1. Spectral classification:
    coarse classification (of about 100 stars per film)
2. Search for unusual objects:
    emission-line stars
    carbon stars
    planetary nebulae
    quasars

A limitation to such purely visual observations is in regard to photometry, where we have to make do with rather rough estimates of magnitude. Measuring the brightnesses or magnitudes of objects is a basic necessity in astronomy, but one that is, ironically, less easy to perform with students than it was ten or twenty years ago. Instruments that were once standard equipment and could be employed on the films - photographic photometers and microphotometers - have fallen into disuse as astronomers receive their data ready processed. For the brighter stars, down to magnitude 13 or 14, magnitudes may be estimated visually to about a fifth a magnitude. This is adequate, however, for our stellar statistics problems (e.g. Fig. 1).

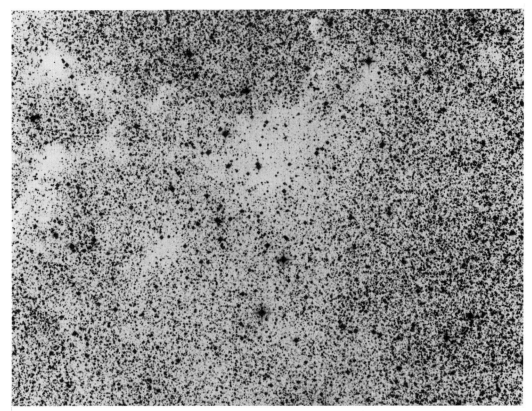

FIGURE 1. A populous region in the galactic plane with an absorption cloud. Star counts yield the extinction in the cloud; photographs of the same region in three different wavebands demonstrate the interstellar extinction law. ©Royal Observatory Edinburgh.

2. The use of enlarged reproductions

Some exercises require the whole field (6.5° × 6.5°) of the films (e.g. Fig. 2). Others in our repertoire do not require a very large area of sky. We have experimented with enlarged photographic prints of smaller areas, incorporated in a book, "Exercises in Practical Astronomy using Photographs"(Brück 1990). Prints, showing all the required detail, were specially produced by the Royal Observatory Edinburgh Photolabs, and excellent half-tone reproductions of these were made by the publishers. Such copies, being enlargements by factors of up to 10 of the originals, mean that linear measurements of images of stars or galaxies can be made with greater ease than from the films.

Some information is bound to be lost in every step of the copying, and in the case of the reproductions in the book, the resolution is limited by the screening dots of 0.2 mm. size. The main loss is in the faintest images, below about 14, when all images tend to be more or less the same size. The reasons for this are twofold: the loss of information due to the copying process, and the fact that we are not able to make full use of the information within the image, as a proper photometer would. The discrimination between stars and galaxies becomes blurred at the very faintest images, but clusters of not-too-distant galaxies show up well. Objective prism spectra, however, are on the whole not suitable for reproduction on paper prints.

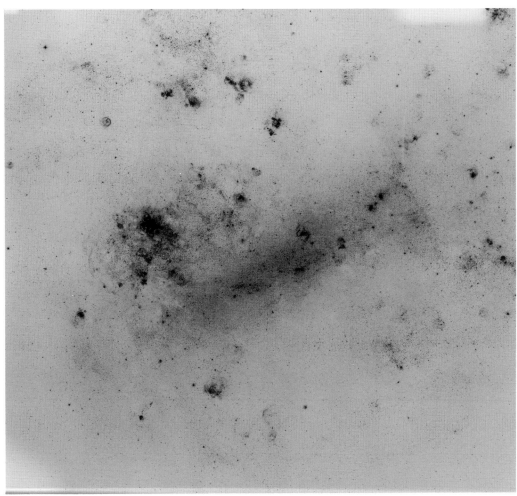

FIGURE 2. The Large Magellanic Cloud in U (violet) light, filling a full field of the UKST. An objective prism photograph reveals emission lines in many of these features. © *Royal Observatory Edinburgh*

3. New Trends.

The title of the conference is "New Trends". As regards our topic, an obvious trend is in the storage of photographs on disk. The entire sky, made up of the Palomar and UK Schmidt telescope surveys, is now available to astronomers on compact disks (The Space Telescope Science Institute 1994). We have done some experiments with this material with a view to its use for teaching purposes. An area of 30 arcminutes square of sky fits conveniently on a 3.5"computer disk and is readily examined on the screen. Many of the subjects of our exercises are accommodated in such a space. Using only basic graphics supplied with a home computer, we have compared the screen images of stars with those of the photographic enlargements and in the printed reproductions in the book, all derived from the same original photograph.

Figure 3 is an area near the south galactic pole containing a photoelectric sequence of stars with magnitudes ranging from 9.5 (the brightest in the field) to 17.3 (barely visible

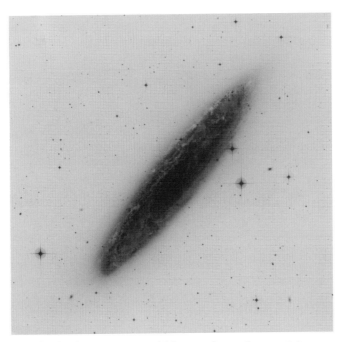

FIGURE 3. A region of half a degree square which contains a photometric sequence ranging from magnitude 9.5, the brightest star, to 17.3 (near the limit of visibility on this photograph). The galaxy is NGC 253. ©*Royal Observatory Edinburgh*

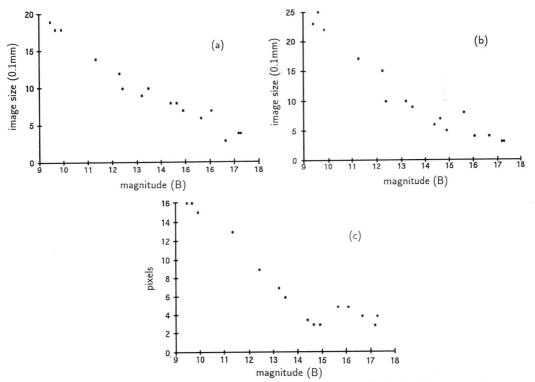

FIGURE 4. Image diameters plotted against magnitude for stars in the field of Figure 3, measured from (a) a photographic enlargement, (b) a printed reproduction and (c) images on the computer screen.

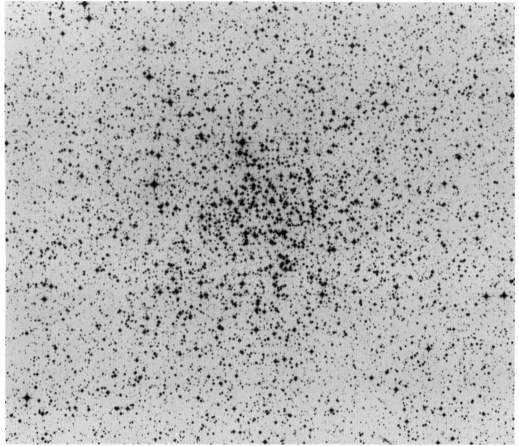

FIGURE 5. An open star cluster: an object suitable for recording and analysing on disk.
© *Royal Observatory Edinburgh*

on the reproduction). Each star image may be enlarged on the computer screen until it shows the individual pixels, so that image diameters may be read off in pixels. The graphs in Figure 4 show plots of image diameters against magnitude for this sequence using the photographic enlargement (a), the printed version of this enlargement (b) and the image on the computer screen (c). It is seen that the diagrams are all similar and that calibration is usable down to magnitude 13 (or better for the first), but impossible for fainter stars. Though the scatter to this limit is perhaps somewhat larger than we achieve using original films, the statistical exercises are viable in all cases. So too are exercises in measuring galaxy dimensions.

This is very encouraging for the prospect of using disks. Personal computers get cheaper, and software becomes more versatile and more accessible to the ordinary user. There is plenty of useful software commercially available with which to analyse the images on the screen (we have used Autosketch 2.1 for Windows): for example, an area may be divided into small parts, individual images may be enlarged, their dimensions and shapes measured, their numbers noted, and so on.

It is possible to fit a great deal of material on a compact disk. This opens up the prospect of more ambitious exercises than we have done until now, involving, for example, a sample of several clusters of stars (Fig. 5) or galaxies. We are not competing here with the experts in the fields of computerised and distance learning but merely showing that useful exercises in astronomy, including some photometry, are feasible on the computer screen. It may well become a favoured method in the future.

Worked examples of some of the exercises may be found in the Proceedings of Colloquium 105 and in "Exercises in Practical Astronomy". Further information on the University and other Packages may be obtained from SBT.

ACKNOWLEDGEMENTS

Our thanks are due to Photolabs, Royal Observatory Edinburgh, for the illustrations, and to Mr M Read and Dr R Hill for technical advice.

REFERENCES

BRÜCK M.T. & TRITTON, S.B. 1990, The Teaching of Astronomy, Cambridge, Cambridge University Press, *IAU Colloquium no.105*, 124-137.

BRÜCK, M.T., 1990. Exercises in Practical Astronomy using Photographs, Bristol, Adam Hilger IOP Publishing.

BRÜCK, M.T.Space Telescope Science Institute, 1994,*The Digitised Sky Survey*, Association of Universities for Research in Astronomy, Inc. USA.

New trends in university education in Russia: teaching Natural History for humanities

By Sergei A. Gulyaev

Astronomy Dept., Ural State University, 620083 Ekaterinburg, Russia

A reform of the content of university education is taking place in Russia today. A restoration of human directed principles, the denial of strict ideological components in education and an improvement in the teaching content of the humanities, are among the most important characteristics of the on-going reforms. An important part of today's activities is the introduction of the basics of natural sciences to the process of teaching humanities. We have gained four years experience in the establishment of natural sciences in humanities at the Ural State University (Ekaterinburg, Russia).

Here I present the methodological strategy of the basic general course of Natural History for humanities. The course is compulsory for undergraduate students of all the humanities (Depts. of Art, Philosophy, Sociology and Politology, Philology, History, Journalism and Economics). It begins from the first year and takes 3 semesters in the Dept. of Philosophy (60 hours of lectures and seminars) and 2 semesters in the other Depts. (40 hours of lectures and seminars). The course is united by a general idea — the History of the Earth. It is divided into three parts: (1) Cosmic period of the history of the Earth, (2) Matter and Energy (only for the Dept. of Philosophy), and (3) Geological and biological periods in the history of the Earth. The first (astronomical) part in turn consists of three chapters: (a) Scientific pictures of the world and their creators, (b) The real Universe (state of art geometry and physics of space), (c) "Genesis" (formation and evolution of the Universe, Sun and the Earth).

The adaptation of this course for different departments is one of the most important methodical problems. We include special topics in the course: Natural Science and Theology for students of the Philosophy Dept., Calendar and Chronology for Dept. of History, Science and the Mass Media for the Dept of Journalism, etc. Additional voluntary special courses were established which have been adapted to the needs of different departments; some of them are: Human Ecology, Legends and Myths of Sky, Synergy for Humanities, Mathematical Simulations for Humanities.

Further work in this field needs the involvement of new educational technology, the preparation of textbooks and methodological materials, the creation of distant-learning courseware and the organisation of conferences and seminars, where experience and ideas can be shared.

International Schools for Young Astronomers, Astronomically Developing Countries, and Lonely Astronomers

By Donat G. Wentzel

Secretary for ISYA and TAD

Department of Astronomy, University of Maryland, College Park MD 20742, USA

wentzel@astro.umd.edu

1. Goals

Early in this century, many cities and universities could support telescopes large enough to do serious research. There were significant observatories even in the less accesable parts of the world. Astronomy was very much an international science, and the IAU was founded to aid this international outlook.

In the middle of this century, astronomy changed in two ways. First, the frontier research turned to new topics. It needed telescopes too expensive for most small and many large countries. Second, physics became a more prominent part of astronomy. That left many of the existing small observatories scientifically isolated, especially in developing countries. The scientifically lonely astronomers there needed new alliances to survive.

Simultaneously, the new prominence of physics led to astronomers appearing in physics departments of developing countries. These new astronomers were also isolated and they also needed to build alliances to survive.

Since the 1960s, the IAU has tried to support astronomy in developing countries, especially the lonely astronomers, by several teaching-related projects, supervised by IAU Commission 46. I shall tell you the more formal aspects of these projects, and then I want to summarize some of the successes and difficulties. But first I want to emphasize an important principle for each of these projects:

<u>Any country or university where the IAU helps to develop astronomy must contribute significantly to this project.</u> The more usual alternative has been tried by other scientific societies. Typically, they donated a piece of research equipment. Far too often several years later, the equipment has been found rusting in some corner. Those of you familiar with tropical countries know that rust destroys neglected equipment very rapidly. The projects of the IAU have minimized equipment and focused on the human contributions. We have selected countries and universities where we are reasonably assured of long term interest because these countries and universities are willing to contribute themselves.

2. The International Schools for Young Astronomers (ISYA)

There have been 22 ISYA: One each in UK, Italy, Greece, Spain, Portugal, Yugoslavia, Nigeria, Morocco, Malaysia, and Cuba; two each in India, Indonesia, China, Egypt, Argentina, and Brazil. The next ISYA will probably take place in Iran during July 1997.

The schools have evolved over the years. During the last dozen years, typically, the schools have been in Asia, Africa, and Latin America. They have been three weeks long. There have been four foreign faculty and one or more local faculty, teaching perhaps a dozen foreign participants and two dozen local participants. Most participants are students with a M.Sc. degree, but some are undergraduates and some are post-doctoral scientists.

The academic degree of the participants is less important than the kind of institution they come from: Many of the participants come from a small observatory or a small university department of astronomy. There the teaching is often quite specialized. The students often have learned one subject well, and they may do some research on that subject, but they have only a basic idea about all the rest of astronomy. Therefore, the ISYA usually emphasize a broad range of topics. The lectures at the schools start out quite basic, but the topics are selected so that the lectures will reach the currently exciting research. The topics are chosen by the host institution. Typical topics are solar physics, stellar atmospheres, our Galaxy, cosmology, high-energy astrophysics, observational procedures and data reduction, and the teaching of astronomy.

The IAU pays only for travel of faculty and participants. In fact, for many years UNESCO sponsored the travel costs together with the IAU. The host institution pays for the rooms and meals of faculty and participants, and it provides the meeting facility including a copying machine. Usually the host institution finds some additional sponsors for these costs. In some places, the rooms at a school or institute are available for very little cost, but in some places we have stayed in hotels. On average, in the last few years, the IAU and the host institution have contributed similar amounts of money. But the amount of time and effort provided by the hosts can be very large. For instance, both in China and in Egypt we stayed one week at an isolated observatory. The logistical effort to supply all of us with food at these observatories was quite significant.

2.1. *What do the participants get out of it?*

A much broader perspective on astronomy and on how science works. Many of the students come from an observatory with just one professor of astronomy. These students have learned a narrow part of astronomy, the topic of their professor. Naturally, they take their professor to be the authority on that subject. They have no idea that the same observations might be interpreted differently at some other observatory. They have never heard a scientific argument. Actually, even students from places where there are several astronomy professors have never heard a scientific argument. Therefore, we encourage faculty at the ISYA to present some common topics but with different points of view. We encourage faculty to ask each other questions, to show the relevance of their topics to different fields. At the school in Malaysia, three of us faculty chose a topic on which we could hold a debate. We found to our own surprise that we had three different points of view. That impressed the students!

Practice to ask searching, challenging questions. A student in a small observatory rarely questions a professor, because the student may appear as criticizing the professor. This is especially true in some Asian countries. Students need some practice to ask questions, and they need practice to phrase a question that will elicit a thoughtful and useful answer. At several schools of the last few years, it took at least a week for students to build up the courage to ask questions of the faculty. Once the questions started, and even once students met individually with faculty, it took another week for students and faculty to find a common scientific language and to communicate effectively. That communication with the students is the main reason we ask faculty to stay at the school for at least two weeks. A faculty member who comes for a few days, gives a few lectures, and disappears again is considered by most students as self-serving: this visiting faculty member comes to show that he or she really knows the subject, but demonstrates no apparent concern for the students.

A new outlook for the students own work. Those students who have done some research are asked to present it briefly. This is often the first time students discuss their topic with a wide audience. They are often shocked that the faculty understand very little of the details they present. In fact, frequently the faculty ask questions during the introduction, because the student does not present the context of the problem. But the students are also often pleased when one of the faculty finds their work interesting, probably in some connection that would not be made at the students home observatory. Quite often the faculty will then suggest other scientific contacts in this subject, leading to international collaborations. I stress again: for this interaction to happen, a faculty member must be at the school long enough to establish a good communication with the students.

Lecture materials. Of course, the participants learn much astronomy and astrophysics. In fact, three weeks of lectures and observing sessions provide much too much material for the students to absorb. Therefore, much of the detailed material is distributed in written form to the participants. When the participants return home, the copied materials often provide a significant increase to the astronomy library. That is why I carefully mentioned the copier machine as an important contribution to the school by the host institution.

Spoken English. Finally, a very important goal of most schools is that the participants learn to talk in English. English is now the international language of science. Indeed, our participants must be able to read English. But most have rarely heard English spoken and most have never spoken English themselves. During the first week of an ISYA, many non-English speaking participants are very quiet. The second week they talk a few sentences. And the third week you cannot stop them talking in English. At the school in China, even the Russians talked to each other in English. In my opinion, learning useful English is the main reason that most schools are three weeks long.

Learning useful English is yet another important reason that faculty should stay at the school as long as possible. It takes much time to establish good scientific communication in spoken English. I know that many participants have stayed in touch for many years, by writing to each other in English. They would not have dared to do this before they attended the school.

2.2. *What does the host institution get out of it?*

Recognition. First, an ISYA may mean official recognition of the importance of astronomy, official recognition from the university and from the government. When the IAU sponsors an event, the word International opens many doors. This practical influence of the IAU may well be worth more to the host institution than the IAUs financial support. In schools such as Indonesia and Malaysia, high government officials have opened the schools. In Nigeria, the school led to the formation of the Department of Physics and Astronomy, the first such department in the developing countries south of the Sahara.

Second, the host institution acquires some international recognition within the astronomical community, because after the school the faculty and participants disperse across some twenty countries.

Start a group of astronomy students. When I first met Bambang Hidayat in Indonesia before the first school there, he was very much scientifically alone at his observatory. Later, I found that this school created a working community of astronomers. Some were working with Bambang, and some were teaching astronomy or working in the space program elsewhere in Indonesia. Similarly, the school in Nigeria led to a great deal of astronomical activity. One of the students from the school in Nigeria has obtained an astronomy doctorate abroad and has now returned as professor at a different Nigerian university.

Broaden the training of astronomy students and young researchers in the host country. This has probably been the main goal for schools in some of the large countries, like China, India, Argentina, and Brazil. The schools in these countries attracted participants from several universities and research institutions spread all over the country. In many such places, there are already some astronomy students, and there are already physics students interested in astronomy. But the courses and the training available in these places tend to be limited and specialized. The ISYA allows the import of several faculty with broad interests who can energize astronomy throughout the country.

Many schools have attracted participants from neighboring disciplines, especially from physics. Some ISYA have also tried to attract amateur astronomers and interested people in substantially other sciences. This has probably been the least satisfactory aspect of the ISYA. These schools had to start with a rather low academic level. Then many participants gained nothing from the faculty's attempt to advance the lectures to some current research-oriented topics.

Do the ISYA really work? I have surveyed the participants from three schools. I received very good reviews from many foreign participants, and in general from those local participants that really were active in the school. We had the least response from the local students who only came part-time. They gained a few scientific facts, but they missed the discussions, they avoided occasions when they had to talk English, and they did not broaden their outlook on astronomy. I believe, the single most important improvement in the ISYA is to insist that local participants attend full-time. Unfortunately, this is expensive for the local host, because the local students should then be given rooms and meals. Therefore, it is difficult to express this request to the host in a diplomatic manner.

I should acknowledge that the first 18 ISYA were planned by J. Kleczek. M. Gerbaldi has shared with me the planning of the last four ISYA.

3. Visiting Lecturer Programs (VLP)

The Visiting Lecturer Programs were instituted to help a university with little astronomy that wanted to enhance its astronomy substantially. Two VLPs were organized, in Peru and in Paraguay. In each case, there was one astronomer, Maria Luisa Aguilar in Peru and Alexis Troche-Boggino in Paraguay. Each worked within the physics department of the national university and was active in training teachers. The goal for the VLP was to establish a small community of astronomers that could be more permanent than one individual.

As a first step, the universities agreed to offer a regular sequence of astronomy courses, as part of the physics curriculum. The first sequence of these courses was to be taught by international visitors. The IAU would pay the travel expenses of the visitors. The host would pay for the local costs. A few students graduating from these courses then were to be sent abroad for further training. After they finish their studies and return home, they would be part of the desired local community of astronomers.

The students in Peru and Paraguay had very little practice in the use of English. Therefore, the VLP courses had to be given in Spanish. We found only very few astronomers both able to speak in Spanish and available for a visit that lasts three months. The astronomers we found could not visit in consecutive semesters. Therefore, only few students were around long enough to attend several astronomy courses. They did not really gain the astronomical background needed to study abroad from these courses. Nevertheless, some of the students from Peru have been able to study in Brazil and are now returning to Peru. Therefore, at least the VLP in Peru seems to have been successful.

Paraguay has been more connected to Argentina. I think there is still hope for Paraguay. It is important to have patience and perseverence.

Since this is a conference on teaching of astronomy, I want to point out that the astronomers in both of these countries have worked hard to help the teaching of astronomy in the schools of their countries. The teaching of teachers was not really a part of the VLP, but I hope that the existence of the VLP helped these astronomers persevere in their teaching activities.

4. Teaching for Astronomy Development (TAD)

The original goal of the VLP was to help countries with little astronomy to develop their astronomy substantially. This goal is still a good one, but the procedures have been made somewhat more flexible under the new program called Teaching for Astronomy Development. We sought applications for TAD from many countries. The IAU Executive Committee has just approved two new programs.

<u>Vietnam</u>. Vietnam has been scientifically isolated for thirty years. We want to help re-establish astronomy in Vietnam. In this, we are joining efforts with astronomers and funds from France and from Japan. The IAU will primarily sponsor a conference to bring current astronomy teachers and physics students up-to-date in several topics of astrophysics.

<u>Central America</u>. We want to help establish astronomy in six cooperating countries of Central America. Primarily the IAU will help fund an annual course in astronomical observations, with participants from all six countries. The course will take place at the new observatory of the National University of Honduras. In this program, we are joining the efforts from the European Commission, which will fund a Central American M.Sc. program in astronomy, somewhat along the lines of our VLP.

It is very important that other international organizations now have similar goals. It is our hope that we can usefully complement the programs supported by the United Nations Office of Outer Space, by the European Commission, by France, Japan and several other international organizations.

5. Does WWW eliminate scientific isolation?

Our international schools and the visiting lecturer programs have been aimed at the scientifically isolated astronomy students and astronomers. But technology is changing. Many developing countries can now receive foreign e-mail for perhaps an hour a day. A few countries have real-time international e-mail. That means they can use programs like ftp and telnet. A very few countries, like Honduras, are connected to the world wide web. In Honduras, at least, once again astronomy has been the driving force for the country acquiring new technology. Of course, once you have the web you have access to the data sets from the HST etc. Therefore, in terms of scientific factual information, many astronomers will no longer be isolated.

But the web does not solve all the problems. I believe that many astronomers in these countries will still be physically isolated. They will still need personal contacts to develop the skills of interacting with other astronomers, they will need practice how to question other astronomers data and their interpretation, and they will need active discussions for placing local astronomy in a broader context. To provide these personal contacts has been the aim of the International Schools for Young Astronomers in the past, and I think it will be a perfectly valid goal for future schools.

India

By J. V. Narlikar AND N.C. Rana

Inter University Center for Astronomy and Astrophysics,
Post bag 4, Ganeshkhind, Pune 411 007

1. Introduction

A summary of work related to astronomy education carried out during the last three years in India is presented here. Since India is a huge country and many educational efforts are made by individuals alone, this report cannot be regarded as complete, but a specific sampling.

2. General Information

India has more than 200 Universities, 8000 colleges, and about 100,000 schools, 33 planetaria, more than 100 museums and about 60 well known amateur astronomers' clubs. Scores of dedicated astronomy oriented school teachers, act as nuclei of astronomy education for the general public and school children. The astronomical almanac, used in a typical household is in some way related to the stars in the sky and the movements of the Sun, the Moon and the planets. Traditionally, a rudimentary knowledge of the celestial sphere is common. The recent developments in space technology have brought a fascination and glamour to modern astronomy for all age groups, and this is noticeably reflected in the number of media coverages of astronomy. There are about 12,000 telescopes of aperture no less than six inches, made by amateur astronomers.

3. Public Awareness

During the past three years there have been at least 300 six inch telescopes made by school children and laymen, under some project or other funded by the government and an equivalent number is also produced from private and individual resources. It takes about two weeks to grind and polish the mirror and assemble it in a suitable mount. After aluminizing the average cost comes out to be in the range US dollars 60–100, for a telescope of size greater than six inches. Small telescopes are acquired off the shelf and the recent eclipse fever has witnessed a sale of about 1000 telescopes by telescope making companies in the private sector.

Training schools in astronomy are also run by several planetaria, science centres and astronomy clubs. These are by and large successful. Some of them also result in a diploma after a rigorous one year course.

On a national scale a Confederation of Indian Amateur Astronomers (CIAA) has been formed in 1993 and registered in 1994, which organises the activities of amateur astronomers throughout the country for the society and the individual. Government of India has also formed a nationwide network of various schemes for science popularisation in which astronomy has played a major role. Such a network has produced more than 10,000 volunteers for educational activities with specific targets as the school children and laypersons who are usually parents of the children.

4. Contribution of IUCAA

The Inter-University Center for Astronomy and Astrophysics (IUCAA) located in Pune has arranged a number of school, college and university programs in the field of astronomy education. Pune has a population of about three million, there are about 200 secondary schools, 40 colleges and 2 universities. We celebrate 28 February every year as the National Science Day. On this day about 100 schools participate in our quiz contest, astro-painting, essay competition and performance of astroballets and astrodramas. We also run summer school programs exclusively for school children of standard VIII, IX and X. Each child typically spends a week with an academic member of IUCAA working on a project. In this one week we introduce them to the basics of astronomy. Usually 70 schools participate every year with each school deputing 2 students as per prior arrangement. Also on the second Saturday of every month we arrange an educational popular lecture or lecture demonstration for school children. One such evening is attended by about 50 schools (each school deputes one science teacher and ten students). Every fourth Friday is reserved for evening sky observation or the star party as it is popularly referred to. Twice during the year the local science and geography teachers are invited for do-it-yourself type science experiments which also include the experiments based on astronomy as prescribed in the text books. About 75 schools join this programme very enthusiastically. Most of them have made a sky globe, low cost planetarium projector, sundial and small telescopes. These items are given away as an incentive to them. The college and university students (BSc/MSc/Mphil) are invited to do their final year project work in astronomy.

Vacation student programmes are arranged for pre-final year post-graduate students wherein they attend introductory review lectures on Astronomy and Astrophysics and carry out projects. At the end of the programme there is a test and an interview. The successful students are offered research scholarships for work in IUCAA after completing the M.Sc. courses in their universities. College and university teacher orientation programmes in astronomy are regularly arranged by IUCAA both on campus and in specific regions of the country. A syllabus is taught with the aim that these teachers can persuade their authorities to introduce a special paper in astronomy. By now 20 universities in India teach astronomy and astrophysics at post-graduate level and often allow complete project work in the field of astronomy.

5. The Total Solar Eclipse 1995

A population of more that 30 million resided close to the path of the totality for the solar eclipse of October 24, 1995. This gave us a unique opportunity to introduce awareness for astronomy to a large population. Scores of seminars, lectures, demonstration, astrophotography, workshops, model making competitions were carried out on a large scale involving not less than three million children. About one million eclipse goggles were sold. Two amateur clubs made sixteen inch telescopes for themselves and took professional type photographs of the eclipse. One amateur club has photographically recorded the shadow bands. Incidentally the same club publishes the sky watchers guide every month at a nominal cost of 0.15 dollar and makes it available to the subscriber before the month begins. Although the articles are not as professionally written, the quality of information is comparable to that of any international magazine.

6. Books

A preliminary survey of astronomy related books in the Indian market suggests that there are about two hundred popular books in astronomy in major languages of the country.

7. Conclusion

In conclusion it may be said that general awareness for astronomy education as a part of science education in the country has been steadily growing both through individual efforts and through government aided institutions as well as non-governmental organizations.

The challenge of teaching astronomy in developing countries

By Lesley Onuora

The Astronomy Centre, University of Sussex, Brighton BN1 9QL

1. Introduction

Having recently returned to England (where I am an Open University tutor) after having spent about 18 years teaching Physics and Astronomy at the University of Nigeria at Nsukka in the Eastern part of Nigeria, I find myself in an unusual position to understand the difficulties of teaching such a rapidly changing subject as astronomy in an isolated place like Nsukka. For example I have seen a great contrast between the OU Astronomy and Planetary Science course material and the few available text books at Nsukka. Although not very mathematical, the OU material includes a lot of the latest research results and theories, whereas at Nsukka the books have hardly changed in the past 20 years.

I am aware that the Astronomy group at Nsukka is not unique. There are other small isolated groups of astronomers (or in some cases only a single astronomer) around the world who are trying to interest their students in astronomy against great odds. These astronomers appreciate the importance of astronomy in awakening interest in science and thus strengthening the basic sciences and developing technological progress. However Governments and even some international agencies often take the view that astronomy is a luxury that is not needed by such developing countries and therefore give little or no support to these efforts.

2. Main Problems

Apart from the lack of teaching materials, extremely limited access to computers and generally poor infrastructure, the one major problem is the extremely poor communications. Often phone, fax and mail do not work reliably, and needless to say there is no e-mail or internet. Access to the World Wide Web is almost taken for granted throughout the world and it is often not realised that the internet is not available universally. It is only since my return to England that I have been able to fully appreciate what is being missed. At the recent United Nations/European Space Agency Workshop on Basic Space Science held in Sri Lanka, discussions were held about astronomical software packages available free of charge on the WWW. Also the availability of astronomical data archives was discussed. This would mean that astronomers in developing countries would be able to do research using the latest data and would not necessarily need their own telescopes. Advance information about scientific papers to be published, conferences planned etc. are all abailable. I am sure we will hear a lot in this colloquium about the plans of the Open University and others to use computers in future for distance teaching. In fact the time may soon come when a lot of information will not be available in hard copy. This is very worrying for astronomers in developmeing countries who do not have access to the internet and will therefore fall even further behind with respect to developed countries.

As I have already mentioned, books are out of date and in any case are not sufficient for the number of students. This places a great responsibility on the lecturer since the only real information that the students get comes from the lecturer. It is a form of distance learning, but in this case the students and lecturer are isolated together!

It is therefore of great importance that the lecturers concerned attend international conferences, workshops, short courses etc. in order to keep up to date. Sponsorship for such activities is a major problem. The IAU has helped a great deal in this regard and have continued to show their concern about the plight of astronomers in some countries.

One might think that the obvious solution to the lack of teaching materials is for the lectureres in the country concerned to write text boks suitable for their students. However, the problem is an economic one of publishing books of suitable quality at prices affordable by the student. The number of students studying astronomy is too few to be able to convince a publisher to produce such a book.

3. Present and future astronomy teaching

The Department of Physics and Astronomy at the University of Nigeria teaches astronomy at both undergraduate and postgraduate levels. All physics undergraduates take an introductory course in astronomy with the option of doing more advanced courses and a project in their final year. It has been my experience that the students are very enthusiastic about their usually first exposure to astronomy, even when taught under such difficult conditions, especially if the lecturer is able to tell them about some of the latest observations and results. Astronomy programmes on the BBC World Service are a useful source of such information. Complimentary copies of some journals, including the Sky and Telescope, which arrive very erratically are immensly useful as are the reprints sent by some observatories.

Availablility of the internet would revolutionize both astronomy research and teaching in developing countries. This has been discussed repeatedly at United Naions/European Space Agency Workshops on Basic Space Science (e.g. Nigeria, 1993, Egypt, 1994).

Astronomy teaching is continuing and in fact gaining momentum in Nigeria. Three other universities are starting astronomy programmes, two of these initiated by doctoral graduates in astrophysics from the University of Nigeria. Recently a collaboration has been started between astronomers in South Africa and the Space Research Centre at the University of Nigeria. It is hoped that this will continue with future plans for some Nigerian postgraduate students to carry out observation projects in South Africa. It is also hoped to foster more interaction between astronomers throughout Africa. The UN/ESA Workshops held in Nigeria and Egypt as well as meetings held by the IAU Working Group on the World Wide Development of Astronomy have helped in this direction.

The MicroObservatory Net

By Kenneth Brecher[1] AND Philip Sadler[2]

[1]Department of Astronomy, Boston University, Boston MA, 02215, USA

[2]Harvard-Smithsonian Center for Astrophysics, Cambridge, MA 02138, USA

1. The MicroObservatory Net

Beginning in 1990, a group of scientists, engineers and educators based at the Harvard-Smithsonian Center for Astrophysics (CfA) developed a prototype of a small, inexpensive and fully integrated automated astronomical telescope and image processing system. The MicroObservatory combines the imaging power of a cooled CCD, with a self contained and weatherized reflecting optical telescope and mount. A microcomputer points the telescope and processes the captured images. Software for computer control, pointing, focusing, filter selection as well as pattern recognition have also been developed. The telescope was designed to be used by teachers for classroom instruction, as well as by students for original scientific research projects. Probably in no other area of frontier science is it possible for a broad spectrum of students (not just the gifted) to have access to state-of-the-art technologies that allow for original research projects. The MicroObservatory has also been designed to be used as a valuable new capture and display device for real-time astronomical imaging in planetariums and science museums. The project team has now built five second generation instruments. The new instruments will be tried with high school and university students and teachers, as well as with museum groups over the next two years.

Though originally designed for use in individual schools, we are now planning to make the MicroObservatories available to students, teachers and other individual users over the Internet. We plan to allow the telescopes to be controlled in real time or in batch mode, from a Macintosh or PC compatible computer. In the real-time mode, we hope to give individuals access to all of the telescope control functions without the need for an "on-site" operator. Users will sign up for a specific period of time. In the batch mode, users will submit requests for delayed telescope observations. After the MicroObservatories complete a job, the images will be e-mailed back to the user.

At present, we are interested in gaining answers to many technical and educational questions including: (1) What are the best approaches to scheduling real-time observations? (2) What criteria should be used for providing telescope time? (3) With deployment of more than one telescope, is it advantageous for each telescope to be used for just one type of observation, i.e., some for photometric use, others for imaging? (5) How much trouble is to be expected in controlling telescopes and shipping images over the Internet in real-time? (6) What access and queuing software should be developed? And, most importantly, (7) What are the real educational benefits of using the MicroObservatories?

The European Astrophysics Doctoral Network

By Thomas Patrick Ray

School of Cosmic Physics, Dublin Institute for Advanced Studies

5 Merrion Square, Dublin 2, Ireland

1. What is the EADN?

In 1986, a group of university astrophysics institutes in eleven Western European countries established a federation known as the European Astrophysics Doctoral Network (EADN). The aims of the EADN, then and now, are to stimulate the mobility of postgraduate students in astrophysics within Europe, and to organize pre-doctoral astrophysics schools for graduate students at the beginning of their PhD research. The network has by now expanded to include about 30 institutes in 17 Western European countries, and ways are being actively sought for expanding the EADN even further to include Eastern and Central Europe. The coordinators have been Prof. Jean Heyvaerts (France) until 1992, Prof. Loukas Vlahos (Greece) 1992-1993 and myself since 1993. The network is financially supported by the European Union "ERASMUS" and the "Human Capital & Mobility" programmes as well as by national funds.

2. The Student Mobility Scheme

The Student Mobility Scheme has been designed to encourage postgraduate, or in some cases senior graduate, students to undertake part of their doctoral or diploma thesis research at an institute which is part of the network. It offers ERASMUS funded grants intended to cover student travel expenses and extra expenses encountered by the student caused by living away from their home institute. The grants are not full grants since it is expected that the student can retain the home grant while at the partner institute. The duration of the visit is usually anywhere between 3 and 12 months and must be preceded by contacts between the student's regular thesis advisor and the network partner advisor. These kinds of interactions often lead to further collaboration and the programme also includes the possibility of financial travel support for teaching staff in connection with the student's thesis defence. A full list of current partners is available on the World Wide Web at the URL "http://www.cp.dias.ie/astro/eadn".

3. The Schools Programme

The schools are essentially intensive theoretical and practical training courses targeted at young astrophysics graduates normally in their second or third year of their doctoral studies. The EADN schools aim to inform postgraduate students in the most modern aspects of astrophysics and to offer them the opportunity of contact with some of the most advanced European researchers in their field. Major attention is paid by the organizers and lecturers to the specialized training aspect of the meetings. Activities designed to personally involve the students are arranged including, for example, discussion groups, talks by students on their thesis research topics, poster sessions and practical exercises in numerical computation. In essence, therefore, EADN Schools follow closely the format of a typical school with intensive courses by experts, contributed papers (both oral and poster from the students) and published proceedings. The content of the courses by the European experts, though advanced, are directed at a broad astrophysics audience. They should, in general, still be understandable by those who are relative newcomers to the themes of the school. This aspect is unfortunately lost in most "professional" conferences and summer schools. At the same time the subject matter of the schools are of a sufficiently specialized nature that *they are often best thought of in a European, rather than a national basis in order to provide the necessary "critical mass"*. Thus the EADN feels it fills an important niche in the training of senior astrophysics Pads and at the same time increasing their awareness of the opportunities being offered in Europe.

To date eight such schools have been organized by the EADN (listed in the Appendix) and the ninth school on the themes of "Stellar Atmospheres: Theory and Observation" is currently being planned for September 1996 in Brussels. The schools themselves are normally of two weeks duration and draw on the talents of around eight European experts to deliver the intensive courses. Typically they are attended by 40-60 European students.

As a rule, each school proposes two closely related themes one being astrophysical and the other more methodological, i.e. in the field of technology or in numerical studies. For example, in Berlin (1992, see Appendix) the themes were "Star Formation" and "Techniques in millimeter and Infrared Astronomy". Here the relationship between the two themes is that technological breakthroughs in millimeter and infrared astronomy have contributed enormously to our understanding of how stars like our Sun form. In the recently organized school in Leiden on "The Structure of the Universe" these 'two streams' of theory and observations again appear, with detailed treatment of the ways in which the work is actually done. Students were given the opportunity to do actual experiments with the observational data. This requires access to local computers and carefully prepared small-scale experiments. We thus aim for a 'hands-on and brains-on' approach, in which the 'streams' do not run sequentially but in parallel.

The format of the EADN schools are intensive with two weeks of lectures, 5 days per week, for 4 hours per day. These 40 hours of lectures are usually given by 8 lecturers so that each individual's contribution on a particular sub-topic consists typically of around 5 lectures. The lecturers normally do not stay for the entire two weeks of the school but typically for 3 or 4 days. During this time as much interaction with the postgraduate students as possible is encouraged. Such a scheme allows time for discussions between lecturers and students and for practical exercises in association with the methodological lectures. In addition each postgraduate is asked to give a short paper on his or her personal research subject irrespective of whether he or she has obtained results, to the other students and lecturers. Facilities for the presentation of poster papers are also made available at the meetings and such contributions are actively encouraged.

The proceedings of the EADN Summer Schools *are published* in the "Lecture Notes in Physics" series by Springer-Verlag. Free copies of the book are distributed to participating students and lecturers and the EADN series has now grown to a size where it is a very useful source of teaching material to astrophysics students (see Appendix A).

4. The European dimension

A clear outcome of our summer schools is that they create amongst the students the feeling that they belong to the wider European scientific community. The network feels that it is important that the European dimension be stressed and implanted at an early stage and this, we think, is successfully done by the EADN both through its schools and its student mobility scheme. Obviously through such schemes we foster movement of astrophysicists within Europe, and we also make them aware of the large ground based and space borne facilities available either within the framework of organizations like ESA

FIGURE 1. The somewhat serious side of an EADN School: Donald Lynden-Bell from the Institute of Astronomy in Cambridge lecturing at the Thessaloniki School on Galactic Dynamics and N-Body Simulation Techniques.

(European Space Agency) and ESO (European Southern Observatory) or in individual facilities such as IRAM, MERLIN, the La Palma Observatory in the Canary Islands, et cetera.

5. The future

With the adoption by the European Union of the SOCRATES programme to replace ERASMUS, the operation of the EADN has entered a period of uncertainty. The emphasis in the past under ERASMUS was on large networks administered by a principal coordinator and the principal coordinator was usually an academic. SOCRATES by contrast is much more clearly aimed at encouraging bilateral exchanges between universities and the administration is taken care of by the university administration through an institutional contract. There has been a lot of heated debate as to whether the latter approach is the best one but in any event time will tell. The new SOCRATES programmes will commence in the '97/'98 academic year and efforts are currently under way to maintain the spirit of the
EADN under these new arrangements.

I wish to thank the organizers of IAU Colloquium 162, and in particular Derek McNally, for the opportunity of talking about the European Astrophysics Doctoral Network.

Appendix A. List of EADN Schools and Book Titles.

Evolution of Galaxies and Astronomical Observations, Les Houches, eds. I. Appenzeller, H.J. Habing, P. Léna, Springer-Verlag, Lecture Notes in Physics, Vol. 333

Late Stages of Stellar Evolution and Computational Methods in Astrophysical Hydrodynamics, Ponte de Lima, ed. C.B. de Loore, Springer-Verlag, Lecture Notes in Physics, Vol. 373

Central Activity in Galaxies and Observational Data to Astrophysical Diagnostics, Dublin, eds. A. Sandqvist, T.P. Ray, Springer-Verlag, Lecture Notes in Physics, Vol. 413

Galactic High- Energy Astrophysics: High Accuracy Timing and Positional Astronomy, Graz, eds. J. van Paradijs, H.M. Maitzen, Springer-Verlag, Lecture Notes in Physics, Vol. 418

Star Formation and Techniques in Infrared and mm-wave Astronomy, Berlin, eds. T.P. Ray, S.V.W. Beckwith, Springer-Verlag, Lecture Notes in Physics, Vol. 431

Galactic Dynamics and N- Body Simulation Techniques, Thessaloniki, eds. G. Contopoulos, N.K. Spyrou, L. Vlahos, Springer-Verlag, Lecture Notes in Physics, Vol. 433

Basic Plasma Processes and Diagnostics of Astrophysical Plasmas, Florence, eds. C. Chiuderi, G. Einaudi, Springer-Verlag, Lecture Notes in Physics, Vol. 468

The Structure of the Universe, Leiden, ed. V. Icke, Springer- Verlag, Lecture Notes in Physics, in preparation.

Distance Learning and Electronic Media in Teaching Astronomy

Distance education in astronomy: at-a-distance and on-campus, a growing force

By Barrie W. Jones

Physics Department, The Open University, Milton Keynes MK7 6AA, UK
b.w.jones@open.ac.uk   http://physics.open.ac.uk

1. Introduction

Distance education has a track record in astronomy and is already making a significant contribution worldwide. It will make an even greater contribution in the future, not only at-a-distance, but through greater use of self-study materials on- campus, where it will liberate staff for more appropriate forms of face-to-face teaching, and help overcome the need to do more and more with less and less resource. Distance education offers huge promise in meeting the educational needs of a burgeoning world population, and because low costs can be achieved there is no need for people in areas of material deprivation to face mental deprivation also. The IAU and The Open University can be proactive in promoting the spread of distance education, and of self-study on campus.

2. What is (successful) distance education?

Distance education is NOT as shown in Figure 1, though its distinctive feature is that the student is remote from the university or college! But in place of a megaphone a mixture of media is used in which printed texts usually carry the bulk of the educational material. There can also be audiovisual and computing media (including use of the Internet and of "multimedia"), and practical work. It is important to play to the strengths of the various media - a current pitfall is that multimedia can turn out to be little more than an expensive book.

It is unusual to find already-published textbooks that are suitable for the distance learner. Common shortcomings are
- inconsistent assumptions about the previous knowledge and skills of the student
- insufficient student activity such as "stop-and-think"questions, and opportunities for self testing
- too much content, leading to the need to omit material that is be essential for the study of later, included topics.

On campus these are less problematical because of the support readily available from teachers, peers, and an academic library. For the distance learner these shortcomings are severe, and therefore specially written texts usually have to be produced.

Though support is less readily available for the distance learner, the student should NOT be unsupported or there will be large drop-out or failure rates. Effective means of support include
- study guides
- a tutor for remote contact (by phone or computer), to mark and comment on assignments, and perhaps to offer a small amount of face-to-face contact
- self help student groups (via phone or computer or by meeting)
- pacing, in which assignment deadlines play a prominent role
- perhaps a residential school (for extra tutorials and supervised practical work as opposed to unsupervised practical work that can be carried out at home)
- perhaps a counsellor for general guidance and advice.

FIGURE 1. This is NOT distance education!

With a modest level of support, and high quality educational materials, over 70% of students will be successful.

To achieve such a high success rate organised feedback from the students is essential, and the institution must be prepared to improve its materials in the light of the feedback.

2.1. *Advantages and disadvantages of distance education*

Though distance education has the disadvantages of a lower level of support, less contact with teachers and peers, and only a small amount of supervised practical work, it does have several advantages, of which a major one is that it reaches people who are unable or unwilling to travel to a campus. Such people include those with job or family commitments. This advantage is reinforced because distance education facilitates study at a time and at a rate that suits the student. A related advantage is that students can support their education without having to find work close to campus. This also enables students to gain work experience alongside their degree, thus improving their career prospects. Full time work is not essential: the cost to the student can be low if the course population is high, and even lower if expensive media are avoided. Related to this is that the cost per student to the institution can also be low.

A different type of advantage is that a small number of astronomers gifted in teaching can prepare materials that reach a huge number of students. Given the widespread interest in astronomy and the small size of the professional astronomical community, particularly in certain countries, this is another important advantage. Local, part-time tutors need not be astronomers and yet can provide excellent student support if the educational materials are well thought out and the tutors properly briefed.

Distance education is complementary to on-campus education.

3. The world scene in distance education in astronomy

There are at least 837 institutions around the world offering distance education. Of the 837, there are 625 universities or colleges, and the great majority of these operate mainly or wholly at tertiary level. Not all of the 625 list their courses in enough detail to tell if astronomy is included, but of those that do, astronomy features in about 40% of them. Applying this fraction to the 625, we obtain a rough estimate of 250 educational institutions world wide offering astronomy courses at university or college level.

The mean number of students per course can only be an educated guess. For a subject as popular as astronomy my guess is several hundred, which gives an order of 100 000 students studying astronomy world wide at-a-distance at university or college level. This includes students that are taking individual courses rather than a complete degree programme. For those in degree programmes, astronomy is typically a component of the degree, frequently a significant component.

Though the 100 000 figure has an accuracy somewhere between that of Hubble's constant and the amount of missing mass, it is clear that distance education is already a force in astronomy.

Lets have a look at the biggest distance teaching institutions, the so-called "mega-universities".

3.1. *The "mega-universities"in distance education*

A "mega-university"in distance education is defined as one that has at least 100 000 students in distance education *degree* programmes. In most cases there are comparable or even larger numbers of students on diploma or masters programmes or taking individual courses. Moreover mega-universities tend to be national institutions, and as such can have a powerful influence on the shape of a country's higher education.

It so happens that all of the mega-universities are devoted almost entirely to distance education. Table 1 includes all ten of them and lists some of their features (Daniel, 1995). Note that three continents are represented, Europe (four of the ten counting Turkey as Europe) , Africa (one), and Asia (five). These huge universities have economies of scale, and as a result the unit cost (expressed as a percentage of the average cost per student for other universities in the country) can be low. In general, the higher cost mega-universities are those where the level of student support is also high.

At least four of the ten have astronomy courses The Open University (UK), The University of South Africa, CNED (France), and UNED (Spain), the courses of the first three being described in these proceedings. The Indira Gandhi National Open University is to introduce astronomy, though Anadolu University, STOU, and the Korea National Open University has no plans to do so. I have no information yet on the other two.

There are several universities that are just below the mega-criterion, but as distance education is growing and institutions are increasing in size, there is no doubt that the number of mega-universities will at least double by the turn of the century. Let's look more closely at astronomy at of one of the mega-universities - The Open University in the UK.

Table 1. The mega-universities (as at the end of 1995).

INSTITUTION (Alphabetical order)	Degree students	Grad'tes per year	Annual budget/ $10^6$$US	% of budget[1]		Unit cost/%
				stud't fees	gov't grant	
China China TV University System	530 000	100 000	??	5	95	40
France Centre National d'Enseignment à Distance	105 000	??	113	65	35	50
India Indira Gandhi National Open University	242 000	8 000	10	30	68	35
Indoneisa Universitas Terbuka	353 000	3 000	2.5	66	34	15
Korea Korea National Open University	196 000	10 000	48	62	38	10
South Africa University of South Africa	130 000	10 000	128	39	60	50
Spain Universidad Nacional de Educación a distance	110 000	1 500	??	60	40	40
Thailand Sukhotai thamm -athirat Open University	300 000	13 000	32	49	23	30
Turkey Anadolu University	567 000	14 000	15	76	6	10
United Kingdom The Open University	150 000	17 000	300	31	60	50

[1] There can be other sources of income other than the two quoted.

4. Astronomy at The Open University

Figure 2 shows the course materials of the main Open University (OU) astronomy course "S281 Astronomy and planetary science". In addition there are all the features of the student support system outlined in Section 1 with the exception of residential schools. There is one tutor per 20 students and an active computer-based self-help group based on the First Class software (see the posters in these proceedings). We also send out a list of astronomy Web pages and astronomy software. However, none of the course materials is computer based, because when this course was being designed in 1992-1994 we were unwilling to place the financial burden on each student of the purchase of computer hardware for S281: computer activities are thus optional. Home-based computing is now

FIGURE 2. The materials for The Open University's main astronomy course, "S281 Astronomy and planetary science".

appearing in several OU courses that S281 students would also be likely to take, and so the hardware cost is spread over several courses. Therefore, when S281 is revised it is probable that we will build in CD-ROM activities.

S281 is the equivalent of two-thirds of a term of full time study (half a semester), and so represents a substantial piece of a three year degree. Like all Open University courses, in addition to assignments marked by the tutor, there is a three hour examination at the end of the course, taken under the usual controlled conditions.

The course is aimed at the equivalent of first year science, maths, or technology students at a conventional university and is therefore somewhat above the level of many of the "liberal-arts" courses in the USA. However, in being aimed at science, maths or technology students it is of broad appeal and accessibility. It is certainly capable of adaptation to the needs of a wide variety of institutions.

An indication of its accessibility is that it assumes zero previous knowledge of astronomy, it uses simple algebra but no calculus, and it assumes a level of physics no more than that acquired by many 16 year old school students.

Table 2 lists some of the characteristics of the 1200 or so undergraduates who take S281 each year a remarkable number for a science-based course, particularly as there are a further 200 students per year not taking it as part of a degree course. About 70% of the students starting the courses obtain passes. Of the 30% who do not, most of them drop out of the course early on: it is a feature of some OU students that they take on one more course than they can handle and then drop the one that, for a wide variety of reasons, is least suited to their immediate needs.

Table 2. Some characteristics of the 1200 undergraduates per year on the OU astronomy course S281 Astronomy and planetary science.

STUDENT BACKGROUND

(on entry to OU)
less than the minimum entry qualifications to conventional UK universities	27%

(on entry to the course)
previously read astronomy, from occasional articles upwards	91%
regularly take more than a passing glance at the sky	65%
regularly watch science fiction at the cinema/on video/on TV	55%
regularly read science fiction novels	39%
have made quantitative celestial observations	9%
have been active members of amateur astronomical societies	5%

PARTICULAR INTERESTS		AGE & SEX OF STUDENTS	
the origin of the Universe	47%	under 30	20%
observing the night sky	29%	30-34	20%
the Solar System	22%	35-39	18%
extraterrestrial life	20%	40-44	15%
history of astronomy	13%	45-49	11%
UFOs	3%	over 49	16%
astrology	2%	female	26%

STUDENT ASPIRATIONS
to enrich their interest in astronomy	73%
to get a job in astronomy or space science	8%
to introduce astronomy examples into their teaching	7%
to go on to postgraduate study in astronomy	6%
to get promotion in a job in astronomy or space science	< 0.5%

More than 70% of the students starting the course give very positive feedback on it, and the course scores well on various in-house quality ratings.

Among the data in Table 2 note that 6% of S281 students intend to go on to postgraduate study of astronomy: this is about 70 students per year, a considerable fraction of the UK total of postgraduate students in astronomy. S281 was first presented in 1994 so it will not be long before this cohort starts knocking on the door of postgraduate study in astronomy at the OU and elsewhere. It is also remarkable that 8% intend to get a job in astronomy or space science - a substantial change in career in most cases.

Another datum worth highlighting is the 20% of the students under 30. Among these there is an increasing number of students in the age-range 18-21. This is the age at which most students are educated on-campus rather than at-a-distance. The increasing number is testimony to the effect of the decline in the UK of maintenance grants and the worsening job prospects for graduates with little work experience. On the positive side it is testimony to the growing realisation of the effectiveness and advantages of distance education.

The results of a survey of a few aspects of the astronomical knowledge of S281 students is given by Broughton (1996). It is important to note that this survey was made BEFORE the students studied S281.

FIGURE 3. Making observations to determine the difference in length between the sidereal day and the mean solar day.

4.1. *Practical work in astronomy at-a-distance: how much can be achieved?*

I noted in subsection 2.1 that only a small amount of *supervised* practical work is feasible at-a-distance. This is because of the huge expense of supervising students spread all over the country. For supervised practical work a residential school is best but this is a rather expensive course component and therefore only a small amount of student time is spent at such schools. It is therefore fortunate that a considerable amount can be achieved through *unsupervised* practical work.

In S281 about 10% of the student time is spent on unsupervised practical work. All of this is achieved with a planisphere being the only item that we send (in addition to a project book and project sheets). Binoculars or a telescope are not essential for the two key projects. Even though these projects are very specific a considerable range of general practical skills are developed, largely because we insist on quantitative analysis of the key projects and a project write up that is marked and commented upon by the student's tutor.

The first of the two key projects is the measurement of the difference in length between the sidereal day and the mean solar day. The measurement procedure is only outlined by us, the details being left to the student. The procedure is hard to distinguish from loitering against a wall, as Figure 3 shows! The student measures the mean solar time at which a star passes behind some obstruction (such as the chimney in Figure 3) and repeats the procedure on several dates over a period of a few weeks. A graph is plotted, with error bars, of disappearance time versus date, and from the gradient of the best fit straight line the difference in length between the two types of day is obtained, along with the observational uncertainty. Extremely accurate values are obtained by many students. They also learn by experience that the night sky is not the same at a given hour of mean solar time on different dates.

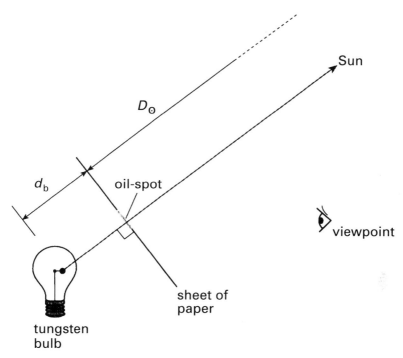

FIGURE 4. Measuring the solar luminosity.

I developed the second of the two key projects from outlines given by others (e.g. Percy 1995). In it, the student improvises what, in essence, is a Bunsen grease spot photometer to measure the luminosity of the Sun. The procedure is outlined in Figure 4. The distance of the electric bulb filament from the oil spot on the sheet of paper is adjusted until the oil spot (illuminated from behind the paper by the bulb) looks to the eye as bright as the surrounding paper (illuminated by the Sun). The luminosity of the Sun can then be calculated from the distances of the Sun and the filament from the paper and the luminosity of the filament. There are many corrections that the student has to estimate to get a more accurate value, the largest coming from the eye being sensitive to a greater fraction of the solar spectrum than the filament spectrum.

Values within a factor of two are obtained (remarkable!) and the students are guided to attach a large uncertainty to the value. The usefulness of this experiment is not so much the value of the solar luminosity obtained, as the general skills of measurement and project write-up that it develops.

There are four other projects, all of which are also carried out successfully in unsupervised mode.

In and around Orion, in which the student, with the aid of a planisphere, familiarises themself with objects in the constellation Orion, and, with home improvised equipment, measures the angular separation of Betelgeuse and Rigel, including an estimate of the measurement uncertainty. Binoculars are needed for some parts of this project.

Limiting visual stellar magnitudes, in which the student estimates the limiting visual magnitude of stars at various zenith angles under various seeing conditions.

Sunspots, and solar rotation, in which the student observes sunspots by projection and uses them to make qualitative observations about solar rotation.

The Moon, in which the student makes binocular observations of the Moon at various phases, comparing and contrasting maria and highlands, and observing crater morphologies and rays.

4.2. *Other aspects of astronomy education at the OU*

There are several other OU courses in which astronomy plays a significant role. The foundation course in science, with over 4000 students per year and equivalent to a whole semester of full time study, describes the Earth's place in the Galaxy and has a major case study, based on popular science articles, on the search for extraterrestrial life. It also has a substantial piece on the large scale structure of the Universe, supported by a "virtual telescope" on CD-ROM (Norton, 1996).

We also have a higher degree programme in astronomy, exclusively by research. There is a conventional aspect to this, with students working full time on campus, but there is also a distance aspect, with students working full or part time but at their normal place of work or at home. In this external programme we have graduates of a variety of universities, not just The Open University.

The external programme in astronomy has proved viable for various reasons. Among them are the following three. First, none-too-modest "amateur" equipment is available to support research on topics such as variable stars, asteroids, and the search for extraterrestrial intelligence at optical wavelengths (OSETI) (Jones, 1995). Second, there are library and computer based data archives, the latter including data from the Hubble Space Telescope, the InfraRed Astronomy Satellite, the International Ultraviolet Explorer, the 100" Hooker Telescope. Third, there are now robotic telescopes with public access (Baruch, 1996).

The possibilities of obtaining a PhD at-a-distance are improving all the time.

5. Expansion of distance education and the use of self study materials on campus

There is growth in the number of institutions largely or wholly devoted to distance education. There is also growth in the size of existing institutions, and it has already been noted that the number of mega-universities will at least double by the turn of the century.

Another important aspect of growth is in the distance education activities of *conventional* institutions. There is enormous potential here, but there are enormous pitfalls too. *Successful* distance education requires

- high quality materials appropriate for the distance learner
- a measure of student support as outlined in Section 2.

The Open University has also seen its materials adopted by other universities in the UK and abroad for self-study *on-campus*. This enables universities to meet the pressures to do more and more with less and less resource. Also, by reducing the need for lecturing it liberates academic staff for more appropriate types of face-to- face teaching such as tutorials and problem classes. The "golden rule" is that you do not leave the students alone with a book: books and other materials must be suitable for self-study, and there must be support in the form of tutorials, problem classes etc. Our experience with self-study on-campus has revealed that

- nearly all on-campus students are comfortable with the use of self-study materials
- self-study is a useful skill for students to develop
- face-to-face contact can be reduced by 20% and yet the quality of education is better

• the flexible study pattern is popular, with its liberation from the lock-step of the lecture course.

The Open University has long provided a consultancy service to other institutions, and this is now being expanded under the name Open University Worldwide. OU Worldwide is coordinating and promoting a variety of existing activity strands, including

• collaborative partnerships with other institutions

• direct teaching of students beyond the UK

• materials sales, including versioned materials and materials specially prepared for other institutions

• provision of a distance education system, a "railroad".

In the particular case of astronomy, considerable benefit would be obtained were the IAU and the OU to act together to promote distance education in astronomy, and also to promote the spread of self study on-campus.

6. Summary

Table 3 is a (whimsical) summary of the main features of distance education.

Table 3. Features of distance education.

Distance education	
• must be <u>O</u>pen	O
• is best on a <u>B</u>ig scale	B
• it is <u>A</u>ffordable	A
• must build on the students' <u>F</u>oundation	F
• it is <u>G</u>rowing	G
• must have courses of high <u>K</u>wality	K
• must make effective use of the various <u>M</u>edia	M
• is part of a growing worldwide <u>R</u>evolution in education	R
• is relatively <u>N</u>ew	N
• must provide <u>S</u>upported self-learning	S

In addition, distance education materials have enormous potential to support self-study on-campus. The IAU and the OU can together help promote distance education and self-study.

REFERENCES

BARUCH, J.E.F., 1996, these proceedings.

BROUGHTON, M.P.V., 1996, these proceedings.

DANIEL, J., 1995, "Proceedings of the IXth annual conference of the Asian Association of Open Universities, *The Government Information Office*, Taiwan, R.O.C., 7-18.

JONES, BARRIE W., 1995, *Astronomy Now*, London, vol. 9 no. 11, 43-45.

NORTON, A.J., 1996, these proceedings.

PERCY, J., 1995, *Proceedings of the fifth international conference on teaching astronomy*, ed Rosa M Ros, Universitat Politecnica de Catalunya, 157-159.

Teaching Astronomy at the University of South Africa

By W.F. Wargau[1], B. Cunow[1] AND C.J.H. Schutte[2]

[1] Department of Mathematics, Applied Mathematics and Astronomy, University of South Africa, P. O. Box 392, Pretoria 0001,
[2] Chief Executive Director: Science, Technology and Informatics, University of South Africa, P. O. Box 392, Pretoria 0001,

1. The development of distance education in South Africa: historical background and the University of South Africa

The University of South Africa celebrates its 50th anniversary this year. Over this period it grew, becoming one of the largest tertiary distance education institutions and the largest university on the African continent.

South Africa always had a mixed racial population with each group having its own culture. This difference between people is further aggravated by differences in the level of "westernisation". Furthermore, South Africa also suffers from an extreme urbanisation problem where on the one hand we find modern cities and on the other tribal groups. All these factors led to a differentiation of the population into a first world and third world component.

In 1858 the government of the Cape Colony decided to institute a board of public examiners in literature and science. The task was to set syllabuses and to set and conduct examinations at college level. In 1864 this board instituted a certificate which was equivalent to the British matriculation certificate. The board only conducted examinations, but offered no training. In 1873 the parliament of the Cape of Good Hope decided to establish the University of the Cape of Good Hope. The University still was an examining body only, which set syllabi, conducted examinations and held graduation ceremonies. Its degrees were recognised by the British Commonwealth.

This institution had to face some very adverse criticism from those who felt that a university can only function in a direct teaching situation, that it was too "foreign" (British) for the country and that it was a mere factory of certificates.

In 1916 the Parliament of the newly created Union of South Africa decided to change the name of the university to "University of South Africa" (Unisa) which started operating in 1918. In the same year the university moved to Somerset House in Pretoria.

Parallel to the University of South Africa as an examining body for colleges etc., a number of residential universities were established over the years. A large number of students who did not want to receive their tuition at colleges or seminaries but were not able to study at residential universities wished to register as external students at Unisa. Finally, the student body grew so large that much discussion started around a concept of distance education at university level. In 1946 the University of South Africa decided to offer distance education. The first director, Prof. A.H.J. van der Walt, laid the ground for the present university. The total student numbers grew from 5500 in 1955, over 40000 in 1975 to 128000 in 1995. The percentages of black and coloured students were 21% in 1955, dropped to 16% in 1975 and reached 51% in 1995. In terms of gender, the present student body is made up of 46% male and 54% female students.

Since 1994 the role of Unisa as a distance education institution has become more important than before in order to provide adequate education for all population groups. The main aim of distance education in the new South Africa is to afford equal education and employment opportunities to all qualified persons within and beyond the borders of South Africa.

2. The astronomy curriculum

Historically the subject of astronomy has been taught in the Department of Mathematics and Applied Mathematics since 1960. During the first decade astronomy was offered at undergraduate level and was seen as a part of the degree in Mathematics or Applied Mathematics. In the 1970s an Honours course was developed which could be selected by the student and constituted part of the Mathematics/Applied Mathematics Honours. In 1986 astronomy formally became a subdepartment. Subsequently, a full curriculum including some practical components at undergraduate and postgraduate levels was developed.

The University of South Africa offers modules, papers and courses which run over a full year. The smallest study unit is the module which is equivalent to a 50 hour work unit. A paper is reckoned as two modules and a course as three or more modules. The Bachelor of Science degree comprises 30 modules, the Honours Bachelor of Science five papers, the Master of Science a selected project and the PhD own research.

As prerequisite for the study of astronomy, mathematics at matriculation level is required. The astronomy curriculum comprises modules at undergraduate levels and papers at Honours level, while on Master and PhD level selected topics including own research can be chosen.

Astronomy can be selected as the only major for the BSc degree or as a double major in the combinations astronomy and mathematics, astronomy and applied mathematics, astronomy and either physics or theoretical physics. For the double major we strongly recommend the combination astronomy and physics.

When Astronomy is selected as the only major, the following options are available:
- 19 compulsory modules
  - 9 astronomy modules
  - 6 mathematics modules
  - 4 physics modules
- 11 remaining modules which can be selected from other BSc subjects, e.g., chemistry, computer science, statistics, geology, geography or biology. At least four modules must be on second-year and four on third-year level.

Astronomy as part of a astronomy/physics double major comprises:
- 27 compulsory modules
  - 9 astronomy modules
  - 6 mathematics modules
  - 12 physics modules
- 3 remaining modules which can be selected in one or more of the other BSc subjects. We strongly recommend that the student enrolls for some chemistry modules.

Students who have registered for a double major can pursue both subjects at Honours level. In our example that would be physics and astronomy.

At Honours level we offer six papers of which the student must at least enrol for three while the other two can be chosen from physics, mathematics, applied mathematics, chemistry or computer science. Of the six astronomy papers, one is compulsory, namely "Stellar Structure and Evolution".

The tutorial material comprises a Study Guide and up to three Tutorial Letters which the student receives on enrolement for a module/paper. The Tutorial Letter contains general hints, assignments, guidelines for examination admission and procedures, recommendation for further reading; information on textbooks, and information about a video tape containing some comments on the curriculum, our facilities and where the lecturers introduce themselves. The video tape can be loaned from the Unisa Library.

Some Study Guides are wrap-around guides, while others are self-contained. A Study Guide contains all the material, such as explanations, tables and appendices which the student may need to solve the assignments. Contact with students is currently established via mail and telephone. Discussion classes are not offered on a regular basis because the number of students is too small. Lecturers invite students to the main campus in Pretoria: local students make regularl use of this, while distant students come only occasionally. For local students evening classes are arranged on an ad hoc basis. They usually cover selected topics from the undergraduate curriculum, but are not meant as regular lectures.

In the following we would like to look at the modules and papers more in detail. On first-year level we offer two modules. One is a general introduction and the other deals with spherical astronomy and Kepler orbits. For the introductory module the student numbers vary between 70 and 100, while for the second one between 30 and 50. The pass rate versus admission to examination is around 75%. On second-year level, we offer three modules. The first one contains basic instrumentation including optics, telescopes, image formation, spectroscopy and photometry; the second one observational astrophysics including magnitude system, spectral and luminosity classification of stars, interstellar medium, our Galaxy and extragalactic objects; the third module is the introductory practical. Student numbers on this level vary between five and ten students per module. On third-year level we offer four modules. The following topics are covered: astrophysical principles including atomic structure, electromagnetic radiation and gas laws, ionization and excitation processes and black body radiation; radiation transport including line formation, analysis of stellar spectra and stellar structure equations; astrodynamical principles including stability of stellar systems, stellar motions and galaxy dynamics. In addition, we offer an advanced astronomy practical. The student numbers vary between two and five students per module.

At Honours level we cover the following topics. A paper on stellar structure and evolution which is compulsory; papers on interstellar medium, radio astronomy, observational techniques and cosmodynamics. In addition we offer a project paper where the student can choose between a theoretical and practical oriented project. "Theoretical" in this context means that the student reviews recent research on a certain topic, while "practical oriented" involves observations at South Africa's national observatory in Sutherland, at the radio astronomy observatory in Hartebeesthoek, or at the Unisa observatory. The aim is to familiarise the student with observing and reduction procedures. Student numbers are typically three per year.

3. The Unisa Observatory

Astronomy is a practical science. It is unimaginable that the entire curriculum is based only on the printed word. Although the situation has improved in the past years due to the advent of electronic media, there is no real substitute for face-to-face contact for practicals. In order to cater for practicals Unisa had set up obligatory practical courses for physics, chemistry, geology and the life sciences in the seventies.

Originally, the tuition for astronomy was entirely based on the printed word, and contact to students was only via telephone, or by occasional visits. As from 1986 the need for face-to-face contact became more and more pressing with the foundation of the Subdepartment of Astronomy and with the introduction of a fully-fledged curriculum including undergraduate and postgraduate levels. The idea of building a small but modern observatory which could be used for training students was conceived in 1987. Although the idea was very challenging, it would be very expensive. On the other hand, such a project was unique to South Africa and for the African continent. The University officials were persuaded to grant the money for building the observatory. On 17 August 1992 the Unisa Observatory was inaugurated by the then Principal, Prof. van Vuuren. In 1993 a practical module on introductory level was launched, which was followed by an advanced practical module in 1994.

The technical equipment of the observatory comprises the following:

- A Compustar C14 computer-controlled 14-inch Schmidt-Cassegrain telescope.
- A fully motorized dome
- An automated PMT-based UBV photometer filters.
- A Daystar 0.6 Angstrom ATM filter and energy-rejection cover.
- Spath-crystal micrometer.
- A two (600, 1200 grooves mm−1) grating spectroscope with reducer, reference source and photographic recording.
- A CCD imaging camera with UBVRI and clear filters.
- Audio-video and photographic equipment including reproduction facilities.
- A 586 Pentium 100 MHz and a 486 DX 33 MHz PC.
- An excellent slide and video collection.
- Complete set of filmcopies of ESO/SERC J and R surveys.

The observatory premises include a library, a lecture room, a kitchen, a restroom and sanitary facilities. In other words, the observatory is self-contained.

The introductory practical, which is part of the second year course, includes thorough training at the telescope, planning and preparation of astronomical observations, astrophotography and solar observations. The advanced practical is part of third year and gives a thorough training in professional observing techniques such as UBV photometry, spectroscopy and UBVRI CCD photometry. Both are compulsory and are scheduled for the South African winter month July, where cloudless conditions are fairly predictable for Pretoria. Each practical runs over two weeks. We offer no practical on first year level because the student numbers are relatively high and an adequate theoretical foundation is needed before a student can benefit from a concentrated practical session. The practicals render face-to-face contact possible for students − which is so important for the practical courses of a distance teaching university.

At Honours level, we offer a project paper which includes work at the observatory. Besides doing a literature study the student can prepare and carry out observations at the telescope. Presently, students are mainly involved in testing the equipment. At Master level the student may choose a dissertation which is linked entirely to the observatory. One of our Master students, for example, wrote a dissertation on the evaluation of the Unisa photometric system. Alternatively the student can do his/her project paper as well as his/her Master dissertation at the South African Astronomical Observatory in Cape Town/Sutherland or at the radio astronomy facilities at Hartebeesthoek. For overseas students special arrangements are made with local universities.

Allocation of observing time at international observatories becomes more and more difficult for graduates. Although they do not lack research ideas, they do not have the required know-how to handle sophisticated equipment. Observatory committees which schedule observing time are reluctant to allocate precious observing time to newcomers. This deficit can be partly ironed out by giving the students sufficient training at a modern but scaled-down observatory. The Unisa observatory is ideal for this kind of training.

4. Future developments

4.1. *Collaboration*

A very recent project includes combining the study of Physics and Astronomy at Unisa (distance teaching) with that of the University of Pretoria (UP; residential). The idea is the following: first year astronomy is offered at UP; from the second year onwards the student may carry on with astronomy at Unisa, while the physics subjects are taught at UP. The student obtains a BSc degree in Physics from UP. It is planned to implement this scheme in 1997.

4.2. *Students on Line*

The communication system which Unisa has developed for the two-way flow of information between the university and the student can be described as dependent upon the printed word. The main carrier of information is the postal system. Over the past decades this has been enriched with modern technology, such as telephone, audio tapes and student centres at locations outside Pretoria. More recently computer-aided instruction and PictureTel interactive video classes were introduced.

As more and more students have access to electronic media, we thought of a more effective way for carrying out Honours and Masters projects in astronomy. At the observatory the student plans the observing run and obtains the data, which will be transferred to a file server and the student can retrieve them via the Internet. The data reduction and analysis can then be performed at home according to instructions supplied by the Lecturers.

In the light of Unisa's total student numbers the Department of Production yearly prints about 500 million pages of study material, the Department of Dispatch yearly handles millions of postal articles and the Department of Assignments annually handles millions of assignments. This was made feasible by implementing one of the best computer management systems in the world which was developed by Unisa. In recent years student numbers and therefore costs increased tremendously so that Unisa is now facing big problems. If more and better study material is to be printed, several million Rand are needed immediately. The handling of assignments which mainly depends on machines and human hands is also getting more and more expensive. A third problem is the slow speed of mail delivery in South Africa which causes severe delays in receiving and sending assignments.

A powerful solution to these problems is the use of the World Wide Web (Internet) service which is fairly easy accessible for those who have access to a PC. We can make excellent use of this source in the following way:

- communication between students and lecturers through e-mail
- the students can download study material and even parts of prescribed and recommended books
- students can submit assignments which are electronically registered and marked by lecturers without ever printing them out; they are then electronically returned to the students
- students can search through the catalogues of the Unisa library and order books and reprints electronically
- students can "talk" to other students through an electronic forum
- students can enrol at the University without having to write or to travel to a Unisa office

In order to achieve these aims, an electronic server at Unisa, called SOL (Students On Line) has been set up. Unisa is now in the process of implementing the server step by step. At the end of the day it will save the University an enormous amount of paper work. The idea is to develop new teaching methods and to redesign our study material for the new medium. It will take some time for the students and lecturers to get acquainted with the new system. But in the year 2000 and beyond the University will probably operate in a different way from that of today. There will always be made provision for students who prefer the paper model and/or who do not have the necessary computer access.

A multi-resource system for remote teaching in Astronomy : its aims, its design, the point of view of the learners

By Michele GERBALDI[1] AND Annie XERRI[2]

[1] Université de Paris Sud - XI
Institut d'Astrophysique - CNRS
98 bis, Boulevard Arago, 75014 PARIS - FRANCE
gerbaldi @ iap.fr
[2] Centre National d'Enseignement à Distance Institut de Vanves
60, Boulevard du Lycée - 92171 VANVES CEDEX - FRANCE
Annie.Xerri @ cned.fr

1. INTRODUCTION

A distance teaching course in Astronomy was developed three years ago by the CNED (Centre National d'Enseignement Distance) in collaboration with professional astronomers from the University of Paris Sud XI.

We wish to present our course with :

- the conceivers and designers' point of view
- the learners' point of view.

2. Creation of the course.

2.1. *Centre National d'Enseignement à Distance (CNED).*

The CNED was created in 1939. It is a public administration under the supervision of the French Ministry of Education. Its first founding mission is to provide teaching and training to those who cannot take courses under usual conditions. But the CNED now operates at all the levels of the educational system from primary up to higher education, in all fields of training, initial, vocational and continuing education.

In 1995-1996, 360 000 students were registered in 2 500 training modules.

Among them, 80% are adults, 190 000 on post baccalaureat level programmes (27 000 registered students reside outside France, in 176 countries).

2.2. *A partnership between CNED and Paris XI University.*

As no such course existed for astronomy, its creation was timely. So, as we did for meteorology in 1990, the CNED which does not deliver diplomas, offered and set up a partnership through an agreement with the University of Paris XI.

We worked with a team of Professors from that university, professional astronomers who are also well-known for working in collaboration with primary and secondary school teachers (CLEA†).Together we decided, conceived and designed a remote teaching course with a multi-resource system.

2.3. *Who is the course designed for ?*

This course has been developed for a large audience, non-specialist, but highly motivated.

Among journalists, Astronomy is one of the most popular sciences ; nearly every week, some "hot news"in Astronomy is offered to the public on T.V. or through the newspaper channels.

† Comité de Liaison Enseignants Astronomes: a non-profit association created in 1978

There are about 27 800 amateur astronomers (1/2000 of the total population) practising their hobby in more than 425 associations or clubs ; there exist also 100 scientific associations run by the municipalities. To augment their teaching at school, the pupils come to visit those places where they can have informal scientific – and astronomical – activities under the guidance of amateur astronomers or staff members. So, astronomy is widely taught outside the schools, in clubs or associations. The amateur astronomers, the staff members, feel the need for some kind of training in astrophysics : usually they have a wide knowledge in the domain of practical observation, but they need to structure their knowledge as well as to receive some kind of training on basic astrophysical concepts.

Astronomy is present throughout the French educational system : the primary school system (ages 6 to 10), the junior high school (ages 11 to 14) and the senior high school (ages 15 to 18). Very few teachers had courses in Astronomy during their university studies. Most of the teachers in the primary schools graduated with degrees either in literature or social science ; very few graduated with degrees in science. So, there is a wide need for teacher training in the domain of Astronomy. To conclude, participants in such a course are expected to have academic qualifications.

2.4. *Which scientific content?*

All the texts of this remote teaching course have been produced with the aim of providing a basic knowledge in Astronomy and Astrophysics. This course offers the opportunity to get an academic credit, *"diplôme d'université"*, at an undergraduate level for those who want it. Astronomy is, not only one of the oldest sciences, it is also a very modern and active field of research, whose discoveries are widely presented to the public by the media. The progress of this science is intimately connected to technological advances, as well as to new developments in physics.

This course (150 hours) has been developed, taking into account the following aspects:

– it refers to history,

– it develops the role of observation,

– it emphasizes the importance of the physical laws, mainly the gravitational law and the radiation laws,

– it presents the phenomena which are the agents of the evolution of the stars and of the Universe itself, showing the strong interaction between observation and theory.

In some cases the historical approach is used, partly because it poses the problem in its original setting and partly because it shows the iterative nature of the scientific reasoning. As far as possible, this course is based on observation: this can be done directly, in some cases by the students themselves (phases of the Moon, sunspots...) or, it is shown on several types of documents.

We would like to insist on the fact that the documents presented in this course have not been selected for their aesthetics but only for the phenomena that they are representing. One of the main objectives of this course is to demonstrate, through astronomical examples, the reasoning methods used in science. In each chapter, the manner in which astronomical phenomena are interpreted in terms of physical laws will be developed:

– this is what constitutes the foundation of astronomical knowledge.

– this goal is not easy to achieve, due to the fact that the participants to such a course have hetereogenous backgrounds.

This course is based on the experience gained during the past twenty years, through the activities developed by the CLEA.

2.5. *Which teaching materials are provided?*

2.5.1. *Printed texts*

3 volumes have been edited instead of only one thick one. They are divided into 14 chapters ; an index, a glossary and a bibliography are included. An appendix will contain a summary of the laws of mathematics used frequently in this course as well as the main results concerning the motion of two bodies.

The printed text contains more information than that is strictly required for the examination. The motivations and the background of the students are heterogeneous, so we have included in the text, typed in small letters, some complementary information which can be read usefully only by part of the students. For example the amateur astronomers will be interested in computational details on Bouguer law, while the physics professors will read complementary computations on spectroscopic binary stars.

A leaflet of exercises with homework experiments is provided, the exercises have detailed answers to the questions. Each exercise refers to a specific chapter. Leaflets accompanying the other media are also provided and, of course, a guide for the whole course, explaining and counselling how to learn with the different teaching materials and services offered. 41 black and white photographs are printed on separate sheets; several of them are used to do some self-correcting practical work.

2.5.2. *Slides*

A set of 60 coloured slides illustrate also the various chapters or represent the result of homework experiment; a replica of a grating is also supplied for use in a homework experiment. Students are urged to make observations by themselves : these observations are helped in several cases by slides or printed pictures.

2.5.3. *Videocassettes*

Three videocassettes are included in the package sent to each student. One is 50 minutes long and is about the solar system which is described from various points of view : comparative structure of the planets, chemical composition... This video is divided into short sequences, each of them on a different topic, so that they can be looked at separately. The two other videos are 20 minutes long each; they are on the Motion of the Earth and Eclipse and Phase phenomena. These videocassettes are also divided into short sequences, all of them having motion-picture cartoons, sometimes in 3D to illustrate the phenomena. Each sequence can be looked at separately and without the sound. A leaflet is given with each of these twenty minute videos and for the one about Eclipses, practical exercices are suggested using the video images.

2.5.4. *Software*

Several programmes for a PC are loaded on to a micro floppydisk to illustrate in an interactive manner, planetary motions, Doppler effect, some of the radiation laws.

2.6. *Which services are offered to the students?*

Learning alone is difficult so the various services described below have been developed in such a way that the students can test their progress.

2.6.1. *Assignments*

In order to test progression in understanding, questions are distributed through each chapter, the answers being given at the end of the chapter. Moreover, 3 long homework assignments are offered. They are sent to a professional astronomer who corrects them individually and sends them back with a detailed critique of the homework, any corrections needed will be completed and sent back to the students within 3 weeks.

2.6.2. *Meeting in observatories*

Twice a year, a full day meeting in a professional observatory is offered to the participants. During each of these days, several activities are offered, not only conferences but also practical activities in some places. All the professional observatories take part.

2.6.3. *Planetarium sessions*

Two planetaria participate: one located in Paris at the Science Museum: "Palais de la Découverte"and one in Brittany (Pleumeur Bodou). Each of them offers two different shows to the participants; these shows have been specially conceived for this course. These programmes have been developed in order to make phenomena clear which are difficult to understand with drawings in 2 dimensions only.

2.6.4. *Telematic service (E-mail with Minitel)*

A direct link with a tutor can be obtained at any time with the French E-mail system: Minitel. To use Minitel, only a telephone line is necessary. The tutor is one of the professional astronomers who corrects the homework exercises. Three services are proposed on-line with Minitel. A series of multiple choice questions on each chapter is offered, the correct answer is given on-line as well as a score. Naked eye observations are proposed for the beginner with a special emphasis on planetary observations. These ephemerides are updated every 3 months. The main role of Minitel is to establish a direct link between the student and a tutor. What is important is that the tutor is always the same person, so through the questions the tutor can acquire an overall feeling for the difficulties of the students.

This year the Minitel link can be used in a new interactive way : there is a general file where all the questions are stored, so every participant can read all of them as well as the answers of the tutor who can also give general advice on difficulties etc... (we call it – the forum). The participants can also communicate among themselves very easily with Minitel : for example some of them asked about people in their geographical area who were learning with this remote teaching course in order to be able to communicate with them directly.

2.7. *Calendar of the course.*

Students can study at their own pace and progress at their own rate (characteristic of the distance learning population). However, a written guide as well as a calendar of the course are given for help (annex 1).

3. Registrations and exam results.

The course is 3 years old. 1 100 students were enrolled in it and 60% of them wanted to register at the university. To enroll in this course, no specific background is required, except for those who want to be registered at the University. For the latter, it is required that they obtained their diploma at the end of their high school studies (the French baccalaurat): if not, their registration at the University can be possible if they have worked for 5 years: this is a general rule of the Ministry of Labour. Of the latter group, nearly 70% sat for the exam and of these 80%, depending on the sessions, passed the exam.

Student Performance

	1993-1994	1994-1995	1995-1996
Students enrolled	644	265	209
Students registered at the University	378 (59%)	162 (61%)	121 (58%)
Students who sat for the exam	251 (66% of the above)	111 (68% of the above)	84 (69% of the above)
Students who passed the exam	208 (83% of the above)	89 (80% of the above)	69 (82% of the above)

4. The learners' point of view.

The feedback from the participants is obtained, each year, through a dense questionnaire covering the scientific content but also their feelings about the multi-resource system available.

4.1. *Which types of information have we looked for?*

We wanted to know who our students really are (gender, age, academic qualifications, occupation, place of residence, motivation...), what are their working conditions and access to equipment; also what are their reasons for taking the course. We wanted to know as well what the students think of the course : overall assessment, assessment of each teaching aid and each service, main criticism and main assets, interest in a follow up course etc...

4.2. *How many students answered the questionnaire?*

58% answered after the first year of the course, 47% after the second year and 34% (up to now) for the third year. We were interested to note that, as we asked for it, some of them answered even when they had to stop learning during the year. Most of them did not remain anonymous - which was a possibility offered - and many of the students wrote numerous and constructive remarks and comments.

4.3. *Who are the students?*

The table shows the great variety among the learners, in every domain: variety of occupations, location, ages....

	1993-1994	1994-1995	1995-1996
Men	74%	73%	83%
Women	26%	27%	17%
Teachers	25%	27%	17%
Staff members of science centers	6%	9%	7%
Engineers	13%	9%	11%
Technicians	12%	6%	16%
Medical professions	6%	9%	6%
Others	38%	40%	43%
LOCATION			
Paris and greater Paris	24%	23%	27%
Provinces	71%	73%	70%
Foreign Countries	2%	4%	3%
Overseas Territories	3%	3%	2%
AGES :			
> 40 years	22%	16%	
< 40 years	53%	61%	
< 30 years	23%	25%	

4.4. *What is the overall assessment of the students?*

Even with this huge variety of participants in term of:

– motivations

– ages

– professional origin

– their points of views are convergent i.e. they found the content that they expected in this course and they highlighted the pedagogical process. In fact, this purpose is met because of the large variety of resources used for this course... Nevertheless, everybody agrees that the written text is essential: it is still the corner stone of such a distance learning course.

4.5. *What do they point out?*

Among this rich feedback, some points stand out.

4.5.1.   Written text, exercises and assignments are essential for all of them.

4.5.2.   To the question *"What is your main criticism?"* most of the answers, at the end of the first year, concerned errors in the text (*"errors of youth"* said some of them!), and also insufficiency in the explanations for self- correction.

4.5.3.   *"What do you most appreciate among the different teaching materials and services?"*.   Each media or service is selected several times, whatever it is and students often justify their choice. These answers confirm that a large variety of resources is useful.

3.5.4.   The Minitel service.

The equipment is available to 85% of the students but is used by 50% at most, depending on the rubrics offered. The non users give all kinds of reasons (*"no need; no question; no reflex to use it; no Minitel available when I need one; the cost; psychological barrier..."*).

The users are satisfied especially by E-mail and forum and explain why they are. *"clear and quick answers; reliable tutoring; availability; user friendliness; feeling of belonging to a group; making it easier to situate oneself in relation to the others; feeling of security; access to the data base of questions/answers of all..."*.

We must point out that despite an uptake rate of 50% at most for Minitel users, there is spin off for every student. Indeed, during the first year errata and addenda – quickly identified owing to e-mail questions from, and remarks by, students – were posted to everyone – and for the second year and subsequent sessions, teaching aids were improved thanks to the students feed back.

5. CONCLUSION

According to the opinions expressed by the participants, the variety of media and services used in this distance learning course fulfill their objectives because of the heterogeneity of the students, as we mentioned before, but also because of the variety of learning methods and approaches to the subject, characteristic, as we know, of each individual.

In spite of the limited use of Minitel by the students (as noted before) and the real limits of this service (difficult connexion from abroad, its cost from the French overseas territories...), we are convinced - and it was confirmed by our students – that it is a precious aid in distance teaching for tutoring as well as improving the quality of the courses.

As for the limits of the service such as it is now, they should be overcome in the near future with different networking using PCs in association with the Minitel system.

One of the major results of this distance learning course in Astronomy is that it has greatly contributed to structure the knowledge of the students, knowledge previously acquired either from long term amateur observations or from personal reading.

Following the wish expressed by numerous participants over the past three years, a second- year course based on the knowledge so far acquired is being conceived.

The following points will be included:

– scientific understanding based on the knowledge acquired in the preceding course.

– using and deepening the knowledge acquired in a perspective to interpret astronomical phenomena with the exploitation of documents, the construction of models, a critical reading of historical and contemporary texts (newspaper articles).

We are happy to share experiences of distance teaching in Astronomy for that kind of public with you. Can we imagine a European at-distance course and certification – created together?

Use of the World Wide Web in astronomy teaching

By Jay M. Pasachoff

Williams College-Hopkins Observatory Williamstown, Massachusetts 01267, USA

I discuss the burgeoning World Wide Web and how it can be used to aid astronomy teaching. I supply a list of a variety of useful Web sites.

The World Wide Web was invented 5 years ago at CERN, which is now translated as the European Laboratory for Particle Physics, as a way of aiding access to information from remote sites. The invention of graphic interfaces, notably Mosaic by a group at the National Supercomputer Center in Illinois and then Netscape Navigator as a private development by many of the original Mosaic people, led to an explosion in use of the Web. Millions of people around the world are now able to access information from over 100,000 Web sites.

There is much astronomical information on the Web, though that information make up only a small fraction of all the information available through this medium. The astronomical information is of many varied types, from images of observations to tables of data to lesson plans to journal articles. The question for us to address here is how best to make use of this information for astronomy teaching. Even with the increased resources available at our desktops, the individual teacher remains an important part of the educational enterprise.

One set of alternatives deals with whom the Web information is aimed at. To present new Web data in class, it is useful to have a means of projecting computer information on a screen, which is most often done with an LCD projector panel. An alternative is to use a color printer to make a transparency, a process that can now be done with a relatively inexpensive and portable setup. Another possible alternative is for the students to have their own direct access to the Web, directing their attention to sites to access from their own rooms or from some central locations.

The Web is most useful when you have access from a high-throughput source, such as the type of access known as a T1 line. Universities will often have such lines, which allow a Web page to appear on your screen in seconds. An intermediate quality line known as ISDN is sometimes available in the US, but is not in much use in the education market. Other people have to use ordinary modems, and even at the current maximum speed of 28,800 bps it can take minutes to download individual Web pages, which makes the job of "surfing the Web"tedious. I think that only when everyone has access to the Web through T1 lines or equivalent will the system be truly widely useful. In the next years, access over cable TV lines may provide such universal high-speed access.

Of course, not all countries have good Internet access as of now. But as such access spreads, the Web will be a mechanism to bring in much information useful for astronomy teaching.

Individual Web sites begin with "homepages,"the top level page of information that shows when someone signs onto your site. These homepages contain both text and graphics; new versions of Netscape Navigator allow motion of features on the pages, which may become unsuitably jazzy as a result. Still newer capabilities, such as those provided by Sun Microsystem's Java, allow control information to be sent out with the standard Web information, allowing still further customization. Web sites allow not only still images but also movies and sound to be downloaded.

A typical Web address, accessible through "Netscape,"might be:

http://www.yoursite.youruniversity.edu/teaching.html.

The address can be parsed as follows, with the "dots" as separators and read as "dot":

- "http://" stands for "hypertext transfer protocol,"the format by which data are transferred; all Web addresses begin with it, though it is not necessary always to type it.
- www is a common beginning for Web addresses.
- yoursite is the "server"computer that is holding the Web information.
- youruniversity is the computer your University uses for access to the Internet

- "edu"is the suffix for U.S. educational sites; other U.S. suffixes are "gov"for government, "org"for organizations (like the American Astronomical Society), and "com"for commercial; sites in other countries end with two letter codes for those countries, such as "fr"for France.
- "/teaching"means that on the computer is a set of files known as "teaching"; you can have many documents within those files.
- ".html"stands for "hypertext markup language,"the computer language used.

I maintain a site at

http://www.astro.williams.edu/jay

at which, in coordination with my textbook Astronomy: From the Earth to the Universe, I maintain updates to a variety of astronomical topics as well as links to other Web sites around the world. Merely clicking on such "hotlinks"sends the computer to the other site. Netscape includes "forward"and "back"buttons to click to allow you quickly to see sites you have looked at recently. You can also enter a list of "bookmarks"that list sites by name rather than by the address; such bookmark lists quickly grow too long for easy use and then can be categorized by subtopics.

Transoceanic downloading of Web pages can take a lot of time, and increasing Web use has slowed transfer times worldwide. Some sites have "mirror sites" set up that will be closer to many users. The popular "The Nine Planets"site, for example, has a dozen mirror sites around the globe at which all the information is downloaded periodically.

A problem for Web use is that sites are sometimes "down"or else inaccessible because of delays or other problems on Internet lines. Thus it is not recommended that you require the Web for a class lecture. But you can download the information before class and store it on the hard disk of your computer for later replay.

Delays in the gratification of receiving Web information also mean that one cannot play the type of video games that are so popular. These delays leave an important niche for CD-ROMs (to be updated in capability in late 1996 by DVDs: digital video disks, with much more capacity). For example, an astronomy CD-ROM like Maris Multimedia's *RedShift* provides the capability of calculating the appearance of the sky from the Earth's surface or from positions elsewhere in the solar system (even close to another planet) for any time or date, as well as many hundreds of still images and a dozen short movies. I have worked with Maris on a further CD-ROM entitled *Solar System Explorer*, with still more capability for information about and images of our solar system. If the role of the Web is limited to providing a limited amount of supplemental information to such CD- ROMs, then there will be many fewer delays than by using the Web exclusively.

Though the works of Shakespeare and other authors out of copyright appear on the Web, contemporary works are less likely to be there because of the need of authors and publishers to be remunerated for their efforts. Thus for the moment, textbooks are not appearing on the Web, though supplemental information is. The whole Web is in a state of flux, and it is not yet possible to know how it will evolve. Some publishers like Time, Inc., with its magazines, and newspapers like *The New York Times* or *The Times (London)* are now available free on the Web, though they are likely to try to attract paying subscribers and cancel free access. Advertising appears on many homepages, but has not (yet?) proved to provide enough income to provide the large expenses of establishing and maintaining elaborate sites. Some attempts are under way to allow billing of small increments of Web access, such as the Clickshare process developed by a company in my hometown of Williamstown. We scientists and educators are used to free access to information on the Internet, but whether that no-cost access will continue is not now known. In any case, the Web provides powerful access to information around the world. Table 1 lists a number of Websites of astronomical interest; they are also available through hotlinks on my own site. Many of the sites also have hotlinks to the same sites, so what results is truly–what else–a web. The Figures show samples of interesting homepages.

At Williams College, we are proud to be the home of an educational tradition: it was put 150 years ago that the ideal education is Mark Hopkins (then our President) on one end of a log and a student at the other end. In some cases, modern technology like the World Wide Web can help teachers maintain one-on-one education though with larger classes and with access to the world outside and the Universe beyond. It is up to us as classroom teachers and as developers of laboratories, lesson plans, and other educational materials to guide astronomy on the World Wide Web and to guide students in how best to take advantage of it.

Table 1. World Wide Web Sites for Astronomy

A.A.O. IMAGES   http://www.aao.gov.au/images.html

AAVSO   http://www.aavso.org/index.html

ADS Astronomy and Astrophysics Abstract Service
        http://adsabs.harvard. edu/abstract_service.html

American Astronomical Society (and access to Astrophysical Journal)   http://www.aas.org/

The American Physical Society   http://aps.org/

The Astronomical Society of the Pacific   http://www.physics.sfsu.edu/asp/asp.html

AstroWeb: list of astronomy departments   http://cdsweb.u-strasbg.fr/astroweb/dept.html

Astronomical World Wide Web Resources   http://stsci.edu/astroweb/net-www.html

Big Bear Solar Observatory (daily solar images)   http://sundog.caltech.edu/

Chaisson Proto   http://tthep2.phys.ttu.edu/dka100/alansill/chaisson/index.html

The Compton Observatory Science Support Center   http://antwrp.gsfc.nasa.gov/

CTI Centre for Geography Home Page   http://www.geog.le.ac.uk/cti/index.html

European Space Agency   http://www.esrin.esa.it/

FUSE Home Page (future spacecraft for the uv)   http://fuse.pha.jhu.edu/

Galileo countdown at Jupiter   http://www.jpl.nasa.gov/galileo/countdown/

Online from Jupiter - Galileo quest   http://quest.arc.nasa.gov/jupiter.html

Gemini 8-m Telescopes Project   http://www.gemini.edu/

General Astronomy Information   http://www.ast.cam.ac.uk/RGO/leaflets/

Global Oscillation Network Group - GONG   http://helios.tuc.noao.edu/gonghome.html

History of Astronomy: General, Historians, Archaeoastronomy, Links
        http://aibn55.astro.unibonn.de:8000/~pbrosche/astoria.html

History of Astronomy: Mesopotamian astronomy
        http://ccwf.cc.utexas.edu/~hope/aneastro.html

Hubble Space Telescope (see also STScI, below)   http://www.stsci.edu

Hyakutake (C/1996 B2) Sky & Tel   http://www.skypub.com/comets/hyaku3.html# top

IAU: Central Bureau for Astronomical Telegrams
        http://cfa-www.harvard.edu/cfa/ps/cbat.html

IAU (International Astronomical Union)   http://www.lsw.uni-heidelberg.de/iau.html

ISO (ESA's Infrared Space Observatory)   http://isowww.estec.esa.nl/

Jupiter/Comet Collision FAQ [Frequently Asked Questions] - Post-Impact
        http://www.isc.tamu.edu/~astro/sl9/cometfaq2.html

Kronk - Comets and Meteor Showers   http://medicine.wustl.edu/~kronkg/index.html

Links from Scott   http://www.keele.ac.uk/depts/po/scott/003.htm

Links from Astronomy Magazine   http://www.kalmbach.com/astro/HotLinks/HotLinks.html

MIT X-Ray Timing Explorer Project   http://space.mit.edu/XTE/XTE.html

NASA K-12 : Live from the Hubble Space Telescope
   http://quest.arc.nasa.gov/livefrom /hst.html
NASA Jet Propulsion Laboratory (including Galileo images)   http://www.jpl.nasa.gov/
The Nine Planets [many mirror sites exist]
   http://seds.lpl.arizona.edu/nineplanets /nineplanets/nineplanets.html
NSSDC home page [lots of data and images]   http://nssdc.gsfc.nasa.gov/
Pasachoff's Astronomy Text Updates and Links   http://www.astro.williams.edu/jay
Royal Astronomical Society   http://www.star.ucl.ac.uk/~jl/mypage.html
SDAC - NASA Solar Home Page   http://umbra.gsfc.nasa.gov/sdac.html
SkyView home page   http://skyview.gsfc.nasa.gov/skyview.html
SoHO - The Solar and Heliospheric Observatory   http://sohowww.nascom.nasa.gov/
Solar Eclipse Images   http://umbra.nascom.nasa.gov/eclipse/images/eclipse_images.html
Solar Image Index   http://www.sel.noaa.gov/images/
SolarNews - Index   http://helios.tuc.noao.edu/SolarNews/index.html
Space Physics Homepage   http://umbra.nascom.nasa.gov/spd/spd.html
STScI Press Releases   http://www.stsci.edu/pubinfo/PR.html
STScI/HST Public Information   http://www.stsci.edu/public.html
Stardust NASA Comet Mission   http://pdcsrva.jpl.nasa.gov/stardust/home.html
StarWorlds - Astronomy and Related Organizations
   http://cdsweb.u-strasbg.fr/~heck/sfworlds.htm
SXT Home page-Lockheed/Yohkoh (X-ray Solar Images)
   http://pore1.space.lockheed.com:80/SXT/
Sun: Granulation movies   http://www.erim.org/algs/PD/pd_home.html
Sunspot number SIDC   http://www.oma.be/KSB-ORB/SIDC/index.html
Sunspot/butterfly graph: Marshall Space Flight Center/Solar Physics Branch
   http://wwwssl.msfc.nasa.gov/ssl/pad/solar/
Transneptunian Object List   http://cfa-www.harvard.edu/cfa/ps/lists/TNOs.html
Ulysses/ESA   http://www.esoc.esa.de/external/mso/ulysses.html
Ulysses/JPL   http://ulysses.jpl.nasa.gov/
USGS's list of astronomy links   http://info.er.usgs.gov/network/science/ astronomy/index.html
USGS/JPL Planetary [access to prints and slides of planetary images]
   http://acheron.jpl.nasa.gov/PIA/PIA/html
Yahoo - Science: Astronomy [a Web Crawler/search engine]
   http://beta.yahoo.com/Science/Astronomy/

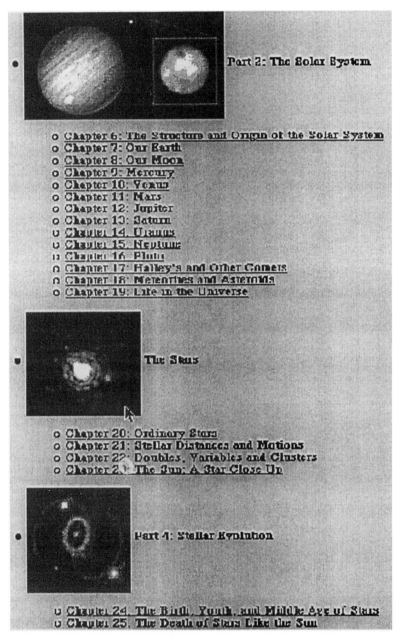

FIGURE 1.   Part of Pasachoff On-Line, http://www.astro.williams.edu/jay.   Updates and hotlinks. As always, underlined text represents hotlinks; clicking on an underlined word takes you to the other site.

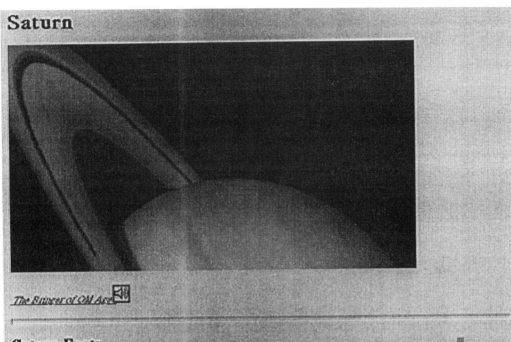

FIGURE 2. The Saturn page from "The Nine Planets":
http://seds.lpl.arizona.edu/nineplanets/nineplanets/nineplanets.html.

On-Line Resources for Classroom Use: data and science results from NASA's Hubble Space Telescope and other missions

By C.A. Christian

Office of Public Outreach, Hubble Space Telescope Science Institute, 3700 San Martin Drive,
Baltimore, MD 21218, USA
carolc@stsci.edu, http://www.stsci.edu/pubinfo/TESTBED/

The data, scientific results, and expertise from NASA's Hubble Space Telescope (HST) and other NASA Missions are being integrated into programs that support innovative and experimental methods to improve content in science and math education. Partnerships with science museums, teachers, other educators, community colleges, universities and other key organizations integrate unique and cutting edge science data and the associated satellite technology into resources which have the potential to enhance science, math and technical learning. The inspiring nature of astronomical data and the technology associated with the HST and other missions can be used by teachers to engage students in many inventive activities. The resources created through collaborative teaming will be discussed, as well as the process for creating partnerships to benefit the education community. Many NASA supported programs encourage electronic access and distribution of multi-media interactive activities and curriculum support materials distributed across the Internet. Space Telescope Science Institute (STScI), in particular, is endeavoring to make the science results announced through the news media and through public information channels particularly relevant to a broad audience, including through resources for pre-college classrooms and informal science centers.

1. Introduction

1.1. Background: NASA Involvement

The science divisions within NASA have a specific charter to provide unique and often new technology instrumentation in orbit for the purpose of conducting top rank science research in the Earth, Space, Planetary and Astrophysical Sciences. The expense of commissioning useful orbiting observatories necessitates that excellent research be efficiently accomplished with the facilities. Therefore, by and large, NASA mission support concentrates on optimal planning and scheduling of telescope use, efficient data acquisition, rapid delivery of data to science researchers, appropriate analysis software, and robust archival services. At the same time, it is realized that public awareness of the purpose and results of NASA missions is frightfully incomplete, and in addition, the collective perception of the return on the investment in NASA is vague. The Office of Space Science (OSS) along with several other divisions in NASA are struggling to initiate and support programs and strategies to improve return on the dollar to the public, and the strategies proposed have notable, but hopefully not debilitating, impact on science research.

It is keenly realized that NASA should not and could not redirect the *majority* of its effort towards education, outreach, and public understanding of science. Therefore, the educational resources being created through several NASA funded programs are dominated by strategic partnerships and cooperatives making leveraged use of NASA funded expertise and resources, without taking on systemic reform of the entire educational system - the latter task is best left to other agencies. The NASA programs for education often emphasize innovation, new approaches, experimentation, and electronic access to digital materials.

1.2. Strategic Partnerships

The strategic partnerships being established by NASA missions should be guided by the principle that mission scientists, engineers and staff must bring to the table unique expertise and experiences along with knowledge and capability to access NASA data, tools, information and related resources. Mission personnel must develop a healthy respect for and understanding of the expertise of educators, informal science (museum) personnel, and other individuals in order for collaborations to be successful . In addition, it is an important goal that NASA missions, which will be shorter lived and less well funded than those in the past, feed products and resources into self- sustaining infrastructure and existing systems which have a charter to disseminate information to the public. Strategic partnerships are built on the insight that access to NASA mission resources is dynamic and that interaction with specific researchers and technical staff, financial support and enthusiasm can dwindle. The HST STScI, for example, has partnerships with informal science education centers (science museums), planetaria, various teacher organizations and community efforts in addition to its fairly good relationship with the press and media, but these connections must be continually infused with new energy, expertise, and enthusiasm to reinforce firm trust between collaborators.

1.3. *Electronic Access and Innovation*

It is no secret that the scientific and technical community has totally integrated electronic access into its functional structure, with the business community shortly to follow. In contrast, in spite of the pressure to obtain electronic connectivity, the public hunger for access to commodities such as posters, slides, lithographs, prints, and mission paraphernalia reaches almost unbelievable proportions, especially with regard to some of the exquisite HST data. However, as users become more technically literate and capable, electronic access to images, graphics, animation and software services is recognized as more affordable and acceptable. In recognition of this, many NASA education/outreach programs must emphasize innovation and experimentation using electronic means to catch the eye and retain the interest of the future target audience. These programs also accentuate prototyping and testing without the obligation (or funding) to scale, replicate or expand specific projects into systemic initiatives, as for example, the IDEA program†. A feasible strategy is to offer highly modular resources that are reasonably de-coupled from the distribution interface – this encourages reuse and longevity.

2. Resources Available

2.1. *Example 1: Hubble Space Telescope Integrated Releases*

http://www.stsci.edu/

The Office of Public Outreach (OPO) at STScI is charged with disseminating information regarding the Hubble Space Telescope to the public. Our response to this directive is to showcase scientific results and technological advancements (brought forward by principal investigators) through high visibility public information releases. The process requires an understanding and delicate treatment of the various, often fickle, clientele (media, journalists, funding agencies, scientists, engineers). Releases demand a resource intensive, "immediate"investment which results in a variety of modular products, such as images, text, animation, graphics, video and audio clips, that have the potential to be used in venues beyond the short lived news, journal articles and electronic communiqus. OPOs new scheme is to integrate the transient process into a longer term integrated production cycle that should result in resources suitable for education, informal science, and "life-long learning"applications. Integrated packages are aligned along themes and topics and serve a fairly broad audience.

The integration initially involves intensive work with the principal investigators (scientist or engineer) and their colleagues to map out the type of information, data, other products and key points contained in a specific scientific or technical release. Image processing, animation, graphics, video, interviews and textual material are co-authored by a team that has a mix of expertise in graphics, animation, writing, video, instructional design, informal science and education. Rarely is the science result initially delivered in a form that can be used verbatim for

† Initiative to Develop Education in Astronomy: http://www.stsci.edu/idea.html

public release‡ , and therefore usually must be reconfigured before it is considered comprehensible. *This situation is symptomatic in projects that do not include representatives of the target audience in the creation process.*

The "educational resources" are created over a longer investment period than required for the initial release. Some materials are fairly simple, but more complex resources, including those available on the Web are created by teams of teachers, scientists, technical staff, informal science experts and production staff. These richer resources are created in the specific context of a classroom activity, a series of lesson modules or a self-guided application. Once the activity is threaded together, the modular pieces are regenerated in a fairly generic form (if possible), and offered independently for use in other contexts. Generally, OPO has not placed emphasis on the creation of stand- alone custom software, but rather, of modules accessible over the Internet through Web browsers†.

An example of a release package which is the headliner topic for several resources and modules is the Hubble Deep Field, observed with HST during 10 contiguous days in December 1995. The data was released from the STScI Archive immediately after calibration, with no proprietary data rights reserved. Teachers and their teams have been designing activities to use HDF for math and science classes, including exercises to classify, measure and count galaxies, test various counting methods, record journals of projects, collect data from other students around the Web, form theories regarding galaxy types, write and present reports, and conduct further inquiry into both historical records on the understanding of the nature of galaxies as well as current research on the HDF.

Other packages in progress or planned are based on themes such as Technology of HST, Technology of Satellites, Servicing HST, Origin of Planetary Systems, Lifetimes of Stars, to name a few. The kinds of resources should include interactive models of satellites, interactive evolutionary sequences of stars, Web based "games", planning and scheduling various satellite related activities, etc. Some of these resources will be directly related to the Smithsonian Traveling Exhibit showcasing HST, so that teachers will have a suite of activities available for classroom use before and after a visit to the exhibit at their regional museum. One additional program, which also supported the "Passport to Knowledge" *Live from Hubble*, program, provides students with an opportunity to collaborate collectively and under the guidance of an advocate scientist, to specify and analyze observations collected during a few HST orbits dedicated to educational use. Numerous other programs are supported by OPO also, but *Integrated Releases* form the core of the OPO/STScI resource creation.

2.2. *Example 2: Remote Sensing Public Access Center (RSPAC)*

http://rspac.ivv.nasa.gov/

Numerous other educational initiatives fill out the ensemble of programs sustained by NASA. One of them, created to enable *Public Use of Remote Sensing Data* (RSD) was initiated to explore methods for Earth Science and Space Science to become useful specifically over the Internet, to the general public, commercial ventures and the education community. This program, initiated under NASAs Information Infrastructure Technology and Applications (IITA) component of the US Federal High Performance Computing and Communications (HPCC) program, was expanded through separate funding to bring Aeronautic content to the classroom.

‡ There are impressive exceptions to this rule however, where scientists or engineers have made a real effort to make modular encapsulations of their science results and tools.

† However, some resources may require special capabilities in the browser such as Java, RealAudio and other products for full impact of the material. Resources also must be only loosely coupled to the interface to prepare for migration to newer information technology interfaces.

Hubble Deep Field (HDF) Products and Resources

Initial Release

Images —Wide view, high res, colors

Text – Image Captions

Discovery, major result – counts, cosmology, implications for galaxy formation, morphologies

General Background – Context of the HDF program, research being conducted, methods of analysis

Video/Audio – Science Team Interviews, Zoom in sequences

Resources-modular

Digital products as above, Slides, Lithographs, Prints

Educational Activities/ Resources

Math – Galaxy Counting: collect results from around the Web; Proportions, area, volumes; Geometric calculations
Math, Research Methods – Galaxy morphology, galaxy color
Astronomy – Nature of Galaxies, Galaxy Trading Cards with facts, Links to other information
Poster – Working poster with educational manual

The RSPAC was commissioned to create access methods for all varieties of NASA data, and also to coordinate the projects specifically funded under the RSD program. The center creates a forum for exchange of information on problems and best practices from the projects, and offers other services such as software testing and porting resource evaluation and coordination of efforts to defend funding. RSPAC provides one of the entries to the projects including:

• *Windows to the Universe – University of Michigan (PI: Roberta Johnson)*: Rich array of Earth and Space science resources for museums, libraries, and student research, Emphasis on historical and cultural ties between science, exploration, and human experience, Multi-level threads for beginner, intermediate and advanced users, Encapsulated self-guided modules.

• *Weathernet 4,- - WRC-TV, Washington D.C.(PI: Dave Jones)*:RSD and state-of-the-art visualization in daily weather reports: Tornadoes, Lightning strikes, weather trends, What is the weather predicted for the Olympiad? etc., Collection of data from school weather stations, Dissemination to 214 other TV stations.

• *The Public Connection – Rice University (PI: P. Reiff)*: Digital museum publicly accessible through four interactive displays of real-time earth and space science data at the Houston Museum of Natural Science (HMNS), Touch screen kiosks, auto-download to schools and other institutions.

• *Science Information Infrastructure - University of California Berkeley (PI: C. Christian - UCB & STScI)*: Described below.

• *Virtually Hawaii – University of Hawaii (PI: Peter Mouginis-Mark)*: Real-time imagery of Hawaii for daily TV weather tourists, residents and students (K-12 and community colleges), Space Shuttle photography, Imaging radar data from SIR-C/X-SAR experiments, NASA aircraft data (visible, thermal and microwave), Aircraft data from private-sector partner.

• *Classroom of the Future – Exploring the Environment -Wheeling Jesuit College, West Virginia (PI: Robert Myers)*: Problem-based course modules for high shool teaching of environmental earth science, Summer in-service teacher instruction.

2.3. *Example 3: The Science Information Infrastructure (SII)*

http://www.cea.berkeley.edu/~edsci/SII/

The SII was created to serve as a demonstration project in response to a specific need for a stable, coordinated framework and infrastructure for dissemination of educational and informal science resources, especially those from NASA, across the US. In the SII model, informal science education centers (i.e., science museums and planetaria) form the pylons for the infrastructure. Science museums in the US are institutions which serve as nodes for informal science exploration, exhibition of technology but also contribute significantly to the professional development of teachers. One of the key success criteria for brokering science information to the education community is the element of trust and credibility. With explicit regard to education, science and math teachers know that the curriculum support, scientific expertise and resources and knowledge concerning educational standards can often be found at science museums. Remarkably, this effective venue for interacting with educators has not been given much general attention by the NASA community, although there are particular examples to the contrary. More detail on the SII is presented elsewhere at this conference.

2.4. *Summary*

NASA researchers are participating in numerous projects aimed at delivering research results, information, materials and technology to a wide audience. Several coordinated programs supported by the Office of Space Science and the Aeronautics Division of NASA are producing innovative resources for education accessible from the Internet. These efforts are characterized by strategic partnerships that involve educators and science museum personnel to be effective. In addition, the Education Division at NASA and individual missions are producing additional products and materials for educational use.

Bringing the Universe into the Laboratory– Project CLEA: Contemporary Laboratory Exercises in Astronomy

By Laurence A. Marschall

Department of Physics, Gettysburg College, Gettysburg, PA 17325

1. The Dilemma of the Introductory Astronomy Laboratory

Were we meeting a century ago to discuss the state of astronomy education, we might have noted that remarkable changes were taking place in our field. The discipline, then regarded as a branch of geometry or mechanics, concerned itself primarily with the determination of positions in the heavens and the mapping of places on the earth. But with the advent of spectroscopy and the construction of large telescopes, astronomy was beginning to probe the how and the why of the heavens as well as the where and when. It was, in short, transforming itself into astrophysics, the study of the physical nature of the universe.

A century ago, we would have called for a change in the things we teach; and in fact there was such a change. When we look at the astronomy of the succeeding century, the material we now offer to introductory astronomy students at most universities and colleges, we see only a vestige of the earlier preoccupation with place and time. Judging by most textbooks, and by the course syllabi I have seen, most of us devote only a small fraction of our courses to astronomical coordinate systems, timekeeping, geodesy, and celestial mechanics. When we teach the solar system, we teach comparative planetology. When we teach the stars, we teach about main sequence and giant branch, about hydrostatic equilibrium and neutron degeneracy, about pulsars and supernovae. When we discuss the universe at large, we teach about the physics of the early universe, the dynamics of galaxies, and the fundamentals of general relativistic cosmology. All these subjects draw heavily on a hundred years of work with large reflecting telescopes, sensitive imagers, photometers, and spectrographs, radio telescopes, spaceborne instruments, and even underground particle detectors.

But have astronomy laboratories kept pace with astronomy classes? Judging by published materials, our labs contain a disproportionately large fraction of 19th century astronomy, along with a smattering of exercises on astrophysics that rely on old data, the measurement of photographs, and the copying of tables and graphs (e.g. Hoff, Kelsey, and Neff, 1992; Bruck, 1990; Ferguson, 1990; Johnson and Canterna, 1987; Culver, 1984).

The reasons for this dissonance between curriculum and laboratory are quite understandable. Many of the results of modern astrophysics have been obtained using highly complex instrumentation, not easily operated by students. The availability of inexpensive CCD cameras and computerized telescopes has alleviated some, but not most of these problems. Many of the processes in astronomy take place at inconvenient times or over inconvenient time scales, and not simply because astronomy must be done at night–adolescent students, after all, are nocturnal creatures. Measuring the height of a lunar mountain requires a particular phase of the moon; observations of a binary eclipse or a stellar occultation must be precisely timed; pulsar variations demand split-second time resolution, while observing the proper motion of a star cluster demands a baseline measured in decades or centuries. None of these procedures fits comfortably into the measured routine of a weekly three-hour lab period, especially in temperate climates, where cloudy nights are frequent.

Yet if the limited options available for the astronomy laboratory are understandable, they are nevertheless regrettable. For many students astronomy is the first and only science course they will have in college, and it can powerfully shape their perceptions of what science is like, not to mention their understanding of the nature of the world around them. Studies have shown that students understanding of science can be most firmly grounded on a base of concrete personal experience. (e.g. Hake, 1996, Thornton and Sokoloff, 1990; Rosenquist and McDermott, 1987).

79

Without a laboratory that complements the astrophysical material presented in textbooks, astronomy becomes a recitation of facts, many of them lacking in meaning to the first-time student. Students complain that laboratory exercises have little to do with the subject they are studying, or grumble that the exercises are make-work, involving little more than measuring photographs someone else has taken, copying numbers from tables to graph paper, and copying phrases from the textbook.

Hands-on experiences are common in experimental (as opposed to observational) sciences like physics, chemistry, and biology. In introductory chemistry, for instance, where the emphasis is on qualitative analysis, students learn quickly that many substances that appear as indistinguishable white powders at first glance differ profoundly in their densities, reactivities, and molecular structure. How many astronomy students come out of a course with the simple realization that most of the objects in astronomy are indistinguishable dots of light, and that, like the chemist, a major task of the astronomer is to tease out the fundamental structure of these objects using a standard array of analytic methods? How many astronomy students, rather, confuse the pretty diagrams of pulsars, black holes, and the spiral arms of the Milky Way with the faint objects astronomers actually see through the telescope.

How we know what we know is, I would maintain, as important as what we know. I do not intend to argue that the laboratory experience is the cure for all the ills of introductory astronomy education. I do believe, however, that the traditional introductory astronomy lab, with its emphasis on simple positional astronomy, elementary optics, and the clerk-like copying of data, lacks an important element of connectedness to the conceptual framework of modern astrophysics and to the methods of modern observational astronomy that impedes understanding and deprives students of the fun of discovery.

2. The Origins of Project CLEA

In recent years, a number of educators have noted that personal computers offer an powerful solution to the problem of the introductory astronomy laboratory. (e.g. Dukes, 1990; Marschall, 1995). Computers can provide simulations of astronomical observations that are extremely realistic, which is not surprising considering that computers, rather than telescopes, are the most widely used astronomical tools. Computers control optical and radio telescopes, interface with spacecraft, and are used in all steps of data analysis in astronomy. Large stores of digital images, spectra, and other data are available on CD-ROM and on the world-wide web.

A few astronomers, in the early days of the PC, began to develop computer simulation labs. Michael A. Seeds (Franklin and Marshall College), and Robert Dukes (College of Charleston), both developed telescope simulators in the 1980s, first on Apple IIs, and then on Macintoshes. Later, a group led by John D. Trasco (University of Maryland) created exercises that ran under early versions of Windows. Some astronomers began to use commercial planetarium programs, like *Voyager* for the Mac, and *Dance of the Planets* for the PC, to produce exercises in positional astronomy (Wooley, 1992).

In 1992, my colleagues (Michael Hayden and Dick Cooper) and I at Gettysburg College set out to develop a comprehensive set of computerized exercises. Realizing that new computers and new operating systems opened up wide possibilities for simulations, we obtained funding from the U.S. National Science Foundation which enabled us to hire a full-time programmer, Glenn Snyder (who has a Ph.D. in astronomy and 20 years experience with NASA science missions), and to establish Project CLEA (Contemporary Laboratory Experiences in Astronomy). Over the past 4 years, CLEA has developed 8 computerized laboratory exercises in astronomy which have met wide acceptance by the educational community. Distributed by mail and by the internet, CLEA materials are currently in use in all 50 states and more than 50 nations worldwide. (Marschall, 1995; White, 1996). Four CLEA exercises won a 1994 *Computers in Physics* award for best educational software.

3. Description of the CLEA Exercises

The CLEA exercises were designed with two princpal goals in mind: to reinforce important concepts in astronomy, and to help students experience how astronomy is done. Each exercise consists of a software disk, a student manual, and a teachers guide with technical notes.

Several principles guide our design of classroom materials.

• Exercises are targeted at non-science students, who make up, by far, the majority in undergraduate introductory astronomy courses, as confirmed by our user surveys. While designed for standard 2 to 3 hour lab periods, CLEA programs have optional settings that make them useful for upper-class science majors, where appropriate, and they have been successfully used in classes from elementary school to senior college astronomy majors. (About 20% of our users are high-school students).

• Exercises are only developed for cases where expensive equipment, inconvenient time scales, or unavoidable complexity would make real observations impractical. Naked -eye stargazing is still the best way to learn the constellations; optics can be taught with real lenses and mirrors.

• Exercises have well-defined learning objectives and goals, which we state in our student manuals.

• Exercises are designed with the cognitive skills and technical sophistication of introductory astronomy students in mind. Some of these design considerations can be anticipated from previous studies in the literature (e.g. Arons, 1990). Others are taken into account by formal and informal feedback from our classes and by user surveys.

• Exercises avoid, as much as possible, rote copying and graphing. They emphasize how data is collected and analyzed, or demonstrate phenomena that can not be examined otherwise.

• Exercises make use of real digital data wherever possible. Our existing exercises use digital spectra, redshift, and radio source data from on-line or CD-ROM data archives.

• Exercises include instructive "real-life"features, such as detector noise, sky background, and weather interruptions, while avoiding complications that might distract from the principal learning objectives of the lab. Some exercises may include optional features that demonstrate the sociology as well as the logic of science, e.g. students must apply to use a large telescope– and may not be granted the time!

• Exercises are carefully designed for ease of use and uniformity. Extensive feedback from both students and faculty at other institutions makes the software easy-to-use and highly portable. This adds greatly to development time, but it insures that the materials we produce will have lasting value to a wide range of users.

• Exercises are independent of one another, and carefully documented so that instructors can pick and choose to fit individual needs. Since only about 20% of introductory astronomy courses are taught by professional astronomers (Michael Seeds, private communication, 1995), we make the exercises easy to install and use, but flexible enough for advanced users.

There are currently 8 CLEA exercises, covering a wide range of techniques and concepts from planetary astronomy to cosmology. The software runs under Windows, (Versions 3.0 and higher, including Windows 95 and NT), although one color Mac lab is available, and a number of rudimentary black and white Mac programs, lacking in many of the features of the Windows programs, are available as well. A brief description of each exercise follows:

The Revolution Of The Moons Of Jupiter

Purpose: *To illustrate the measurement of the mass of a planet using Kepler's third law.*

The software provides a view of Jupiter at four magnifications (actual Voyager images are used), along with a highly accurate ephemeris program that draws the four Galilean satellites in their proper positions relative to the planet at any time. Students make observations of Jupiter and its satellites at regular intervals over a period of several weeks and, by graphing the separation of each moon from Jupiter versus time, they measure the period and radius of each satellite's orbit. This is sufficient information to derive the mass of Jupiter. Students use the mouse cursor to identify the moons and to measure distances. A cloudy night feature provides some incompleteness to the data. Instructor-settable options set the percentage of "cloudy"days, suppress the automatic moon identification, and suppress the automatic calculation of distance from the planet. (Windows and color MAC)

The Rotation Of Mercury Using Doppler Radar

Purpose: *To illustrate the measurement of the rotation rate of a planet using the Doppler shift of a returning radar pulse.*

Students are given control of a simulated radar telescope. They point it toward Mercury and send off a narrow pulse of radio waves. During the time required for the pulse to go to Mercury and return (10 to 20 minutes, depending on the aspect of Mercury), the screen displays an animation showing the inner solar system and the position of the wave front of the pulse and its echo. The received echo, frequency versus intensity, is displayed for 5 time slices, which correspond to echoes from surface elements progressively further from the subradar point. By clicking the mouse cursor, students can measure the Doppler shift from the approaching and receding limbs of the planet, and from this determine (with appropriate geometrical corrections) the equatorial rotational velocity of Mercury. This, in turn, gives the period of rotation of the planet. Students use off-line calculators to reduce the data.

The software features on-line data recording, a printing option, and an on-line checker, which tells students if their calculations are "in the ballpark". Options include the ability to speed up the return pulse, reducing the waiting time. (Windows and b/w Mac)

The Flow Of Energy Out Of The Sun

Purpose: *To illustrate the statistical nature of radiative transfer in stellar interiors and stellar atmospheres. To show how photons diffuse in a random-walk pattern from the core of a star . To show how spectral lines are formed by random processes of absorption and re-emission in its atmosphere.*

This software contains several animated simulations that show photons in slow motion as they interact with matter and are redirected due to random processes of scattering, absorption, and re-emission. Students view the animations and measure the statistics of these processes. Each simulation can be run independently, or the entire set can be used as a full class exercise. One set of simulations deals with the scattering of photons in the interior of a star. The program demonstrates the random walk pattern of a photon generated in a simulated star as it makes its way to the surface. Students can vary the number of layers in the star to see how the number of interactions and the total length of the photon's journey vary in response. They can also shoot a pulse of up to a thousand photons all at once to see how the original spike diffuses out in an extended period of time.

Another set of simulations illustrates the transfer of radiation and the formation of spectral lines in the atmospheres of stars. In these programs, students can investigate the random nature of absorption processes, noting that photons whose energies correspond to electronic transitions of the atom seldom travel straight through a gas of those atoms. Students can experiment to see how many "line"photons make it through undeflected, as opposed to "continuum"photons, which travel straight through most of the time. They can experiment with samples of different gases. For a given gas, they can vary the photon energy, and by shooting photons through the gas and plotting how many make it through as a function of energy, they can learn how the statistics of a large number of photon interactions produce the dark absorption lines we see in the spectra of stars. (Windows and b/w Mac).

The Classification Of Stellar Spectra

Purpose: *To introduce students to digital spectra and to the process of classifying different spectra by the relative strengths of lines. To familiarize students with the sequence of spectral types. To teach how spectra are obtained. To show how the distance of a star can be estimated from its spectrum and a measurement of its apparent magnitude (spectroscopic parallax). To illustrate the need for large-aperture telescopes for the observation of faint objects.*

This exercise incorporates two separate but interrelated features:(1) a set of tools that enables students to view digital spectra and compare them with standard spectra from an atlas of representative types. (2) a simulated telescope that enables students to obtain spectra of unknown stars using a photon-counting spectrograph. In the first part of the exercise, students are asked to examine a number (typically 2 dozen) spectra of main-sequence stars and, by comparing them with a stored atlas of representative spectra, determine their spectral type. The software allows the digitized spectrum of the unknown to be displayed, flanked above and below by standard

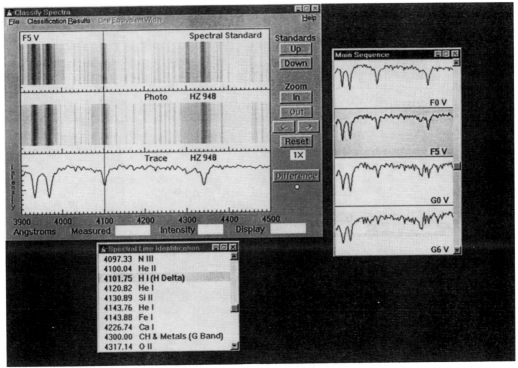

FIGURE 1. A screen from the Classification of Stellar Spectra Exercise.

star spectra. Zoom and pan features are provided, along with a measuring cursor. An atlas of spectral lines, coupled to the cursor, allows easy identification of prominent features. We provide nearly 200 spectra for use in this lab, including spectral standards for several luminosity classes as well as the main sequence.

Options include the ability to display the digital difference between an unknown and a standard spectrum, the ability to display the spectra both as graphs of wavelength versus intensity and as gray-scale photographs, and the ability to measure equivalent widths. Students can record their measurements on paper or disk files, import their data into spreadsheets, and run spreadsheets from the program shell. Since all the files read by the program are ASCII text files, the user can supply alternate digital spectra and additional line identifications.

In the second part of the exercise, students are given control of a simulated telescope. They can view a star field (a randomly generated field is supplied, but again, alternate fields can be specified using text files), center the spectrograph slit on an unknown star, and record spectra using a simulated photon-counting spectrograph. Spectra can take from several seconds to several hours to record sufficient signal-to- noise to identify faint features. The recorded spectra can be saved to disk and imported into the first part of the program where their spectra can be compared to standards. This information, along with the observed apparent magnitude of the star, can be used to derive a spectroscopic distance to the unknown star.

For faint stars, students may want to use a larger telescope. Three telescopes, with 0.4 meter, 1.0 meter, and 4 meter apertures, are available. To use the large telescopes, students must apply for time. Instructors can set the probability of success for using the larger telescopes (or make them open to unrestricted use), and the amount of time granted. A utility called GENSTAR is available from CLEA which generates customized files of star fields for use in this program (Windows only)

Photoelectric Photometry Of The Pleiades

Purpose: *To familiarize students with the technique of photoelectric filter photometry and* *counting statistics. To acquaint students with the use of a computer controlled telescope. To illustrate the use of equatorial coordinates for finding stars in a cluster. To introduce the use of H-R diagrams for analyzing the age and distance of clusters.*

The software for this exercise puts students in control of a computer- controlled telescope with sidereal tracking (the "tracking"must be turned on or stars drift out of the field of view). Two fields of view are provided, a "finder"view of 2.5 degrees, and a magnified "instrument"field of view of 15 arc minutes. In the instrument mode, students see the outline of an aperture superimposed on the field. This marks the position of the entrance aperture of a photon-counting photometer. Controls are available to select one of three filters (U,B,V), to set the length of integrations and the number of integrations. Once these controls are set, students can take readings of the number of photons received from either a star in the aperture or from the sky (if no star is in the aperture).

In the configuration most suitable for introductory students, the program automatically calculates stellar magnitudes from the count rate, as long as a sky reading has been taken through a given filter sometime before the star is observed. For upper-class students the magnitude calculation can be suppressed and students can use the raw star and sky counts to calculate magnitudes. Information on the signal-to- noise ratio is also displayed after each series of integrations. A student guide supplied with the software describes how to use B and V photometry of cluster members to plot an Hertzsprung-Russell diagram of the Pleiades and to determine its distance.

We provide a database of stars in the Pleiades, but the data files for the stars are text files which can be modified by the user, so that data on any cluster can be substituted for the Pleiades, and exercises on the comparative ages of star clusters are possible. A utility called GENSTAR is available from CLEA which generates customized files of star fields for use in this program.(Windows and b/w Mac)

The Hubble-Redshift Distance Relation

Purpose:*To illustrate how the velocities of galaxies are measured using a photon- counting spectrograph. To show how this information, along with estimates of galaxy distances (from their integrated apparent magnitudes) yields the classic Hubble redshift-distance relation. To determine the value of the Hubble parameter and the expansion age of the universe.*

At the controls of a simulated telescope, students view distant clusters of galaxies and obtain their spectra with a photon counting spectrometer. The telescope offers two fields of view, a wide field view of 2.5 degrees, and a magnified field ("instrument view") of 15 arc-minutes. Stars are represented by realistic point spread functions scaled to magnitude, and galaxies by images from actual CCD frames. In the instrument mode, students can position the slit of a spectrograph on the galaxy and take spectra. The photon counting spectrograph simulates actual Poisson statistics and contains both a sky background and a galaxy spectrum. The relative contribution of the galaxy depends on how much "light"from the image is included in the slit, so that the highest signal-to-noise is obtained when the slit is positioned on the brightest part of the galaxy, just as with a real spectrograph. Students are advised to obtain spectra with signal-to-noise of about 10, so that they can see and measure the Ca H and K lines, which are used to determine the redshift of the galaxy.

Wavelengths can be measured using the mouse cursor, and recorded for further analysis. The spectrometer also records the integrated apparent V magnitude of the galaxy, which is used, along with an assumed absolute magnitude, to determine the distance of the galaxy. With this information for five or six galaxies at various distances, students can plot out a Hubble diagram, determine the Hubble parameter, and estimate the age of the universe.

A wide variety of instructor-settable options are available. Instructors can construct their own galaxy fields using GENSTAR, a utility supplied by CLEA, and can even install their own image files to represent galaxies. The integration time to reach a given signal-to-noise can be set to conform to the needs of the class and the speed of the computer. Even the value of the Hubble parameter can be specified by the instructor; the default is 75 km/sec/Mpc. (Windows and b/w Mac)

Radio Astronomy of Pulsars

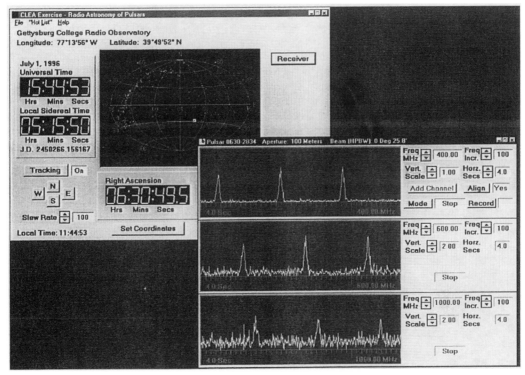

FIGURE 2. A screen from the Radio Astronomy of Pulsars lab currently under development.

Purpose: *To understand how radio telescopes are used to detect pulsars. To understand the fundamental observational characteristics of pulsars. To understand noise in observations. To understand how dispersion can be used to get a measure of the distance of a pulsar.*

This exercise, now in alpha-testing form, presents students with a radio telescope whose operating characteristics (beam width, receiver noise, steerability) can be set by the instructor. The telescope can be pointed at a source in the sky and the output of the receiver displayed on a graphic display (like a chart recorder or digital oscilloscope). The receiver can be tuned to a range of frequencies between 400 and 1400 Mhz, and additional independently tunable receivers can be added and displayed on the screen. Background and instrument noise are simulated along with noise intrinsic to the source. Data from each receiver can be saved for later analysis. Measurement windows, on which timing data can be recorded using the cursor, are provided as a default feature. Additional analysis tools (e.g. power spectra displays) can be called up at the instructor's discretion. Sound features are available to "listen"to the signal.

Students first view the signal and measure fundamental characteristics such as period and pulse width. They then view the signal at different frequencies and use the dispersion phase delay to estimate the distance of the pulsar. The radio telescope has optional features which can be used for additional labs, including a "transit"mode and a map of the galactic background noise. (Windows only).

4. Prospects for Future Development

Project CLEA's plans for the next two years include the development of at least 4 more exercises, selected from over a dozen outlines we have drawn up over the last few years. The list includes exercises on astrometry of asteroids, cepheid variables, radial velocity curves of binary stars, and a "capstone"lab where students identify an unidentified object using standard analytical techniques.

At the same time, CLEA plans a systematic evaluation of its materials to find out how students and instructors regard the material and how effective it is in the classroom. Since the beginning of the project CLEA materials have been tested at Gettysburg College, and a wide range of informal responses have been received from users around the world. Though invaluable in fine-tuning the software and hardware, evidence of pedagogical effectiveness is anecdotal at best. Beginning in January 1996, however, an independent evaluator, Dr. Marcus Lieberman, began a formal evaluation process, including a program of pre- and post-testing of students in classes at Gettysburg and the University of Wisconsin, Madison. First results show that students using CLEA materials, and other hands-on exercises score significantly higher than students in lecture-only courses, an outcome we intend to explore more fully with a large sample of students and improved testing procedures. An extensive phone canvass of users is also being conducted.

Further work at CLEA is contingent, of course, on continued funding. But we note with pleasure that others are beginning to follow our lead. Translations of CLEA manuals into Spanish and Italian have already been made; French, Hebrew, and Chinese, we understand, are in the works. Instructors have been sending us copies of alternate student exercises and manuals they have developed, and alternate databases they have incorporated into our software. We note also the variable star exercises of Claud Lacy, of the University of Arkansas; Binary Maker, by David Bradstreet of Eastern College, Pennsylvania; the TS-24 telescope simulator by Craig Young, a New Zealand amateur; and a series of exercises on the web by Greg Bothun of the University of Oregon.

We look to the future with optimism. As astronomy education changed a century ago, so it is changing now. Astronomy classes in the future will offer a more experiential approach to the subject, and we take some satisfaction in knowing that Project CLEA has played a role in bringing this about.

5. Obtaining CLEA materials

CLEA materials are available over the world-wide web at URL
http://www.gettysburg.edu/project/physics/clea/CLEAhome.html
or can be downloaded from the anonymous ftp io.cc.gettysburg.edu (138.234.4.10). Manuals are available on-line as postscript files, but users who are unable to print these files can, or who cannot download the software, should contact us by mail: Project CLEA, Gettysburg College, Gettysburg, PA 17325 USA, by telephone, 717- 337-6028, or by email: clea@gettysburg.edu.

Acknowledgements

Thanks go first and foremost to my co-investigators at Project CLEA, Glenn Snyder, Rhonda Good, Mike Hayden, Dick Cooper, and Mia Luehrmann. Thanks also to the students who have helped us over the years, especially Shawn Baker, Michelle Vojtush, and Akbar Rizvi. Thanks to our evaluators, Marcus Lieberman and Michael Chabin, to the crew who participated in the 1994 summer development workshop, and to the many astronomers and teachers around the world who have offered criticism and encouragement. Project CLEA has been supported by grants from the National Science Foundation and Gettysburg College.

REFERENCES

BRUCK, M.T., 1990, *Exercises in Practical Astronomy Using Photographs*, (Adam Hilger, Bristol, UK).

CULVER, R., 1984, *An Introduction to Experimental Astronomy*, (W.H. Freeman and Co., New York).

DUKES, R.J., 1990, "Microcomputers in the Teaching of Astronomy", in *The Teaching of Astronomy*, IAU Colloquium 105; Pasachoff, J.M., and Percy, J.R., ed. (Cambridge University Press, Cambridge, UK).

FERGUSON, D.C., 1990, *Introductory Astronomy Exercises*, (Wadsworth Publishing Co., Belmont, CA).

HAKE, R.R., 1996, *Interactive-engagement vs. traditional methods: A six-thousand- student survey of mechanics test data for introductory physics courses*, preprint, submitted to the American Journal of Physics.

HOFF, D.B., KELSEY, L.J., & NEFF, J.S., 1992, *Activities in Astronomy, Third Edition*, (Kendall-Hunt Publishing Co., Dubuque, IA).

JOHNSON, P.E., & CENTERNA, R., 1987, *Laboratory Experiment for Astronomy*, (Saunders College Publishing, Philadelphia, PA).

MARSCHALL, L. A., 1995, *Virtual Professional Astronomy*, Sky and Telescope, 90, #2 (August), 92.

ROSENQUIST, M.L., & McDERMOTT, L. C., 1987, *A Conceptual Approach to Teaching Kinematics*, American Journal of Physics, 55, 407.

THORNTON, R.K., AND SOKOLOFF, D.R., 1990, *Learning Motion Concepts using Real-time Microcomputer-based Laboratory Tools*, American Journal of Physics, 58, 858.

WHITE, J.C., 1996, *Have Observatory, Will Travel*, Mercury, 25, #3 (May/June), 16.

WOOLEY, J.K., 1992, *Voyages Through Space and Time*, (Wadsworth Publishing Co., Belmont, CA).

Project LINK: A Live and Interactive Network of Knowledge

By Barry Welsh AND Isabel Hawkins

EUREKA Scientific Inc., 2452 Delmer Street, suite 100, Oakland, California, 94602, USA

Project LINK (A Live and Interactive Network of Knowledge), is a collaboration of Eureka Scientific, Inc., the San Francisco exploratorium Science Museum, and NASA/Ames Research Center. Project LINK has demonstrated video-conferencing capabilities from the Kuiper Airborne Observatory (KAO) to the San Francisco Exploratorium in the context of science education outreach to K-12 teachers and students. The project was intended to pilot-test strategies for facilitating the live interface between scientists and K-12 teachers aboard the KAO with their peers and students through the resources and technical expertise available at science museums and private industry. The interface was based on Internet/macintosh video conferencing capabilities which allowed teachers and students at the Exploratorium to collaborate in a live and interactive manner with teachers and scientists aboard the KAO. The teachers teams chosen for the on-board experiments represented rural and urban school districts in California. The teachers interfaced with colleagues as part of the NASA-Funded Project FOSTER (Flight Opportunities for Science Teacher Enrichment).

Teachers from Project LINK participated on two flights aboard the KAO during the Fall of 1995. Lesson plans, classroom activities, project descriptions and lessons learned are currently being disseminated through the World Wide Web. Further details of this Project LINK can be found at: http://www.exploratorium.edu/learning_studio/link.

Computer as a tool in astronomy teaching

By Francis BERTHOMIEU

Lyce Jean Moulin, Place de la Paix, 83300 DRAGUIGNAN, FRANCE

As yet, astronomy, the most ancient of all sciences, surprisingly is not included in French secondary science classes. Recent trends in favour of a more attractive and motivating scientific education have taken it up.

Astronomy has, at all times, been arising curiosity, and now provides a privileged field to scientific approach :

- **Observation** of the vault of heaven and its peculiarities
- **Description** of its general appearance and of the specific movement of stars and planets
- **Measurement** of distances, coordinates and angles.
- This will make it possible to define successive **models** , which will be ever closer to the observed reality.

The obstacle of mathematics must be avoided or bypassed : many devices and demonstration models allow for a simplified and convincing approach. Computers may be valuable tools. My purpose is not to go through the multimedia version of an encyclopaedia but to follow some new trails.

DIGITAL IMAGES are efficient tools for first experiences : observation can be adapted to a specific public and digital images can guide pupils through observation. They facilitate measuring operations : interaction will incite users to creativity and discovery, and numerical models will be exploited much more easily.

The movement of planets is a quite convincing example. I use for that purpose a series of digital images of the sky : each photograph represents the constellation of Taurus, all taken during the 1990-1991 winter. My software allows pupils to recognize the characteristic stars of that region and to locate the moving planet Mars among them.

With a mouse click, the position of Mars is transfered to a map of the sky, that was conceived in order to represent the constellation of Taurus such as it would appear on a photograph. Beside that, the map includes the celestial coordinates, so that the pupils can also perform some double measurements. In addition, the software constructs, point by point, the apparent trajectory of the planet and the pupils become convinced that this movement really takes place near the Ecliptic plane (see Fig. 1).

Another example can be that of the Moon. Many people confuse the explanation of its phases with that of eclipses. Our younger pupils can discover the difference playing with a ball and a projector. A formal explanation involves some difficult 3-D geometrical construction, which can be easily presented by computers. Digital pictures of the Moon were taken in the course of one month, and introduced into other software. The user can easily draw the circular border of the Moon. Then the program allows you to choose the position of the Sun so that the drawing of the Moon looks like the photograph. You obtain also the size of the circle and the value of the angle between the axes Moon-Earth and Moon-Sun. Nothing remains but to find the reason why the size of the Moon apparently changes during the month, and how it moves in relation to its celestial neighbours, the Earth and the Sun! (See Fig. 2, 3.)

MOVING PICTURES are another means for visualizing astronomical concepts. Even though underlying mathematics are sometimes out of the users reach, moving pictures facilitate the understanding of the phenomena. They are a solution particularly well- adapted to geometrical models.

Unfortunately Figure 4 shows a static copy of the computer screen. The pupil, indeed, can observe and analyse the moving picture. He notes that six moving points determine two triangles. These triangles get distorted but remain identical one to another. As one of their vertices does not move, the program draws the trajectory of the others.

It is a great pleasure to detect the pupils pleasure when they suddenly discover that these

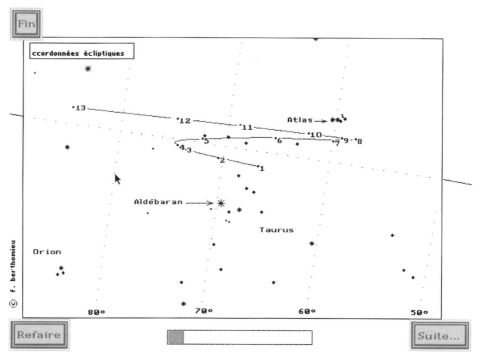

FIGURE 1. The track of Mars in the constellation of Taurus.

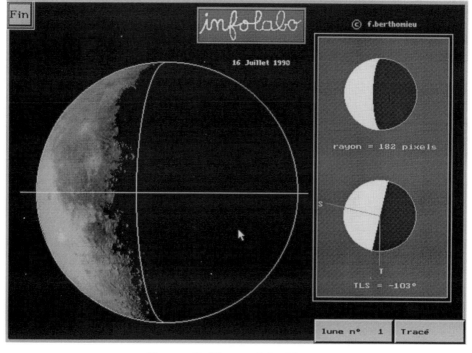

FIGURE 2. Phases of the Moon.

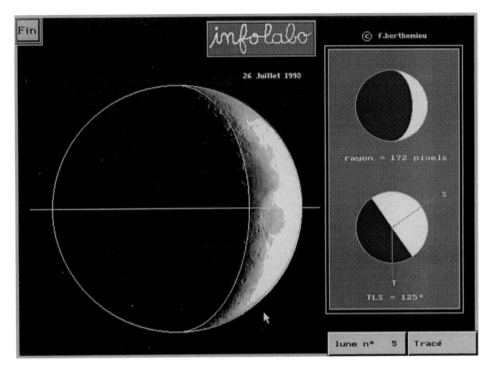

FIGURE 3. Phases of the Moon.

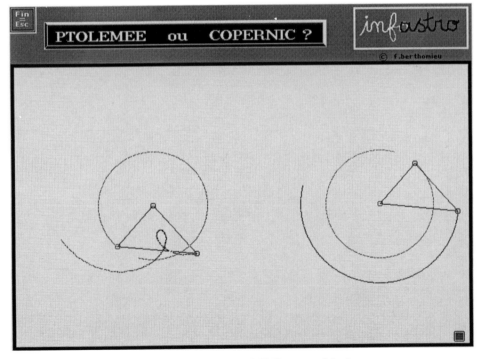

FIGURE 4. Geocentre and Heliocentre Motion.

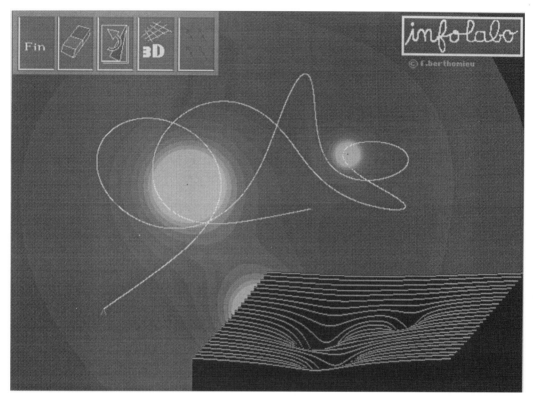

FIGURE 5. Illustration of Orbital Motion.

points on the screen can be identified with the Sun, the Earth and Mars : it is the exact moment when they understand for themselves the difficult problem of Solar System models.

While scientific research has been using it for decades, numerical simulation is coming (too shyly) in our teaching methods. Figure 5 shows an example of what can be done in that area .

When studying movements in a complex gravitational field, my software allows you to launch then a virtual spacecraft and to observe its erratic movement all around the gravity wells. A 3-D representation may induce some visualization of the space curvature...

I am now carrying out such researches. My purpose is to produce such easy to use software so that most teachers may rapidly put computerised methods into practice in their classroom.

Mathwise Astronomy and the TLTP: aiding or degrading education?

By E.M. McCabe

School of Mathematical Studies, University of Portsmouth, UK

1. Background

The Teaching and Learning Technology Programme (TLTP) in the UK was launched in 1992 to "develop innovations in teaching and learning through the power of technology". Increasing numbers of students with mixed abilities and backgrounds were entering into higher education. Flexible course structures and the need for remedial teaching added further motivation in the search for methods of improving productivity and efficiency.

Since 1992 over 33 million of funding has been awarded to 76 projects spanning the university curriculum. When support from host institutions is taken into account, overall funding for the TLTP is estimated at 75 million. TLTP materials are now becoming available to assist institutions in maintaining and enhancing the quality of their teaching provision. The successful implementation of this new technology is requiring each institution to rethink its teaching and learning strategies (Laurillard, 1993).

Approximately one quarter of the projects are based on a single institution and are concerned with the culture change, the integration of technology and staff development. The remainder are consortia concerned with courseware development and involve staff from between two and fifty universities.

Astronomy is represented within one of the largest consortia, the UK Mathematics Courseware Consortium (UKMCC), which has received 1.3 million of TLTP funding. Other projects include Software Teaching of Modular Physics (SToMP) and Statistics Education through Problem Solving (STEPS).

2. UK Mathematics Courseware Consortium

Mathwise, the product of the UKMCC, is an exciting new computer-based learning environment for students of mathematics in the sciences and engineering (Beilby, 1993 and Harding, 1996). A set of fifty modules in foundation mathematics and its applications are being developed. The first thirty, including the astronomy module, were completed by December 1995 and the remaining twenty are scheduled for completion by December 1996.

	Phase 1	Phase 2
Foundation	20 modules, e.g. Standard Conics	10 modules, e.g. Vector Calculus ($\times$ 2)
Applications	10 modules, e.g. Astronomy	10 modules, e.g. Sports Maths

First year undergraduates in science and engineering are the target audience, although topics at higher and lower levels are covered. The foundation modules provide full coverage of the SEFI† core curriculum in mathematics for the European engineer (Barry, 1992). Astronomy was identified as an exceptional application for motivating science and engineering students to explore and understand its underlying mathematics. As a result, the astronomy material considerably exceeds the nominal five hours of study time allocated to each module.

The project has five main support sites for the project development, namely the universities at Birmingham (lead site), Cambridge (Hypercard), Coventry (Toolbook), Heriot-Watt (Authorware and assessment) and Keele (multimedia). While Mathwise Astronomy has been developed for the PC using Toolbook, other modules have been developed using Authorware

† Society for the Education of the European Engineer

for the PC/Mac or Hypercard for the Mac. Conversions between Toolbook and Hypercard have been completed for some modules. In all, around forty authors at thirty different universities have developed the courseware. The academics have been responsible for writing, designing and authoring the software to a common specification using templates and tools developed by the support sites.

Universities within the UK can use all Mathwise modules freely and at minimal cost. The first evaluation CD-ROM was released in March 1995. An extended version is currently being distributed by the Numerical Algorithms Group (NAG Ltd.) of Oxford. Several international publishers have expressed their interest and it is expected that the material will become more widely available on CD-ROM in 1997.

3. Mathwise Astronomy

Mathwise Astronomy (McCabe, 1994, 1995a) is a spectacular demonstration of the powerful features of Mathwise. Many other astronomical CD-ROM products, such as Redshift 2, are the software equivalent of coffee-table books. They are extremely attractive and interactive, but do not offer a deeper understanding (Howarth, 1996). Several astronomical textbooks offer practical exercises using photograph measurements (Brück, 1990), supplied datasets or suggested observations, but are less interactive. Mathwise Astronomy provides outstanding opportunities for exploring and understanding the Universe, by combining the better features of the interactive CD-ROM and the practical textbook.

The astronomy module is divided into six learning units (Grand Tour of the Universe, Distances to Planets and Stars, Mass of the Sun and Stars, Luminosity and Temperature of Stars, Star Formation and Stellar Evolution) and thirteen leaflets (e.g. Population Density of Stars, Reflecting Telescope Mirror, Escape Velocity, Small Angle Approximations, Energy of a Gas Molecule, Stellar Magnitudes, Derive Kepler's 1st / 2nd / 3rd Law, Powers of Ten).

The learning units have a predominantly linear structure, likened to lectures. Leaflets 'bud' from and return to the main learning units, providing opportunities for further investigation and explanation. Leaflets can also be explored from within the learning units of other foundation or application modules and are likened to drop-in tutorials.

module	lecture course
learning unit	lecture
leaflet	tutorial
resource	library books and programs

The resources available include References (glossary, maths handbook, biographies and bibliographies) and Tools (calculator, grapher, star maps, simulations, and experiments).

Screen-based observations of the motions of the planets and stars lead on to an understanding of how distances, masses, temperatures and luminosities can be determined. Accurate simulations of astronomical phenomena, such as solar and lunar eclipses, the phases of Venus and stellar parallax, are frequently exploited to provide personalised data. Collapsing interstellar gas clouds and the changing appearance of stars in the H-R diagram allow a basic mathematical understanding of star formation and stellar evolution to be developed.

There are over two hundred screens of highly interactive multimedia. Video clips recorded by Dr. Patrick Moore introduce each learning unit and animations illustrating the astronomy or mathematics appear on most screens (Figure 1).

4. Computer Aided Virtual Experiments (CAVEs)

The Mathwise astronomer is encouraged to make observations, take and record measurements, analyse the measurements, derive empirical results, put forward hypotheses, develop mathematical theories and derive analytical results. This idealised scientific method guides the discovery and helps to guarantee success on a reasonable time-scale. The use of a simulation in

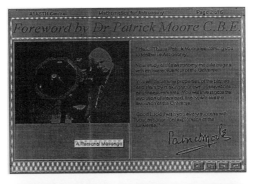

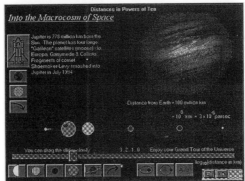

FIGURE 1. Introducing the Grand Tour of the Universe

this context is described as a Computer Aided Virtual Experiment or CAVE. Resources include a wide range of CAVEs for projectiles and celestial orbits (Figure 2).

The dimensions and separations of objects on the screen are measured by 'rubber- banding'. The measurement is recorded continuously while the mouse is dragged and is fixed when the mouse is released. For example, the technique has been used to allow screen measurement of angular separations and diameters, distances of planets from the Sun and parallaxes. Errors occur naturally when making the on-screen measurements, although their origin is quite different from 'real measurement'errors. Data gathered in this way can be used subsequently for the application of mathematical and statistical techniques, which deal with the experimental error.

Time intervals during experiments can be measured by a digital stopwatch with start/stop and reset buttons. The speed of some simulations are controlled by the user, while sensible speeds are fixed for others. An option to allow the use of standard datasets as an alternative to making measurements is generally provided.

Mathwise astronomers can investigate the parameters which govern telescope mirrors, celestial trajectories, stellar properties, star formation or stellar evolution, establish mathematical relationships from their screen measurements and derive theoretical results.

5. Flexible Study Modes

The stages in any calculation encourage thought and offer appropriate feedback, in such a way that even a mathematical proof becomes visually exciting! Interactive calculations and examples are illustrated numerically, algebraically, geometrically and graphically in both two and three dimensions.

Within many of the mathematical pages, flexible study modes are permitted (McCabe, 1995b). A reader can choose between a Study mode (cf. a manually driven car), a Revision mode (cf. an automatic car) or a Skim mode (cf. a car in overdrive). The screen is displayed either

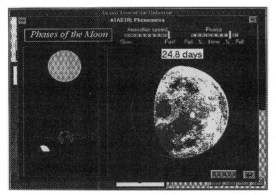
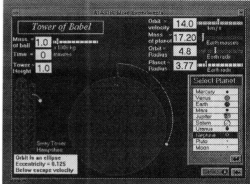
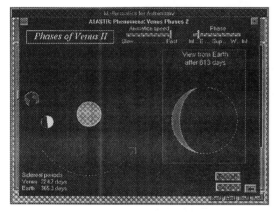
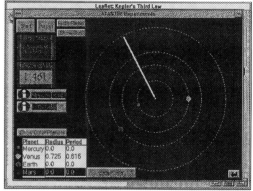

FIGURE 2. A selection of Computer Aided Virtual Experiments (CAVEs)

step-by-step under user control, automatically at a fixed speed or with full details displayed immediately. These options are especially useful for presenting algebraic derivations or proofs, where a careful explanation is required at first reading, but only the final results need to be accessed at a later stage.

Figure 3 shows how a mathematical result is broken up into stages by using a set of menu items. By clicking on each step, the next stage in the reasoning is shown and different parts are highlighted or animated. Any step can be reviewed while following the proof. This technique even allows the reader to investigate the "Grand Evolution of the Universe"from the Big Bang up to the present day!

6. Evaluation: First Aid or Lemonade ?

Foundation Mathematics modules are intended to provide remedial materials, encourage revision, extend basic concepts originally taught in the classroom and raise the confidence of the learner. Applications modules are intended to motivate students to acquire the skills and understanding and to provide worked examples which extend the student's technical knowledge to applied problems. A student can receive 'first aid' outside timetabled teaching hours without the fear of failure or embarrassment. Under supervision the teacher becomes an ally rather than an opponent. Trials have demonstrated the effectiveness of the techniques used, but with some reservations.

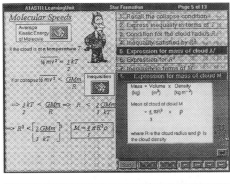

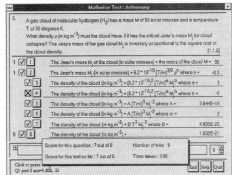

FIGURE 3. Flexible Study Modes and Structured Response Assessment

There is a natural tendency for students to focus attention on visually attractive simulations and the parts with "fizz". Considerable effort has been made to provide well specified activities with clear objectives. The structure of Mathwise encourages a logical progression, while allowing the option to skip leaflets.

Informal self-assessment is provided within units. A test mechanism for more formal summative assessment has been developed at Heriot-Watt. In common with other Mathwise modules, computer-based questions can be randomly selected from a bank of multiple- choice or structured response questions, which may include several random parameters. Tests may be completed in either a practice or exam mode. In a structured response question a sequence of key parts must be answered, with the option to reveal intermediate steps (Figure 4).

7. Further Developments

When used prudently, Mathwise Astronomy and the TLTP are aiding rather than degrading education. They are not an all-embracing panacea, which can eliminate the need for teachers, but they do provide an exciting new dimension to learning, which can significantly improve the quality of education.

Although Mathwise astronomy is designed to be self contained, it is planned to write a series of worksheets to accompany the module. These will provide additional paper- based exercises, which will suggest alternative activities. In some instances it may make the recording of complicated results from screen based simulations more straightforward.

Another TLTP consortium, Software Teaching of Modular Physics SToMP, is also planning to add an astronomy component to its existing open learning materials. The project, involving nine UK universities, uses a flexible multimedia database system, Microcosm, to present and cross-reference information. Adaptation for the World Wide Web and linking to robotic telescopes are further possibilities, but is not yet straightforward for such highly interactive material.

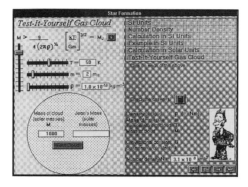

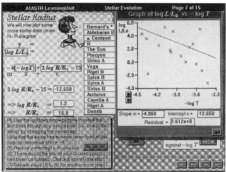

FIGURE 4. Interactive Stellar Evolution

Acknowledgements

The support of authors and staff working on the UKMCC Project is gratefully acknowledged.

8. Further Information

TLTP products, including Mathwise and SToMP, can be used freely in UK universities.
UKMCC Project, Centre for Computer-Based Learning, University of Birmingham, Edgbaston, Birmingham, UK. B15 2TT
Tel: 0121 414 4800   Fax: 0121 4146267   email:mathwise@bham.ac.uk
URL: http://www.bham.ac.uk/mathwise/homepage.htm
Software Teaching of Modular Physics, Department of Physics, University of Surrey, Guildford, Surrey, UK.
Tel: 01483 259414   Fax: 01483 259501   email: stomp@surrey.ac.uk
TLTP email: tltp@hefce.ac.uk URL: http://www.icbl.hw.ac.uk/tltp
Author email: mccabe@sms.port.ac.uk

REFERENCES

BARRY, M.J.D., 1992, A Core Curriculum in Mathematics for the European Engineer, *SEFI (Society for Education of the European Engineer)* document 92.1 .

BEILBY M.H., 1993, Mathwise, CBL Environment for Mathematics, *CTI Maths and Stats Newsletter*, Vol. 4 , No. 4 .

BRÜCK, M., 1990, Exercises in Practical Astronomy Using Photographs, Adam Hilger.

HARDING, R.D., 1996, Cooperative Cross-Platform Courseware Development, ALT-J, *Association for Learning Technology Journal*, vol. 4, no. 1, 22 - 27.

HOWARTH, I., 1996, Review of Redshift 2, *Observatory*, vol, 116, no. 1132, 194 - 6.

LAURILLARD, D., 1993, Rethinking University Teaching: A Framework for the Effective Use of Educational Technology, Routledge.

McCABE, E.M., 1994, I in the Sky: Interaction with the Universe in the Mathwise Astronomy Module, *UK Toolbook User Conference 94 Proceedings* ISBN 0 86292 427 8, 37 - 49.

McCABE, E.M., 1995a, Sliding into Mathematics: The Creative Use of Widgets for Engineering the Universe, *Eurographics UK '95 Conference*, Loughborough University, ISBN 0 952 1097 2 7 233 - 247.

McCABE, E.M., 1995b, The Incorporation of Flexible Learning Modes into Computer Based learning of Mathematics and Astronomy, *Hypermedia in Sheffield '95*, Sheffield Academic Press, 525 - 532, ISBN 1 85075 5728.

A Virtual Telescope for the Open University Science Foundation Course

By Andrew J. Norton[1] AND Mark H. Jones[2]

[1]Astronomy Group, Department of Physics, The Open University,
Walton Hall, Milton Keynes MK7 6AA
[2]Centre for Educational Software, Academic Computing Service,
The Open University, Walton Hall, Milton Keynes MK7 6AA

1. Introduction

The Open University is the UK's foremost distance teaching university. For over twenty five years we have been presenting courses to students spanning a wide range of degree level and vocational subjects. Since we have *no* pre-requisites for entry, a major component of our course profile is a selection of foundation courses comprising one each in the Arts, Social Science, Mathematics, Technology and Science faculties. The Science Faculty's foundation course is currently undergoing a substantial revision. The new course, entitled "S103: Discovering Science", will be presented to students for the first time in 1998.

The University has always aimed to make use of appropriate technologies for delivering its teaching material. For the first time, this new version of the Science Foundation Course will make extensive use of fully integrated CD-ROM based activities. One of these is a "Virtual Telescope" package designed to give students an appreciation of what is required to measure the expansion of the Universe.

2. S103: Discovering Science

The four science disciplines of biology, chemistry, Earth sciences and physics each contribute in equal measure to the course. Whilst parts of the course are deliberately multi-disciplinary, in order to give students a feel for science as a whole, other parts of the course reflect the very different natures of the four component disciplines. The course will be studied over a thirty-two week period and is accredited at 60 CATS points at level one. (CATS stands for Credit Accumulation Transfer Scheme and is the national scheme within the UK for classifying higher education courses.) A degree is awarded for an accumulation of 360 CATS points, split between levels one, two and three. An outline of the Course is given in Table 1.

The Course begins with a multi-disciplinary opening that is strongly skills led. After another combined science block which considers the question of 'A Temperate Earth?', the course narrows down to some more discipline specific content. Within 'The Earth and its Place in the Universe' one week out of the four is spent on a Cook's Tour of the Universe, travelling from clusters of galaxies down to the solar system. The remainder of the block covers aspects of Earth science. Following this opening, the students encounter discipline specific blocks in biology ('Unity within Diversity'), physics ('Energy'), and chemistry ('The World and its Atoms'). They then complete their tour from the very large to the very small with two weeks of physics in 'Atomic and Subatomic Structure'. In this final block of the first half of the Course, students learn about electrons in atoms, nuclei in nucleons, and quarks in nucleons from the point of view of a quantum description. This is balanced by a largely experimental section about the dual wave and particle nature of light.

Table 1. The Structure of S103: Discovering Science

Block	Title	Weeks	Disciplines
1	Entering Science: Water of Life	2	Multi
Taking the world apart...			
2	A Temperate Earth?	4	Multi
3	The Earth and its Place in the Universe	4	Physics / E.Sci
4	Unity within Diversity	2	Biology
5	Energy	1	Physics
6	The World and its Atoms	3	Chemistry
7	Atomic and subatomic Structure	2	Physics
Putting it back together...			
8	Building and Recycling with Atoms	3	Chemistry
9	Continuity and Change	4	Biology
10	Earth and Life through Time	3	Earth Sci.
11	Universal Processes	2	Physics
12	Life in the Universe	2	Multi

With the second half of the Course, students begin to rebuild up to the large scale structure of the Universe. Several weeks of chemistry ('Building and Recycling with Atoms'), biology ('Continuity and Change') and Earth science ('Earth and Life through Time') culminate in another physics block, but this time with a cosmological slant ('Universal Processes'). It is this block that is home to the Virtual Telescope, of which more in a moment. The Course concludes with another multi-disciplinary case study on the theme of 'Life in the Universe'. As can be seen, there are six weeks of study devoted to each of the four science disciplines, coupled with eight weeks of multi-disciplinary activity. Of this, around two-and-a-half weeks of study are on themes that may be broadly classified as Astronomy based.

3. Course Components

Each of the twelve blocks of the Course is based around a Book text that is specifically written for the purpose, by authors at the OU. Accompanying this is a Study File that contains notes to the other activities within the Course and provides a place for the students to keep track of their own skills development and learning process.

The other Course components include twelve video programmes, ten broadcast television programmes, about twenty-six CD-ROM based computer activities and eight practical work activities to be carried out at home. The students will attend a one-week residential school which incorporates three full-day laboratory sessions on inter-disciplinary activities, a day-long field trip, a project to be worked on in groups throughout the week, and tutorial sessions each evening. Students are also allocated a personal tutor-counsellor who hosts regional tutorials throughout the year and is available for guidance and support at any time.

Assessment of the course is via eight tutor marked assignments (50%), plus a centrally marked assignment based on the final block of the Course (10%), and an end of year examination (40%).

4. Universal Processes

The penultimate block of the Course is entitled 'Universal Processes', and an outline of its structure is given in Table 2. The block links together the two most exciting areas of modern physics, namely the quest for a theory of everything and an investigation of the early Universe. The particle physics and cosmology are presented in a way that is accessible to the students with minimal mathematics, and yet builds on the background knowledge of subatomic structure and fundamental forces gained earlier in the course.

Table 2. The Structure of Block 11: Universal Processes

Section		Title	Other components
1		Introduction	
2		The evolving Universe	
	2.1	The expanding Universe	+CD-ROM
	2.2	The cooling Universe	
3		Four universal forces	
	3.1	Electromagnetic interactions	
	3.2	Strong interactions	+CD-ROM
	3.3	Weak interactions	+CD-ROM
	3.4	Gravitational interactions	
	3.5	Four forces in the Sun	+Video
4		Unified theories	
	4.1	Electroweak unification	+TV
	4.2	Further unification?	
5		The history of the Universe	
	5.1	From big bang to the present	+CD-ROM
	5.2	The future	

The block begins by investigating the two pieces of evidence which allow us to conclude that we live in an evolving Universe, namely the Hubble expansion and the cosmic microwave background. This is where the 'Virtual Telescope' sits in the Course, as discussed below. After establishing that the Universe is expanding and cooling, the question is posed as to how the hot big bang and the very early Universe can be understood. This requires an appreciation by the students of the four fundamental forces of nature. After these are each described in turn, with the help of further CD-ROM based activities, the processes within the Sun are used as an example of the four forces in action. This section is accompanied by a video programme shot at a solar observatory and a neutrino detector experiment. The unification of these fundamental forces is discussed next, with the help of a TV programme filmed at CERN. The students are now equipped to embark on a tour from the big bang to the present day, and far into the future. Again this is done with the help of a CD-ROM based activity.

5. A Virtual Telescope

The aim of this activity is to provide the student with a virtual experiment in which they measure the expansion of the Universe. Whilst the emphasis is on making the experience as 'real' as possible, we are *not* dealing with astronomy students. So we have no mention of magnitudes when discussing brightness or luminosity, and there are no coordinate positions given in terms of right ascension and declination, for instance.

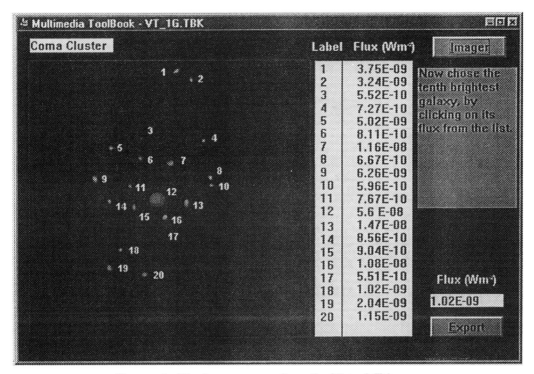

FIGURE 1. The imager screen from the Virtual Telescope

From a study of the text, the student brings to this exercise a knowledge of optical spectra and the measurement of speeds using a redshift. They also appreciate the inverse square law relating the brightness (in W m^{-2}) to the absolute luminosity of a galaxy and its distance, and an understanding of the statistical arguments behind the adoption of 'standard candles'.

5.1. *Imager*

The package opens with a visible light image of the night sky, in which eight clusters of galaxies are highlighted. The clusters (Perseus, Coma, Hercules, Ursa Major I, Leo, Gemini, Boötes, and Hydra) are chosen to be representative and at a range of redshift between about 0.02 and 0.2. On selecting an individual cluster, the virtual telescope provides an image of that cluster. The images used are from the digitised sky survey, available over the internet from STScI. Regions of sky about 10 arcmin square are used and the images have been digitized at a resolution of 1.6 arcsec per pixel. A typical view of the imager screen is shown in Figure 1. The student then has the choice of selecting either 'spectrometer' or 'photometer' to carry the investigation further.

5.2. *Spectrometer*

Selecting 'spectrometer' and choosing an individual galaxy within the cluster, the student obtains a spectrum of the galaxy in question between about 350 nm and 700 nm. These galaxy spectra are generated internally from a template spectrum of a galaxy in the Coma cluster, adjusted for the required redshift and noise characteristics. On calling up a 'standard spectrum' (actually that of a K-type star) the student is invited to

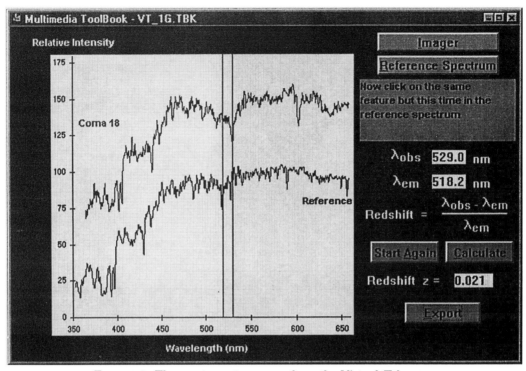

FIGURE 2. The spectrometer screen from the Virtual Telescope

identify common features in the two spectra, from which a 'redshift' is calculated. This is converted into a recession speed in km s^{-1} and exported to the on-line notebook. The procedure may be repeated for other features in this spectrum or for spectra of other galaxies in this cluster – all of which should give similar results for the recession speed of the cluster in question. An example of the spectrometer screen is shown in Figure 2.

5.3. *Photometer*

Returning to the 'imager' screen and selecting the 'photometer' option, the virtual telescope performs aperture photometry on (up to) 20 galaxies in the cluster and identifies which galaxy has which brightness. The student selects the 10th brightest galaxy in the cluster from the information presented. By comparing the brightness with that of a standard 10th brightest galaxy at a known distance, the distance to the cluster in question is determined. This distance is converted into Mpc and exported to the notebook. An example of the photometer screen is shown in Figure 3.

5.4. *Results*

The whole procedure of speed and distance measurement is repeated for up to eight clusters of galaxies. The distance/speed data that have been obtained are plotted and a straight line fit to the data is found. The gradient is measured – this is the Hubble constant in km s^{-1} Mpc^{-1}. The answers that the student obtains at each stage can be checked (i.e. the redshifts and distances for each galaxy, and the final values of Hubble constant). If deviant values are obtained the computer will prompt with re-try messages.

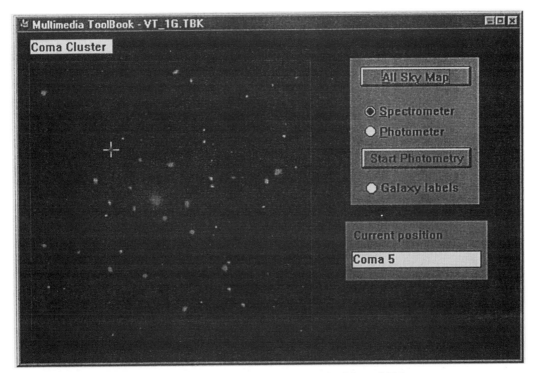

FIGURE 3. The photometer screen from the Virtual Telescope

6. Conclusions

The S103 Virtual Telescope is still under development; however, it is hoped that it will provide our students with an exciting and challenging experience of making real cosmological measurements. We hope it will demonstrate that CD-ROM based multimedia activities have an important role to play in teaching astronomy concepts as part of a distance taught, broad based, introductory science course.

The presence of multimedia in astronomy teaching

By H. J. Fogh Olsen

Copenhagen University Observatory, Denmark

1. Introduction

Sometimes I find my self in a society in the middle of The Global Village and sometimes in a society in a little state with a large number of computers not speaking the language I usually talk. When a prevailing part of the population are working in one area the society is named after that area, allthough a lot of other things can characterize the society. The latest societies are:

- Agricultural society
- Industrial society
- Information society

The agricultural and industrial societies have come to an end. When a society comes to an end, it is usually because the efficiency of production reaches a level higher than necessary, to keep all the workers busy. Many of the workers are attracted to other kinds of work, which gives rise to the next culture.

We must imagine a similar over production of information, so that the number of people occupied by producing information will start to decline. Some say that we have reach the end point already, because we have access to information from all over the world through computers, Internet and World Wide Web in an amount larger than we can handle. But that may not be true because we are waiting for large numbers of the population to learn to utilize all that information. The demand may increase for some time to come.

We can see the extremly high impact computers have on politicians, compared to their previous interest in libraries. In Denmark more money has been put into school computers during the last five years, than have ever been used for school library books.

2. Media carrying information

Up to a few decades ago it was difficult for the general public to get substantial information about more complicated subjects such as science. Although the books have existed for quite a long time they were not for the general public but only for people with more academic interest.The period starting with Isaac Newton, can be described by words and mathematics, and so it has been for 300 years. But it has also been dominated by demonstrations of the physical laws. Many sophisticated models demonstrating the physical laws at work among our planets have been produced over the years.

2.1. Planetary models from Ole Romer.

The Jovilabium made by Ole Romer demonstrates how the moons of Jupiter can orbit Jupiter and look the same as seen in a small telescope from Earth. The Eclipsarium demonstrates when either a lunar or a solar eclipse will occur. Only a few solar eclipses from the polar regions are not "discovered"by the machine. The remaining eclipses are all included at their precise time of appearance.

The planetarium, modelling all the planetary orbits, was also produced by Ole Romer. Several examples of a royal transportable model were made in Paris, and a more public model was made for The Round Tower in Copenhagen. All were hand driven. The latter now exists as a copy in Tycho Brahe Planetarium in Copenhagen. The other models made by Ole Romer exist in copies at the Ole Romer museum in Copenhagen. Since the first utilisation of electricity, a projection planetarium was soon introduced and has developed during the following 100 years to the point where we now have CRT projectors.

2.2. *Graphics and Images*

There is no doubt that the human brain is very efficient, when it receives impressions from the eyes either directly from nature or from images, graphics or drawings. Images are very valuable, but have not been used much in physics. Images from the real world are compared with nature, so the images is always related in size, colour etc.

Astronomical images made with telescopes differ from terrestrial telescope images, in the sense that you can not make a comparison of image and reality, because we cannot perceive the Universe with the naked eye.

The disappointment of observing a star through a telescope has happened to almost everybody who has the chance to do so, because the star looks the same and does not show the detail expected. The beautiful images from the sky are not available to the unaided eye with or without a telescope. They must therefore be explained in more detail than would be the case for a terrestrial image. Not many teachers in Denmark have introduced the use of astronomical images.

2.2.1 Globular Clusters

Graphic representation is so much easier, but should also be taken with care.

A classical example is the presentation of The Milky Way Galaxy. The Globular Clusters are often shown distributed uniformly around a line, occupying a rather large area. The Globular Clusters are located far from the Galactic Plane and the Central Bulge, which is not the reality. Does such display add anything to understanding of the structure of the Galaxy?

2.2.2 The H-R diagram.

If you try to find

- The location of the Sun
- The luminosity of white dwarfs
- The lowest effective temperature of a star

you may reach very different results. In two pages of a text book you must learn to understand why astronomers use absolute magnitude, luminosity, apparent magnitude or spectral type, effective temperature or colour to show the same linear relation but with the plotted line placed in a funny way because astronomers use opposite signs for the axes to normal.

2.2.3 The Moon

The last example of that sort is also a classical one. Since it was very expensive to produce illustrations for books, it was an advantage to put as much information as possible into one illustration as possible. When you want to illustrate the phases of the Moon astronomers are able to do that in one figure. From what point on earth would you be able to observe the moon as illustrated?

Who can blame the illustration of the more popular literature when they play with images to give a more fascinating background. But how is it possible for the innocent reader to distinguish between that and a scientific image produced to describe new scientific areas?

3. Cyberspace and virtual reality in a digitized culture.

- teachers' explanations and blackboard work
- prose and mathematical formulas
- graphics and images
- models and experiments

Taking up a new name like multimedia, it can be hard to imagine what it should be. After all it is nothing new, except that it is just everything from everywhere brought together at a time you wish from The Global Village.

All these media have been digitized for the purpose of being electronically handled by wire from the storage medium to the CRT. The storage medium is usually a hard disc or a CD-ROM but can be one of the many other systems in use today or new systems to be developed in the future. One advantage and important thing is the way it has been established almost free of charge and available to an overwhelming and rapidly growing number of people.

One disadvantage is the extremely fast way it changes so what you may have found interesting one day is not available the next and it happens to be published in no more than a single copy. Very often you only know the date it appeared, you do not know who wrote it or published it. Sometimes you may find the previous text etc. but it does not look the same. You cannot really rely on it for reference. Since you receive information from all over the world and from all levels of sources it is difficult even for professionals to judge the value of the data received. School pupils value is crucial because the most exciting data is often not appropriate for teaching.

The resulting consequence of this new access to information is already with us; in the future it will be more complicated to teach but maybe easier to learn.

In the old simplified model the teacher is able to control all information. The teacher selects the books to be utilized.

In Denmark, you cannot find any books describing science in the Danish language other than the teachers choice of text book. Books in foreign language are usually not available and certainly not at the high school level or popular science level.

Up to now the teacher keeps control of the curriculum since the pupils are forced to follow the set guidelines. In the near future everybody will have access to The World Wide Web and they will be able to collect information of all kinds and levels in many languages. Teachers need to change their job from carrying information to the pupils–trainers, or maybe word coachs will be more appropriate in the future.

Probably the most obvious thing is to start at university level but very soon Internet connection will be available to many schools too. We should not hesitate to take the initiative to set up schemes to evaluate the quality and value of information on the Internet. How can we ensure that the information obtained from the Internet is of the necessary quality? At least in Denmark, the astronomical or probably physical curriculum must include some new items:

- evaluation of source material
- testing individual students progress
- recording the transient Internet sources
- reviews of www information

It is not the time to find a standard, but it is the time to discuss how we ensure that the previous high standard of information at school level is maintained when information of all kinds of quality continually arrives from the World Wide Web.

The Student Learning Process

What to Cover and When

By Philip M. Sadler

Science Education Department, Harvard Smithsonian Center for Astrophysics and Harvard Graduate School of Education, Cambrdige MA, USA

Imagine trying to teach reading to students who do not know the alphabet or driving to someone who does not know the purpose of the brake. As teachers, we have a view of what the fundamental ideas that our field are and make decisions about their coverage and order in our courses. Yet, research shows that students rarely have the foundation that we expect; they hold misconceptions about the physical world that actually inhibit the learning of many scientific concepts. Moreover, the metaphors that we employ for building student understanding: reliving the historical development of the field, journeying from the closest to farthest reaches of the universe, and observing the objects in the sky, are only based on our own beliefs in their effectiveness. Empirical evidence shows that they are of little value; there is rarely any lasting change in students' conceptual understanding in science. Yet, by testing large populations, one can tease out the relative difficulty of astronomical conceptions, which misconceptions inhibit understanding of scientific ideas, and which concepts are prerequisites for others. These relationships allow the determination of an intrinsic structure of astronomical concepts, the way in which novices to experts appear to progress naturally through to an understanding of the field. Such a structure has application in the classroom. Certain ideas appear to be so fundamental to understanding light, scale, and gravity that no headway can be made until they are mastered. If we learn to set realistic goals for our students and teach the prerequisite notions prior to the more exotic ones, we may be able to optimize student learning and build understanding that outlasts the final exam.

Alternative frameworks amongst University of Plymouth Astronomy Students

By Mike P.V. Broughton

Centre for Teaching Mathematics, University of Plymouth, Plymouth, PL4 8AA

1. Introduction

In recent years much research into conceptual understanding of science has been carried out. Oddly, Astronomy (one of the smallest sciences in terms of pupil numbers) is possibly one of the most widely studied subjects, with numerous papers being produced revealing the intuitive ideas of (usually) young school children. Within these papers it is generally recognised that if students cannot assimilate the fundamental concepts of a subject, then their own initial frameworks are altered accordingly, producing mis-conceptions.

Much of this research into pre/mis-conceptions, alternative frameworks etc, has been concerned with the knowledge of gravity or the shape of the Earth, the Sun and other such bodies. Another area heavily researched is that of phases/eclipses, and how the young children of today perceive these phenomena.

The research presented here takes the findings from earlier papers and extends it by assessing astronomy students at the University of Plymouth. The experiment probed two areas, the phases and eclipses of the moon and Sun and the ability of students to de-centre.

2. Previous Studies

It has been known for many years now that children usually start to think of the Earth as flat (Vosniadou *et al* (1989)), with age usually removing or adjusting initial frameworks. This may be demonstrated by assuming we have two children, A and B, which both hold the notion of a flat Earth. From the flat Earth model, child A may 'leap' to the concept of a spherical Earth straight away; the child's flat Earth conceptions have been removed and replaced with a model which the child is able to associate with 'space' and thus a spherical Earth. Child B may only have an adjustment of models however, and would 'add' to its current framework the notion of a spherical Earth. This adjustment has two consequences. The original model of a flat Earth is reinforced in the child's memory, not because the child has been told that flat Earths exist, but because the child has not been told. The second, and more damaging† in terms of knowledge is that the child now falls into the category of 'Dual Earths'.

A 'Dual Earth' situation is one in which the flat Earth is the object which Humans, and all life, live on. The second Earth, a spherical body, is floating in the 'sky' and is usually unobservable by people on the ground. The only way in which we may observe this second Earth is by going 'up'. This happens to be true; the only way we can view the whole Earth is by being in space. Pictures which have helped cause the child to transfer between notions are all of the Earth taken from space, and since it is not mentioned that this is the **one and only** Earth, the child constructs an additional branch to the current model enabling it to again fit with observable data.

It is apparent in all research that the acquisition of notions, culminating in the correct scientific view, is gained with age. The age spread is typically from about 8 years (still in flat Earth mode) to 15 years old (Correct notion). Mali and Howe (1979) and other researchers have shown that not only does age play a part, but also the culture in which the child is brought up. The results of an experiment conducted by Mali and Howe showed that children in America were approximately one year ahead of their neighbours in Nepal, even with similar schooling. (Approximately is

---

† The term damaging is used because the model allows the child to tackle quite complex astronomical events and surface with the correct answer, such as phases, whilst holding an incorrect model.

used because at 15 years old, the end of the study, a large proportion of Nepali children still believed in a flat Earth, which Mali and Howe linked to local religion).

When a child has reached the stage of understanding space and approximate spatial awareness, researchers have switched their main focus to that of how the Solar System moves. Through objects such as polystyrene spheres and balls, researchers ask the subjects to animate the Solar System and roughly indicate what orbits what.

3. Advancement of previous surveys

Not much research has been carried out on students after they have left school, although this is probably due to the subject of Astronomy being small and normally only offered at University level. University Astronomy is not concerned with the shape of the Earth, the Moon's orbit (although it does normally cover phases and eclipses) and how our Solar System works, but is aimed more at mathematical and historical events, such as Dopper shift, relativity and Galileo's role in the advancement of science. This assumes that the students who take the subject have a firm grasp of the 'basic' astronomical events.

The University of Plymouth, and its predecessors, have taught astronomy for 100+ years and it was decided that the students who participate in the course should be tested for their understanding of these 'basic' events. The course is a three year minor pathway on a honours degree, with approximately 106 students in total, although around 80% of these students are in the first year.

All three years were to participate in the same questionnaire survey which was designed to test for lunar phases/eclipses and also for their de-centering ability. To test the students' phases notions, three questions were presented in the form of a pictorial multiple choice. Each question featured a picture of a lunar phase, with six possible answers each showing a different configuration of the Earth, Sun and Moon.

Two eclipse questions were asked in a format identical to the previous question, each requiring knowledge of lunar and solar eclipses respectively.

The last two questions were designed to test for de-centering ability directly. The first was an invented countryside map with a small village in the middle. Given two views the student was required to place him/her-self on the map. This is a concept which most students are familiar with and so it was expected that the positive results of this question were to be high.

The second de-centering question was similar in style and approach as to the map question, with two views and a map being provided. The map was however a chart of the Solar System at a given time and date, with the two views of various planets. The student was to use this information to locate themselves within the Solar System.

The initial de-centering problem is very similar to those employed by Piaget (1929) in the three mountain problem. Some of the barriers which Piaget faced were overcome by use of the printed questionnaire rather than one-to-one questioning, and by the benefit of using older students who do not require prompting in any way. The second question is again similar but on grounds which students are not familiar with. In terms of method, both should be identical and thus a student who completes the first question correctly should also complete the second equally well.

4. Results from the questionnaire

The questionnaire was administered during September 1995, with all students taking the question paper within 3 days of each other. In this paper, the actual results are not going to be studied or analysed in any great detail but rather the relevant comments the students placed in the space at the end of each question for justification and some reflections from the one–to–one interviews.

Questions one through three were asking for answers to problems involving the phases of the moon. Questions four and five asked about lunar and solar eclipses and the last two were 'map' locating exercises. The percentage of correct responses to each question and a brief outline of the written responses follow.

Question 1:   Link a picture of a thin new moon to its orbital position
It was evident early in the marking that the students were answering correctly yet were giving the wrong reasons for the event. The scores for years 1,2 and 3 are 44%, 50% and 88%.

Question 2 :   Link a picture of a 1st quarter moon to its orbital position
Scores for year 1, 2 and 3 were 75%, 75% and 88%. This question often raised the response that it had to be 'that one' because half of the moon was visible which implied that a 90° angle must be present.

Question 3 : Link a picture of a moon just after full to its orbital position
Scores for years 1,2 and 3 were 46%, 62% and 88%. It should be noted that the third year results are slightly tainted by one person who got every phase question incorrect. Also, the candidate was the only student who justified the correct reasoning yet the wrong picture.

Notes on the phases question
Following is a table showing a summary of the justifications of the students and whether the student got the question correct. (Remarks that were repeated frequently are indicated by an asterisk).

Table 1 - Student Justifications

Justification	Years	Solution
The Moon blocks the light	1	Correct
Earth obstructs light	1	Correct
Earth silhouettes moon*	1,2	Incorrect
Shadow of Earth obstructs	2	Incorrect
Earth blocks light*	1,2,3	Incorrect
Earth casts shadow	2	Correct
Only part of moon lit*	1,2	Correct
Light side points away from Sun (!)	1	Incorrect
Moon is at same distance as Sun	1	Correct

The Table shows an alarming trend which was seen throughout the questionnaire. A number of correct solutions were supplied, but the justification was often widely incorrect, a situation which is commonly termed dual-perspective (Berry *et al* 1991). Table 1 was created mostly from the first years' answers as years 2 and 3 were not as co-operative and often left the justification blank. Of the nine 2nd years, 4 answered with mis-conceptions such as the Earth casting a shadow. One third year remarked that the closer the moon is to the Sun, the brighter and thus more 'visible' it becomes.

The dual-perspective may be equated with a framework similar to that of the Dual Earth model held by children and mentioned earlier in this paper. The concept of the Earth- Moon-Sun system and what phases look like when is 'programmed' in, but the underlying conception of a flat Earth style framework exists in memory.

Question 4 : Match a picture of a Lunar eclipse with its orbital position
Score for years 1,2 and 3 were 53%, 38% and 78% respectively.

Question 5 : Match a picture of a Solar eclipse with its orbital position
Score for years 1, 2 and 3 were 75%, 38%, 78%. The low score for this and the question before are due to reversals in both the answer and justification; i.e. a solar eclipse became a lunar eclipse and vice versa. The high percentage of correct answers in year one was also a surprise

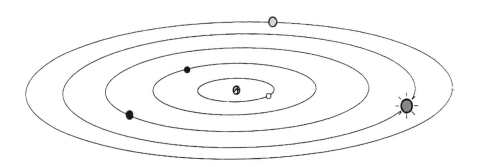

FIGURE 1.

when compared to the previous phase results and may be attributed to the amount of media coverage eclipses obtain.

Question 6 : Locate oneself on the map provided given the views due north and south
This question was, unsurprisingly, answered very well by all years with scores of 95%, 100%, 100% for years 1, 2 and 3. The general opinion of this question was that they (students) could not find a reason for asking it.

Question 7 : Locate oneself on the starchart provided : two views given, 180° apart Year one did not like this question at all with 36% of the students not answering and of those that did, 35% got it correct. All second and third years answered the question with scores of 38% and 44%.

5. Conceptions of our Solar System

It quickly became apparent that the question on position finding in the last de- centering question was posing quite interesting problems. As an extension to the question paper, one–to–one interviews were employed which discovered that the students at the University of Plymouth have pre-conceptions about the Solar System; some of which may only be described as pre-Copernican.

Following are the results of the extension study. The results, which also included drawings by the students, were able to be classified into five notions. Notion five is the most advanced and is the current scientific view of the Solar System.

Figure 1 is a pre-Copernican view or our Solar System. It was found that 8% used this model when talking about our Solar System. Of these students, 4% thought that the Earth was the centre of the Universe. Nearly all agreed that the order of bodies was Earth, Moon, Mercury, Venus, Sun, Jupiter etc. This framework has been dubbed Notion One.

Notion Two, shown in Figure 2, has all but the Earth in its correct position. All planets orbit the Sun except the Earth which is the centre of the Solar System and thus has the Sun orbiting it. The Moon also orbits the Earth and is inside the orbit of the Sun. 6% of students indicated this model.

Notion three (Fig. 3) finds the Sun being placed firmly in the centre of our Solar System. 8% of the students indicated that this was their framework with 40% of those thinking that the Sun was not the centre of the Universe. This is the first notion in which one student positively states that the Sun is moving in a direction away from the Big Bang.

One student (<2%) indicated this model which is included as Notion four (Fig. 4) purely because it demonstrates the method of additions to a concept. Apart from the moon's orbit (dotted) the model is correct. On questioning, the student replied that the moon went round

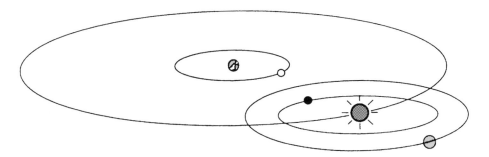

FIGURE 2.

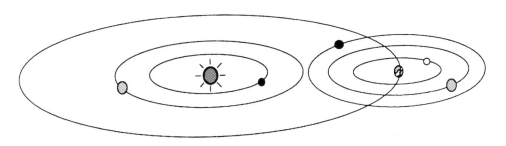

FIGURE 3.

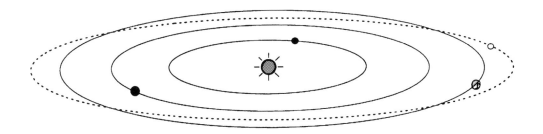

FIGURE 4.

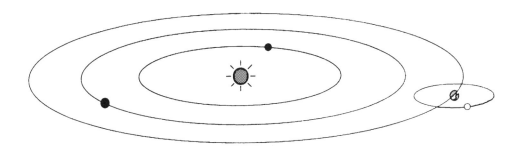

FIGURE 5.

the Sun, and its distance from the Sun made the phases. When asked to draw the model, the student drew the above picture very quickly and explained that the outer lobes are when the moon is 'growing' (waxing) and the inner when it is 'shrinking' (waning).

The correct model, notion five (Fig. 5), was chosen by the remaining 73% of the students. 11% of these students stated that the Sun moves in some manner, often away from the Big Bang. A further 28% said that they were almost certain that the Sun was static with the remainder either happy to abstain or not fall categorically either way.

6. Further study

The results found in this paper were slightly worse than expected. Further analysis and more students were thought necessary and so the Open University was approached. The Open University Astronomy course S281 is the largest single course in Europe (student numbers) and they kindly agreed to the University of Plymouth testing all the students *prior* to their taking the course. (S281 is a 2nd level programme).

After the University of Plymouth questionnaire, which is now seen as a pilot project, the last two questions were removed (the map questions on de-centering) and replaced with questions dealing with orbits of the Solar System and also on the wider universe. The questions asked both for movement and diagrams to be given. One phases question was also removed leaving two similar.

In total 1200 questionnaires were sent out to open University students, with the current number returned at 347. It is expected that not many more will be returned. Initial results are appearing to be very similar to the results gained at the University of Plymouth, and are listed below in Table 2.

Table 2

Question Overview		Common Justifications for phases
Correct phases response	78%	Moon Blocks light
Correct eclipse response	91%	Earth casts a shadow
Correct Solar System drawn	81%	Moon becomes more reflective
		Moon moves away from Earth

7. Conclusion to the surveys

The findings presented here are shown in a compacted form, yet still show the general trend of 'basic astronomical models' held by undergraduate astronomy students. Although no other research exists in this field, the data does follow from other researchers' findings such as Acker and Pecker (1990), who quizzed the general public, and also those of Durant (1989).

The research shows that phases are often able to be correctly positioned with respect to the Earth-Moon-Sun system, yet the underlying knowledge of the events is still incorrect.

Evidence of altering frameworks exists, allowing the student to fit new models to existing ideas allowing events, such as phases, to again 'work'.

The result of the full survey of Open University students is still being analysed but will soon be complete. Preliminary results of the survey and also those of the University of Plymouth survey suggest that Astronomy courses need to take account of the students' 'basic astronomical conceptions' before attempting to teach more advanced techniques.

If this is not adhered to, we may well find students applying relativity, albeit correctly, to Ptolemy's Universe.

REFERENCES

ACKER, A. & PECKER, J.C., 1990 *Public misconceptions about Astronomy* Cambridge, Cambridge University Press, IAU Colloquium no.105.

BERRY, J. & GRAHAM, E., 1991, *Using concept questions in teaching mechanics* Int. J. Math. Educ. Sci. Technol., Vol 22, No 5, 749-757.

DURANT, J.R. 1989, *Public understanding of Science*, Nature, Vol 340, 11-14.

MALI AND HOWE, G.B. & HOWE, A., 1979, *Development of Earth and Gravity concepts amoung Nepali Children*, Science Edu, Vol 63, #5, 685-691.

PIAGET, J., 1929, *The Childs conception of the world*, Sarreod and Sons Ltd, ISBN 0710030681.

VOSNIADOU, S.& BREWER, W.F., 1989, *Mental Models of the Earth*, Cognitive Psychology, 035-564.

Identifying and Addressing Astronomy Misconceptions in the Classroom

By Neil F. Comins

Dept. of Physics and Astronomy, Univ. of Maine, Orono, ME 04469 USA
GALAXY@MAINE.MAINE.EDU

1. Origins of Misconceptions

Students come into our classrooms with many misconceptions about science in general and astronomy in particular (see numerous papers and references in Novak, 1993 and Pfundt & Duit 1993). These beliefs evolve from a variety of sources throughout childhood and adolescence (Comins, 1993a, 1993b, 1995) . I have found that directly addressing these incorrect beliefs in the context of their origins helps my students replace them with correct knowledge. By understanding the origins of their misconceptions students can screen information more effectively, i.e., they learn to think more critically. My purpose in this paper is to briefly identify origins of misconceptions and classroom techniques for replacing them.

I define misconceptions as deep seated beliefs that are inconsistent with accepted scientific information. Unless we directly address these incorrect ideas at their roots, most students cannot replace them with correct knowledge. Most students retain correct material only long enough to pass tests, and then lapse into believing their prior misconceptions.

In previous works (Comins, 1993a & 1995) I identified a heuristic set of origins that account for all the misconceptions I have identified. It is well worth noting that such a list is by no means unique and, given that I have since added another category, nor is it complete. Nevertheless, this set of origins is an extremely practical one, providing a significant set of tools for understanding and dissecting misconceptions and how these beliefs are used by different people. In an effort to make this set more tractable, I have now revised it to an even dozen (see Sections 1.1–1.12). This was done by combining several sources, such as grouping "Over-generalization", "Belief in Object Permanence", "Belief in Object Uniqueness"and "Choosing the Simplest Explanation"together under the rubric of Common Sense. The new category is Anthropomorphism.

1.1. *Factual Misinformation*

We have all received incorrect information from sources we normally trust and, therefore, whose statements we do not question or analyze very deeply. These sources include teachers, our parents, other adults, textbooks, and knowledgeable peers.

Factual misinformation is data that has been heavily pre-processed by others before we receive it. Therefore this material often contains unstated assumptions and conditions (such as limited ranges of validity) that would affect our understanding and use of the information if we knew them. Children and others who have not yet developed techniques for critically examining information from others are especially susceptible to incorporating factual misinformation into their understanding of nature.

Some examples of misconceptions that often originate this way are that there are 12 zodiac constellations; that the North Star, Polaris, is the brightest star in the night; that the asteroids in the asteroid belt were once part of planet that was destroyed; and, that the Earth's core is liquid.

1.2. *Media Minimalism*

The news provides information that raises our interest and keeps our attention. As a result, the news about a scientific discovery often covers the most sensational, rather than the most important, elements. We incorporate the filtered, biased, incomplete, and often inaccurate information reported in the media into our view of the world and so the conclusions we draw or extrapolations we make from this knowledge are often incorrect.

1.3. *Cartoons and Science Fiction*

Cartoons are a major external source of misconceptions about science for young children. Animators have done considerable damage to young intuitions about nature. For example, a common cartoon ploy ("The Roadrunner" comes to mind) is for the bad character to run off a cliff. When the bad guy rushes off the cliff, he stops moving forward, pauses for several seconds until the impact of what has happened is clear to both him and the young audience, and then he falls straight down.

At an adult level, people willingly suspended their disbelief in order to be entertained by exciting, but physically implausible, science fiction. Unfortunately for the general perception of nature, much science fiction actively violates physical laws. Misconceptions created by science fiction have increased dramatically in the past few decades because of the fantastic animation capabilities now available, making science fiction much more realistic than ever before. A virtually universal belief created by science fiction films is that asteroids are closely clustered together.

1.4. *Mythical Concepts*

Mythical concepts are those we accept "on faith," without sufficient evidence or without a desire to make a rational decision. Literal interpretation of the Bible, for example, requires accepting that the Earth and the rest of the universe formed in six days. Making this a foundation of one's belief about the creation of life leads to numerous misconceptions and conflicts with observations of the natural world.

Similarly, believing in magic as a supernatural power, rather than as sleight- of-hand, or in ghosts or witchcraft, requires accepting that there are rules governing the universe that are outside the realm of science. Such beliefs cause many people to interpret events in non-scientific ways that can then lead to actions contrary to those implied by scientific explanations. For example, believing that prayer can cure illnesses has led to many needless deaths and much suffering.

Related to belief in the supernatural, there is a fascinating layer of thought that everyone experiences called "magical or wishful thinking." It begins in childhood, before experience shows children how the natural world operates. Children frequently invoke magic as a means to an end they cannot otherwise achieve. By the time we are adults, such desires have evolved into "wishful thinking." At some level most of us know that wishing does not cause things to happen, but that does not stop us from trying.

1.5. *Language Imprecision*

In normal usage, most words have several meanings. We overcome imprecision in our language by taking into account the context in which words are used, as well as by adding verbiage to clarify matters or by asking questions. In science, however, every word has a single, rigorous usage. Often a scientific definition of a word is not one of the common ones in everyday usage. Among the innumerable confusing words are "black" in black hole; comet "tail"; "shooting star"; "spring" tide; and Saturn's "rings". To understand science, students must learn the correct meanings of words and not let common usages color perceptions of scientific meanings.

There are also many words that have limited meanings even to non- scientists, but most people never get the meaning correct in the first place. Common among these are "Sun" (often interpreted as a special kind of object, rather than as an ordinary star), "pulsar," "meteor", "meteoroid", and "meteorite" (the latter three often taken to have the same meaning).

1.6. *Erroneous Personal Cosmology*

Cosmological concepts and misconceptions are those that deal with the origins, sizes, locations, motions, and ages of the universe, the solar system, the stars, and the Earth. Within our "personal cosmology" we each have explanations for some, if not all, of these issues. When we encounter a question about the universe that we have never addressed explicitly, our personal cosmology is where we look first for an answer. As an example of this, ask your students where the Earth is located in the Milky Way galaxy. Referring to their personal cosmologies, students will often respond (incorrectly) that we are at the center of it, outside it completely, or in a spiral arm.

All of the elements of personal cosmologies lead to misconceptions. Needless to say, theology often has a bearing on these misconceptions. Notable among the non–theological beliefs are that the distances between stars and galaxies are much smaller than they actually are; that the Sun and other stars last forever; and that the Sun is immobile in space. Many of these misconceptions are based on a lack of specific information, coupled with the use of human-sized distances.

1.7. *Incomplete Understanding of the Scientific Process and Scientists*

The process of doing science includes many features that are not well understood by the public. These include expressing science in mathematical terms, making predictions from these mathematical theories, testing predictions, modifying or replacing theories that make incorrect predictions, using computers to explore the complexities of modern scientific theories, making accidental discoveries, and creating new technology to help understand nature, among others.

Most people also envision scientific information as complete, certain, and unchanging. Therefore, when they learn something about science from a "reliable source," most people incorporate that information into their world view. Thereafter, it is very hard to change their belief about that subject. Furthermore, science is incomplete in that it does not explain everything in any discipline.

Not knowing what most scientists do, many people envision us all as white- haired men and women, spending long, lonely nights looking through telescopes in hope of making new discoveries.

1.8. *Incomplete Information/Reasoning*

Intelligent, well-educated non- scientists often reason correctly about science as far as they go, but they do not include all the scientific facts related to the subject at hand. As a result, they draw incorrect conclusions that lead to misconceptions. Furthermore, when asked about something we have never considered, we all tend to fill in the blanks in our knowledge "on the fly" and without sufficient information.

For example, a student told me he believed that the larger a star is, the more mass it must contain. His reasoning was that the greater the volume, the greater the mass packed into it. What he did not realize is that densities vary from star to star.

1.9. *Misinterpreting Sensory Information*

Our senses pre-filter incoming information before it enters our brains. For example, our eyes are not uniformly sensitive to all the colors of the rainbow, being most responsive to yellow light. Sensitivity to color falls off towards both red and violet. Therefore, what we perceive as the color of each object is not its true color, because the intensities of all the colors from it are changed in our eyes before the color information enters our brain. The same applies to the other senses, as people who differentially lose sensitivity to pitch know all too well.

Furthermore, even when we see a true representation of events, our uncritical interpretations of them are often incorrect (see also Common Sense, below). Examples of misconstrued perceptions include mirages, observations that the Sun, planets, and stars orbit the Earth (they certainly appear to do so); that the Earth is flat; that the Sun is yellow; and that stars twinkle.

1.10. *Inaccurate or Incomplete Observations*

People observe things inaccurately for a variety of reasons. These include not knowing what to look for (a person believing that all the stars in the sky are white is ill-prepared to notice that many of them are colored); not being sufficiently interested in making careful observations; being distracted by certain features of what they see; mis- categorizing what they see; and being biased by what they have been told beforehand.

Other examples are that the same constellations are seen at night throughout the year, that no planets are visible with the naked eye, that the Moon is only visible at night, and that the sword of Orion is just composed of stars.

1.11. *Anthropomorphizing*

Among the myriad things every child must learn is that they are separate entities from their care givers and from the rest of the world. This realization takes several years and the earlier stage of seamless attachment children have to parents and other care givers leaves traces in the anthropomorphizing that virtually all of us do as adults. We attribute human motives and desires to creatures who function solely by instinct and autonomic response to the outside world, as well as to plants and trees, and to inanimate objects like the celestial bodies. By responding to the rest of the natural world as if it had awareness and other human traits, we develop misunderstandings about how things work.

1.12. *Common Sense*

Common sense, defined as "sound practical judgement that is independent of specialized knowledge or training"(Webster's College Dictionary), is an essential facet of our interaction with the world. While it is often useful in everyday life, common sense frequently leads us to believe concepts inconsistent with the laws and behavior of nature. The problem is that so much of what goes on in the natural world defies "common sense."For example, common sense tells us that heavier objects will drop faster than lighter ones. The difference between common sense and scientific fact has led to a growing divergence between what most people believe and what scientists know to be correct.

Common sense overlaps a variety of other sources of misconceptions. Also, the "common sense"activities we each use in reconciling new information with current beliefs depend on learning patterns we developed as children. Here are some specific error-prone common sense activities:

1.12.1. *Over-generalization*

The invalid generalization from the properties of objects we know to ones we do not know causes many misconceptions. We make these generalizations both as we are learning new things and as we are asked about things we have never thought about before. For example, when asked to describe the surfaces of the other planets, most people generalize, from the Earth, that they have solid surfaces. When asked to describe the other moons in the solar system, many people say that they are all spherical, like our Moon.

1.12.2. *Uniqueness*

Incorrectly assuming that objects are unique has the opposite effect to over-generalizing. For example, because Earth is the only body in the solar system with life on it, many people believe that the Earth is also the only body with water. Many people also believe that Earth is the only planet with an atmosphere.

1.12.3. *Permanence*

Most children develop a belief that their social world is static, meaning that the people and objects dear to them will be around forever. This belief is motivated by the fear of change (desire for permanence), a feeling that often continues into adulthood and is therefore applied to new situations and concepts by adults. A belief in permanence can also originate in the lack of perceived change in the Earth or its cycles. In most places on Earth the planet's surface does not change during our lifetimes; the day is always twenty-four hours long; and, the same patterns of stars are visible each year. We know, of course, that while these cycles are reliable over the span of a human life, belief in permanence on astronomical time scales is completely unfounded.

1.12.4. *Choosing the Simplest Explanation*

Most people choose the first, and therefore usually the simplest, explanation that they think of for a new situation. Referring to that most ubiquitous of misconceptions, most people assert that summers are hotter than winters by analogy with the fact that the closer you are to a fire, the warmer you feel.

2. Classroom Remediation of Misconceptions

Remediating misconceptions is made difficult by several factors. First, of course, people are resistant to the idea that their beliefs are incorrect. As a result, we tend to take new information and distort and modify it to make it consistent with our current understanding of any topic. Second, many misconceptions are used to help explain a variety of other things, so removing a misconception can undermine many other ideas we hold. Third, different people give different weight to different common sense beliefs, such as the uniqueness of objects and ideas, or the power of generalization. As a result, different people often hold diametrically opposite misconceptions on a given topic. For example, those people who focus on uniqueness often assert that our Moon is the only moon in the solar system, while those who over-generalize often believe that each planet has one moon.

2.1. *Make students aware that they harbor misconceptions*

This is hard because nobody likes to be told that their beliefs are wrong. The most effective ways I have found of doing this are by using counterintuitive physics demonstrations (levitating a beach ball in a tilted air stream works well) and by taking on some common astronomical misconceptions at the beginning of the semester. That changing distance from the Sun does not cause the seasons and that the Sun is not yellow (i.e., that its most intense radiation is not yellow, but blue- green) are useful.

The process of awakening misconception awareness is important. Invite students to think about (and if they are brave or confident enough, tell) what they believe to be correct before doing a demonstration or giving a correct answer.

"How many of you think that the seasons are caused by the changing distance from the Earth to the Sun?" This type of question is valuable because it makes the students active participants in the process. It also invariably creates anxiety, which should also be acknowledged and explained to the students as useful in confronting incorrect beliefs.

Less effective are questions like "What causes the seasons?" "What color is the Sun?" "What is the shape of the Earth's orbit around the Sun?" "What planet is farthest from the Sun?" The problem with this format of questioning is that if a student confidently gives the correct answer, it can dissuade other students with incorrect beliefs from getting involved.

2.2. *Assure students that misconceptions are unavoidable and that they do not imply stupidity*

Corollary: Students (indeed, everyone) look for someone to blame for their own misinformation. In this case, that is often justified. Anger and frustration often focuses on teachers, religious leaders, parents, friends, the media.

2.3. *Teach students the origins of misconceptions*

The importance of knowing where misconceptions come from is that it helps students more readily accept that they have received or processed misinformation in the past and it helps them evaluate their sources of information more critically in the future. Changing ones' beliefs in the face of irrefutable evidence is hard enough; doing so without knowing how and where the incorrect information that we have originally came from is much harder. Knowing, for example, that our senses are not "ideal" makes us more sensitive to other possible interpretations of what we perceive.

2.4. *Assure students that with suitable effort, they can unlearn many misconceptions*

As discussed above, the process is not easy, but misconceptions can be replaced with correct knowledge. I have found that the more students understand the process of replacing incorrect information, the more apt they are to try and change.

2.5. *Teach the scientific method and critical thinking*

As noted above, most students do not understand what science is, how it functions, or how scientists operate (see, e.g., Lett, 1990). It is worth introducing students to the relationship between theories and observations/experiments, as well as to the predictive power, falsifiability, repeatability, and simplicity (i.e., Occam's razor) of scientific theories.

2.6. *Ask "What If?" questions about misconception-prone topics and explore the results*

"What if?" questions allow you to show students the implications of both incorrect and correct beliefs (Comins, 1993c). By following misconceptions to their logical contradictions, you can forcefully demonstrate why some incorrect ideas must be changed. An example of this is assuming that the Sun is shining by burning gases. Assuming that the Sun's entire mass was composed of fuel and oxidizer, it would have completed combustion billions of years ago and "gone out." In that case, of course, we could not be here.

2.7. *Keep students focused on this issue throughout the course*

Point out common misconceptions throughout lectures. I often offer extra credit to students who provide me with lists of misconceptions (along with a statement of the origins of each belief) that I corrected for them during the course. This incentive keeps them thinking about their prior beliefs, the origins of these beliefs, and the difference between what I say and what they thought they knew. I also include pedagogy designed to keep students focused on misconceptions throughout my new textbook (Kaufmann and Comins, 1996).

REFERENCES

COMINS, N.F., *Misconceptions About Astronomy: Their Origins*, 1993a, in, Proc. of the 3rd Inter. Sem. on Misconceptions in Sci. and Math. ed Novak, J.D. (see below).

COMINS, N.F., *Misconceptions About Astronomy: Their Origins and Effects on Teaching*, 1993b, abstract in Bulletin of the Amer. Astro. Society, 25, #4, 1430.

COMINS, N.F., WHAT IF THE MOON DIDN'T EXIST? 1993c, New York: HarperCollins.

COMINS, N.F., Addressing Common Astronomy Misconceptions in the Classroom, 1995, in Astro. Soc. of the Pacific Conf. 89, "Astronomy Education," ed. John Percy.

KAUFMANN, W.J.III, & N.F. COMINS, 1996, *Discovering the Universe 4th*, NY:Freeman.

LETT, J., A Field Guide to Critical Thinking, in Skeptical Inquirer, V14, Wint.1990.

NOVAK, J.D., 1993, Proceedings of the Third International Seminar on Misconceptions and Educational Strategies in Science and Math, Cornell University, Homepage: http://meaningful.education.cornell.edu/miscon/homepage.htm

PFUNDT, H. & REINDERS D., 1993, Bibliography: *Students' Alternative Frameworks and Science Education*, Kiel, Germany: IPN at the Univ. of Kiel.

Learning Effectiveness of Lecture Versus Laboratory: are labs worth it?

By B. Hufnagel[1,2], E. Loh[1] & J. Parker[2,3]

[1]Department of Physics and Astronomy, [2]Division of Science Education, [3]Department of Teacher Education, Michigan State University, East Lansing, MI USA

1. Introduction

Michigan State University (MSU) serves a large and diverse student population, $\sim$ 1000 of whom take the astronomy course for non- science majors each year. Significant resources are also invested in the related astronomy lab, enrolling about half the lecture students. Although this lab is optional, the students are required to complete one lab course for their degree. In the fall of 1995, we undertook an extensive assessment of student learning in these astronomy courses.

2. The Student Population

Unlilke most astronomy research, information about the entire population under study (403 students) was available. This included name, major, grade earned, and concurrent enrollment in lab and lecture. Fig. 1(a) shows that the shapes of the grade distributions differ for the day and evening classes, and that neither is Gaussian. Therefore the day and evening classes will be analysed separately, and statistics such as *mean* and *standard deviation* are good descriptors for only the day-class students receiving a 1.0 lecture grade or above. The lab grades were also plotted for the day and evening classes separately, and no difference in the shapes of the distributions were apparent (Fig. 1(b)). This indicates that the different grade distributions of the day and evening lectures are lecture-dependent, rather than rooted in the nature of the students taking day versus evening classes. The lecture and lab grades were also plotted for males versus females (gender information was not available for eleven students), and no gender bias was evident.

One way to assess the effectiveness of the lab curriculum is to determine if those students taking the lab plus lecture concurrently have higher grades in the lecture than students taking only the lecture. Figs. 2 and 3 show that the lecture grades for students talking the lab are better. However, this is statistically significant for the women in the day lecture (Pearson's $r = 0.34$), but not so for the men (Pearson's $r = 0.94$) (see Press et al. 1986 for a discussion of Pearson's Coefficient of Correlation r). Another test of effectiveness is to look for a correlation between a high grade in lab and a high grade in lecture. A student's lab grade was found to correlate weakly with his or her lecture grade, with a Pearson's r value of ~ 0.51 for both sexes.

The population statistics can also test the hypothesis that a computer-intensive lab course is gender biased. No gender bias either in avoidance of the lab by women, in retention of women in the lab course, or in lab grades earned by women was evident. There were 48% and 50% women lab students at the beginning and end of the semester, respectively.

3. Introduction to the Survey

The assessment instrument for this fall 1995 study was an entrance and exit survey developed by us to test the explicit goals of the lab course. Of the 403 students enrolled in three lecture sections, 271 entrance and 192 exit surveys were obtained. About half of these students also enrolled for the optional lab course. Anonymity of the students encouraged candid feedback, and the entrance surveys were administered by faculty other than the assigned professor.

The survey consisted of two parts, the first being background questions, and the second twelve

Population Grades: Day versus Evening Classes

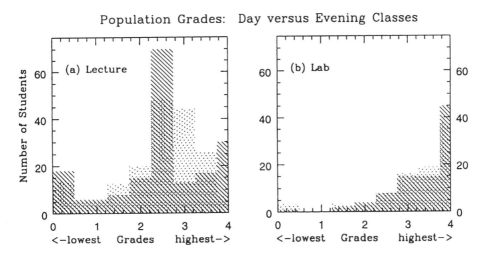

FIGURE 1. The distribution of grades for day (lighter shade) and evening (darker shade) (a) lecture and (b) lab classes. The day-lecture data have been scaled by 0.80 and 0.67 (lecture and lab, respectively), to match the smaller evening-class size. The grades for the day and evening lecture classes are distributed differently, and none of the four are Gaussian.

Effect of Lab on Women's Lecture Grades

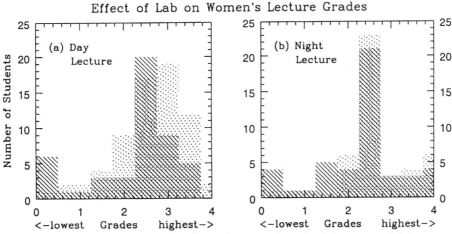

FIGURE 2. Comparison of lecture grades for women taking both lecture and lab versus those taking the lecture only. Panel (a) is the day-lecture grades, and Panel (b) the night-lecture grades. The dark- shaded histograms are grades for women taking the lecture only, and the light ones for women taking both the lab and lecture.

content questions. The twelve questions had two parts, the first a bimodal (e.g., yes/no) choice, and the second an open-ended essay question probing the student's level of understanding.

A concern with any sample is if it is representative of the population. The entrance survey included more women than men (58 and 40% respectively), while the exit survey was representative of the population's almost even split. Other questions, however, such as math background and why they chose astronomy rather than another science course, were answered in similar percentages in both the exit and entrance surveys.

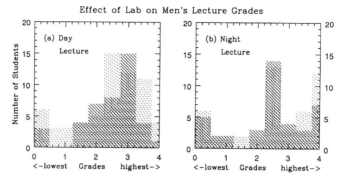

FIGURE 3. Comparison of lecture grades for men taking both lecture and lab versus those taking the lecture only. The dark-shaded histograms are grades for men taking the lecture only, and the light ones for men taking both the lab and lecture.

4. The Survey Results

Was the lab more effective than the lecture? This depended on the lab. For example, the lab on variable stars was less successful in changing a widely-held ($\sim$ 75%) misconception that space-based observing is necessary to observe variable stars, as 59% of lab students vs. 50% of non-lab students still held the misconception at the end of the course. (Only a handful of students were willing to explain their answers.) However, the second lab on measuring mass using dynamics shifted the lab students' partial understanding to a complete one (71% of lab students). The non-lab students' understanding at the end of the course resembled that of the lab students after the first dynamical-mass measurement lab. Almost all of the students began the course knowing that it was possible to measure mass dynamically, but the *lab students where able to explain how to do it.*

Did the lab course enhance retention of knowledge? The "entrance"survey was given after several of the labs had already been taught. The two surveys then measured, for these four labs, whether the students retained the knowledge to the end of the semester. The results are mixed, with the lab-only students doing better in retention on some labs but worse than the non-lab students on others.

Are the students taking lab different from those taking only the lecture? Both lab and lecture students gave "The course sounded interesting"as their most common reason (45%) to choose astronomy. Slightly more lab students expressed a prior interest in astronomy (28%) versus those not taking the lab (20%). The maths backgrounds were similar, but the most significant difference was that lab students were far more likely to be first-year students (40% vs. 14%).

For the exit surveys, we added an open-ended question about the source of their knowledge of astronomy. The most popular response was television (53%), followed by school (33%).

5. Discussion

When the title of a paper is phrased as a question, the answer is usually "No."However, in this case, the answer to the question "Learning effectiveness of lecture versus lab: are labs worth it?"is "It depends."It depends on the lab itself, the gender of the student, and how it inter-relates with the lecture. For example, the second lab on the same topic, such as our two labs on dynamical mass, shifted the typical partial understanding to a complete one. Although lecture grades were in general better for students taking the lab and lecture concurrcutly, only for women in the 250-student day class was the difference statistically significant. We suggest that this is more strongly lecturer- than student-dependent, though it is possible that women in extremely large classes respond particularly well to the cooperative learning format of our labs.

The question as to where students get their understanding of astronomy is an interesting one. We intend to explore this further as guidance for our community outreach efforts.

We recommend that an institution considering changes to its astronomy lab curriculum first define the purpose of the lab. For example, is the lab intended to complement the lecture course with active learning of the same topics? Alternatively, a lab course might be designed to add breadth, such as hands-on use of the analytical tools used by astronomers or familiarity with the night sky. If the lab is meant to complement the lecture, the second step is to compile population statistics to test if this purpose is already being met. This can inexpensively answer the question posed by this paper for current and past lab courses, identify gender or other biases, and discriminate between the need for drastic change in the curriculum versus maintenance. A student survey can then be designed to further investigate which labs work and which don't, to better understand the student population, or to assess the effectiveness of changes as they are introduced.

Acknowledgments

We appreciate the cooperation of the two instructing faculties and their students who cheerfully participated in the survey. We thank the teaching assistants, and also R. Wilhelm for his comments on this paper. BH also thanks C. Suelter for conceiving of and promoting her postdoctoral position in astronomy and science education research.

REFERENCES

PRESS, W.J., FLANNERY, B.P., TEUKOLSKY, S.A., & VETTERLING, W.T., (1986) Numerical Recipes, New York: Cambridge University Press, 630.

Robot Telescopes: a new era in access to astronomy

Department of Industrial Technology, University of Bradford, Bradford BD7-1DP

1. Background

For the teaching of astronomy there can be no alternative to the hands-on experience of using instruments on a real telescope observing on a clear dark night. Such experience is not possible for millions of students who are excited by the ideas of astronomy. It is not merely one of cost. The logistics of assembling a class of students after school hoping for clear skies destroys the possibilities of real observing for the majority of students. Robot telescopes change all that.

In educational terms a robot telescope can provide a range of experiences of observational astronomy. The development of CD-ROM and the Internet to support classroom learning have produced the concept of REAL(Dunlap 1996): a Rich Environment for Active Learning as an appropriate framework on which to develop the classroom response to these technologies. The Bradford Robot Telescope has demonstrated student centred experiences to generate a Rich Environment for Active Learning(REAL), for astronomy. It is based on a massive extension of the library and experiential resource available to the teacher over the Internet, the opportunity for the student to develop and answer questions associated with the learning programme and access to a robot telescope which provides two modes of operation: service observing and eavesdropping. In the concept of REAL the students are:-

• Allowed to, and taught to, determine what they need to learn through questioning and goal setting

• Provided with sufficient scaffolding in the environment to help students with prompts, examples, modelling and collaborative support

  • Enabled to manage their own learning activities

  • Enabled to contribute to each others' learning through collaborative activities.

The objective of the schools programme of the Bradford Robot Telescope is to support REAL environments in the classroom.

2. The Bradford Robot Telescope

The development of robot systems for astronomy has a long history. In 1955 Bart J. Bok (Bok 1955) called for the development of a *"small automatic monitoring telescope which would provide at all times exact extinction information and provide valuable yearly information concerning the photoelectric properties of the sky"*. Vincent Reddish (Reddish 1966) built an automatic system to provide standards for photographic plates and in the mid sixties the US National Academy of Sciences produced a ten-year programme for ground based astronomy with a central proposal that significant funds be allocated to the development of automated telescopes and instrumentation such as plate measuring machines (Whitford 1964). In the end these programmes were suffocated through technological problems (Baruch 1992) which have now been solved with the Bradford system.

Towards the end of the Eighties the UK Science and Engineering Research Council (SERC) produced a Ground Based Plan for Astronomy 1990 (Mitchell 1989). This proposed the construction of small robot telescopes. By the early '90's there were a growing number of problems that robot telescopes could make a significant contribution to solving:

• Scientific investments in satellite observatories operating in wavebands outside the visible were being compromised by the inability to follow up observations and more importantly to perform concurrent observations in the visible wavebands. Important science was being lost

• The limited ability to respond to targets of opportunity e.g. comets, novae, supernovae and cataclysmic variables, was a loss to science

• The award of observing time on telescopes was largely based on the potential returns measured by the number of publications arising from the observations. The science which requires the frequent and long term monitoring of variable objects was not being pursued

• Surveys of particular types were limited and patrols searching for particular events were virtually non-existent

• There were few if any searches for planetary systems outside the solar system

• Within the solar system there was much uncertainty about the science of the satellites of the outer planets and of the asteroids and comets.

Arising from the Ground Based Plan the SERC initiated the development of a robot telescope by Baruch and his team.

The telescope was constructed at the Bradford University Experimental site near Haworth in the Pennines about 15 kilometres from the University Campus. It consists of a local network of telescope control computers at the observatory, linked to the main campus via an ISDN telephone line. The control computers are called Control, Point, Image, Weather and Catalogue each with the appropriate functions. The telescope is an alt-az Newtonian with the two major foci taken through the altitude bearings. The telescope is primarily for CCD photometry and supports a CCD photometer with broad band BVRI&Z filters. A major objective of the telescope design was to reduce telescope costs to a small fraction of the current cost of a 1 metre telescope. The telescope is designed as a robot which can be assembled by two people on the top of a mountain without heavy lifting gear and this limits the aperture to about 1.2m with a hextek type mirror.

The Bradford Robot has a 46cm mirror. The CCD has a field of about 12 by 20 arc minutes. The telescope design allows for two main instrument locations for large heavy instruments and two subsidiary instrument locations. A mirror is rotated to move the focus between the different locations. The major software components are identical with systems used on large telescopes or are commercial software packages modified where necessary to meet the less demanding requirements of 1m aperture robot telescopes.

The telescope is accessed over the Internet via the base station (Baldrick) which is located at the main University campus. The address of Baldrick and the telescope is http://www.telescope.org/ Access to Baldrick on the World Wide Web is via any one of the Mosaic or Netscape type of browsers currently available for a number of platforms including Suns and PCs. The user is then presented with a list of hypertext choices. They can read an on line guide to the telescope, find out technical details of the hardware and software, learn more about stars and galaxies, read weather reports, and control the telescope, all from the same interface.

The detailed operation of the telescope is described by Baruch (1996).

Baldrick presents visitors with a set of links which include: Access to the telescope, the Nuffield Research Project, Selected Internet References for Schools, Stars and Galaxies a CD-ROM conversion with graphics, pictures and hypertext links to provide a basic information resource in astronomy. Apart from the astronomy resource material the telescope has demonstrated two significant active classroom resources: service observing and eavesdropping.

The normal access to the telescope provides service observing. The user can requests an observation by selecting it from a browsable list with hyperlinks to automatically set up the request. The lists include solar system objects, a list of stars by name, the Messier Catalogue and IC, SAO and NGC catalogues. The user can also make a request for an observation of a specific RA and dec. The interface is designed to be helpful and supportive with comments and default values always available. For the least sophisticated user clicking on the mouse when pointing at the name of the desired object will set the system up to take an image of this object and inform the user by email when the image has been taken.

Users are requested to register by entering their username and password. In this way the system can allocate them user space on the system, return a notice that their observations has been made and the users can look at the progress of their observing request on the data base. An important aspect of the observing process is the generation of quality indices. Each observation which is normally a CCD image in FITS format has associated with it a header file which includes a complete set of data associated with the telescope system when the observation was made. It also includes an automatically generated quality index for the observations.

The quality index was originally designed for CCD photometry of stars. The programme searches for stars on the CCD image. It looks for objects which appear above the sky background and have a defined profile and roundness within certain limits. This works well with stars but most of the many tens of thousands of observing requests that the telescope has received are for planets, the moon, nebulae and galaxies. All of these receive a poor rating from the quality index even if they are excellent images.

The system has also demonstrated eavesdropping. This mode of operation allows the user to access the screens of the computers controlling the telescope. In this way the telescope control computers including Point and the image processing computer Image can be monitored over the Internet.

This mode of operation can be linked to detailed support pages on the net which can prepare students for the observations. A project on the velocity of light could provide details of early ideas and theories with a discussion of the size of the solar system to give an introduction to Roemer's method for measuring the velocity of light. The students could be referred to earlier observations of the occultations of Io which enable the period to be determined and predictions of the time of occultation. They could then eavesdrop on observations of Jupiter watching for Io to appear or disappear. Similar programmes could be used for variable stars of different types or for quasars.

Eavesdropping has a further application which is relevant to professional astronomers. Large telescopes could encourage astronomers who win observing time on the telescopes to provide home pages detailing their research programme and the implications of the results they are hoping to get. The Internet can then be used to introduce students to the research programmes and allow students to eavesdrop on the instrument screens as the data comes in. In this way students could be involved with the excitement of professional observational astronomy. Some programmes would not be suitable, but of the 20 large telescopes in the world it is likely that on any one night a few of these would be suitable for Internet eavesdroppers.

The telescope can also be controlled directly over the Internet. This is mainly used for engineering developments by the Bradford Group and for Television programmes which illustrate its use by students from anywhere on the planet. With the poor weather at the Haworth site, direct control or eavesdropping are not viable options

3. The Use of the Telescope for Education

The Nuffield Research Project contains six on-line projects in astronomy for schools and a host of support material for class-room activities. The projects are associated with the UK National Curriculum and link into the Bradford Robot Telescope. The projects were originally provided in paper and magnetic form and are now available on the Internet and consist of about 300 pages.

The projects for years 7 to 11 cover:
• An Introduction to Astronomy
• The Earth in Space
• The Planets
• The Solar System
• The Moon
• The Galaxy and the Universe.

Research projects for years 12 and 13: there is teacher support material which includes:-
• The project student learning objectives associated with the NC
• Teacher guides for the assessment of the projects linking them to NC attainment
    targets
• The objectives associated with the use of the Internet for teaching
• The objectives associated with using a robotic telescope for teaching

- Technical details of the computer set-up and access to the Internet
- Guidelines for the teachers in the use of time.

The Six Projects cover the range of student age and ability consistent with the National Curriculum. The projects are in a 'least-work' for the teacher form and directly relate to the National Curriculum with notes showing how and where they fit in with the NC. They include a variety of support including suggested activities, worksheets and a level of differentiation.

Each project is totally self contained; where necessary it uses service observations by the Bradford Robot Telescope to demonstrate an aspect of the learning objectives of the project. The materials supporting the project are complete, illustrated, ready to use and copy free.

There were ten schools with IBM compatible PC computers which were given software, technical support and funding to link them to the Internet. About 150 schools were circulated with the material and provided with a degree of technical support, and thousands of schools from all over the world have come in to use the projects and the telescope over the Internet.

Our conclusion from running the project was that no schools had the necessary bandwidth to enable a whole class to use the Internet together and the most effective use of the resource was to download it from the net to a local server. There the projects could be printed out on paper for use by the class or if the school had a networked classroom of IBM compatible PC computers they could simulate access to the Internet using the resource on the server. Where the students could access the Internet in their own time then each could submit their own observing request; normally the requests were class requests.

In spite of the telescope being on a very poor site, which with the English weather was unable to give the turnaround necessary for class observations, many students had a very positive experience from the project and attended the final pupil researcher discussion meeting clearly excited with the science and technology they experienced with this project.

4. The Future Plans

The Bradford Robot Telescope is a fully functional system built to evaluate and develop the technology. It is now clear that this technology can provide access to observational astronomy for millions of students. The telescope operates in response to requests submitted over the Internet from anywhere on the planet. It can operate in real time responding to the requests of a single observer or it can respond to many observation requests scheduling them at the most appropriate time. In either case the Internet allows for eavesdropping on these observations with no limit on the numbers of eavesdroppers.

In order to expand this vision it would be necessary to install robot systems at good sites in the northern hemisphere and in the southern hemisphere. For Europe, optimum observatory sites would be in the USA (Arizona and Hawaii), Japan and Australia. For the world, sites in the northern and southern hemispheres which are distributed as evenly as possible around the planet would give access to the world's students at any time of the day or night. It is not just a facility for the developed nations of the world. Many of the sites should be in the developing nations and with true partnerships it would introduce them to the advanced technologies of the robot systems and communications and enable some of their students to take advantage of these systems.

The Bradford Robot Telescope development has illustrated that in a modern robot system 80% of the cost is in the software. The Bradford Group wish to build partnerships with observatories on good sites, with funders who wish to be associated with the projects, with colleges who wish to give their students access to classroom observational astronomy. The Bradford Group wish to share their expertise and software. They have the design for a 1.2 metre telescope that can operate as a robot. They have the software and systems designs to convert most computer controlled telescopes into robot operation. They are currently updating their system to run with a Meade LX200. They can provide a customised Internet interface for any partner and provide for them a daily schedule for their telescope and email to their users when the observations have been made.

There are 25 million school students in Europe over the age of 11 who in the next few years will have access to the Internet. In the developed world there are about 70 million such students. A commercial partner will be associated by these students with an exciting programme linking them into a futuristic system to observe the Universe. It is well worth the few millions that it will cost to copy the Bradford systems and provide them for the school students of the world.

The Bradford Robotic Telescope demonstrates that practical astronomy is within the reach of students of all ages, directly from the classroom. No longer will it be necessary to wait for the dark and the weather. The link to the Internet can transport the students to telescopes on good sites on the dark side of the earth and provide the impression of being in the control room 10 metres from the telescope. It is not the same as having your own telescope outside at night but for millions of young people it will inspire them with astronomy and the sciences, engineering and technology that support it.

REFERENCES

MITCHELL, E.W.J., 1989, *The Ground Based Plan - A Plan for Research in Astronomy and Planetary Science by Ground Based Techniques* SERC Swindon ISBN 1 870669 10 X.

DUNLAP, J.C.& GRABINGER, R.S., (1996) *Rich Environments for Active Learning* FTP://ithaca.icbl.hw.ac.uk/pub/nato_asi/rsq1.rtf.gz

BOK, B.J. 1955 *Size and Type of Telescope for a Photoelectric Observatory* The Astronomical Journal (Astronomical Photoelectric Conference - Flagstaff Arizona 1953) Vol 60. pp31-32.

REDDISH, V.C., 1966 Sky and Telescope Vol 32. pp124.

WHITFORD, J. 1966 National Academy of Sciences - National Research Council Publication No 1234. *Ground Based Astronomy - A Ten Year Programme*.

BARUCH, J.E.F. 1992 *Robots in Astronomy.* Vistas in Astronomy Vol 35. pp399-438.

BARUCH, J.E.F. 1996 *A Robot Telescope on the Internet*, 1997 Year Book of Astronomy. Edited by Patrick Moore, Macmillan - London.

Teaching/Learning Astronomy at the Elementary School Level

By Nicoletta Lanciano

Dipartimento di Matematica, Università "La Sapienza", Rome, Italy

1. What kind of astronomy can be taught to children between the ages of 6 and 11?

There are those who argue that children have little familiarity with the sky and that the study of astronomy should be put off until they're older. We believe, on the other hand, that children have an intimate daily rapport with the sky, the sun and moon especially, based on genuine affection for these celestial bodies which is often expressed in their fantasies, reminiscent of ancient mythology and present-day primitive cultures. Their initial conceptions of celestial objects and phenomena bring to mind ancient philosophical conceptions and the kind of erroneous thinking induced by present-day culture and mass media, and make us aware of how difficult it is to develop personal perceptions and of the powerful emotions that prevent or inhibit us from building new ones. The kind of astronomy we present to young children, with which we have been experimenting for years, is not the kind usually taught in schools and cannot be broken down into various different topics. We have children observe nature, do real life drawings of it, concentrate on it and listen to mythological stories so as to sensitize them to the rhythm of sounds, song, motion, numerical calculation and geometric representation.

This kind of astronomy only deals with what can be seen and recorded with the naked eye: the Earth, Sun, Moon, Venus, Mars, Jupiter and Saturn, the constellations and the sky, a theater of celestial bodies in motion.

ASTRONOMY FOR CHILDREN

- IS NOT A SUBJECT MATTER AS SUCH
- IS NOT BROKEN DOWN INTO TOPICS
- ONLY CONSIDERS CELESTIAL BODIES VISIBLE IN THE SKY TO THE NAKED EYE

RECITING MYTHS   ASTRONOMY
        GEOMETRY

WE CAN TELL FROM THE NAMES OF CONSTELLATIONS WHETHER THEY ARE IN THE NORTHERN OR SOUTHERN HEMISPHERE.

2. Research Methods

We will present:

a) methods for researching children's initial conceptions;

b) activities for interacting with these conceptions, and a comparison with other teaching methods;

c) how these conceptions evolved in groups of children between 6 and 11 under our observation.

TEACHING ACTIVITIES DESIGNED TO IDENTIFY INITIAL CONCEPTIONS,
but which also
ENCOURAGE THEIR EVOLUTION
and
TEACHING ACTIVITIES DESIGNED TO INTERACT
WITH INITIAL CONCEPTIONS AND CONCEPTION BLOCKS,
but which also
IDENTIFY NEW INITIAL CONCEPTIONS

The following are different activities designed to elicit different kinds of perceptions:
- working with clay involves the hands
- doing theater involves the whole body
- dialogue makes use of words which are spoken and listened to (oral)
- individual and group writing makes use of words which are written down and read.
- drawing makes use of sight and hands.

Information is thus gathered in the form of:
- objects produced
- theatrical actions or gestures on the part of individual children
- written texts produced by individuals or the group
- individual drawings

While thinking in terms of products, special attention is paid to the processes which groups or individual children go through in their work.

2.1. *Children's Initial Conceptions of the Sun*

The following are some of the conceptions children have of the Sun, expressed in conversation, mime and drawings.

Many children say that: "The Sun has rays."

Small children think that "at night the rays go back in": they envisage transformations in the natural world with great nonchalance and sense of dynamics. Some of them think that "at night the Sun turns into the moon". Leaving aside what is incorrect, the important thing is to maintain this sense of motion, of dynamics, of transformation: the Sun, being a star, is a body in continual transformation. The problem in terms of teaching is to get this initial concept to evolve in the desired direction.

In terms of these rays, there is a new obstacle:
- we see diverging rays leaving the Sun's disk
- we have the children consider this for a moment and then say that "they are parallel to each other when they reach the Earth"

There is no one correct response. It is not that one of them is right and the other wrong, so on the one hand it is a question of combining reflection on macrospace (the Earth and Sun in which we see diverging rays coming out of the Sun's disk), and mesospace (the Earth's ground where these rays hit and can be said to be parallel to each other), and on the other, we need to introduce the fact that light moves in straight lines and is propagated in all directions, implying that we can intercept a ray, thus introducing the question of diffused light. In this context megaspace is the great celestial space which is inaccessible to us and which we can see only in part, and mesospace is the space we move in, totally accessible to us both physically and visually. There is also what we refer to as microspace, which is close, totally accessible to our eye, and manipulable, and macrospace, which is immense, partially accessible to the eye and inaccessible to us in terms of movement.

MACROSPACE OF THE EARTH AND THE SUN
IN WHICH WE SEE THE SUN'S DIVERGING RAYS COME OUT OF ITS DISK
MESOSPACE OF THE EARTH'S SURFACE
WHERE THE RAYS HIT (WHICH CAN BE CONSIDERED) ALL PARALLEL TO EACH
OTHER

CRUCIAL IMPORTANCE OF STATING IMPLICIT FRAMES OF REFERENCE

PROBLEMS OF LIGHT: children think
- THE RAYS GO CROOKED or THE RAYS GO STRAIGHT

Light:
- GOES IN A STRAIGHT LINE
- RADIATES IN ALL DIRECTIONS
- IT IS POSSIBLE TO INTERCEPT A RAY
- EXPERIENCE WITH DIFFUSED LIGHT

EXPANDING OUR FIELD OF VALIDITY
THE SUN IS A DISK
THE SUN IS A SPHERE THAT EMITS LIGHT
I SEE A DISK — I IMAGINE A SPHERE THE SAME AS FOR THE MOON

2.1.1. *Example of a conception that evolves with age (regardless of formal education)*

Elementary school children from 6 to 7 often maintain that "there are a lot of suns", of different colors, sizes, positions, that can be seen in different parts of the world, etc. They often draw more than one sun in the sky and more than one moon. At the symbolic, unconscious level, and not as a simple expression of knowledge, this could mean that the Sun is important, that it symbolizes festivity.

MANY SUNS   JUST AS THERE ARE MANY EARTHS (THE PLANET EARTH/
             THE EARTH WHERE WE LIVE) JUST AS THERE ARE
             MANY MOONS
to
ONLY ONE SUN

2.1.2. *Activities designed to interact with conceptions, encourage their evolution and avoid certain teaching obstacles*

Our shadows: shadows are not just on the ground but occupy space. To make this clear, we give examples to show that we are all aware of this fact in our daily lives and, for example, park our cars out of the Sun and sit under beach umbrellas. The difficulty, in this case, is to get people to consider everyday experiences when they are thinking scientifically, in a "scholastic" situation in which, due to erroneous teaching, shadows are usually considered in terms of projections on horizontal planes (the ground) or vertical planes (a wall). So what we do is to fill our shadows with branches or with other people's bodies, we divide them into sections with white sheets of cloth. This way we can see that shadows are in space, that they occupy three-dimensional space, and that they can have longer or shorter sections which are deformed if we tilt the cloth.

Throughout this work, in keeping with what will follow concerning deep and emotional elements, we work with the shadows of our own bodies and not the usual shadow of a telephone pole. We feel that as a first approach to the Sun and shadows, this is much more advantageous, not only because it allows students to study the volume of a shadow, but also because it is attached to them. We always have our shadows with us, even when we're not in school, on the beach for example, so we can continue to make discoveries, on our own, comparing them with others, experimenting, etc.

A monument of bamboo canes to trace the Sun's path: during an entire day of hands-on work, observation and reflection, we construct a three-dimensional track in space of the Sun's path in the sky. At the end of the day we have an arc, which may be of varying height, covered in colored disks, showing the path the Sun took that day. The fact that this group activity takes a long time gives everyone the possibility of understanding the sense of what is being done and of formulating questions more and more precisely and to the point. From elements observed in the "here and now", we can form mental images of what can be seen elsewhere or at other times of the year. This gives us a foundation of direct observation and an imaginative and precise record of it with which to reason about things we can see only in our mind's eye.

THE SUN'S PATH WITH BAMBOO CANES and not GNOMON
- TO AVOID ERRORS ABOUT THE DIRECTION OF THE SUN AND THE
  DIRECTION OF SHADOW
- TO AVOID PROBLEMS OF INDIRECT VISION: SUN FROM EAST TO WEST
  SHADOW FROM WEST TO EAST
- TO MAKE SURE THAT THE WORK IS CARRIED OUT OVER A LONG
  PERIOD OF TIME, WITHOUT ANTICIPATING, WITHOUT SPEAKING
  INSTEAD OF OBSERVING, WITHOUT EXPLAINING INSTEAD
  OF EXPERIENCING

2.1.3. *Evolution of conceptions as the expansion of our field of validity*

By evolution of conceptions in students we consider the fact that their conceptions go from a simple level to a more complex one in which they are capable of considering multiple aspects. This does not mean that the new conception will be more "correct"in the scientific sense; on the contrary, greater complexity often leads to confusion, but the conceptual field of validity will be wider.

3. Training courses for teachers

We address the question of initial teacher training and updating courses for teachers at the elementary school level in Italy by presenting several examples of activities carried out in residential courses or in the university, and the problems encountered. Elementary school teachers in Italy are not required to have a degree, and in any case almost none of them has one in a scientific subject. The only scientific background they have is what they receive in high school.

In these courses we work on:

• the geometry of the sky: the great celestial and terrestrial circles: the equator, the horizon, meridians, ecliptics, etc., the height of their angles, etc.

• music, to find connections between what we observe and study about the sky and music, sound, rhythm, song and how musical instruments are made

• motion that helps us have a perception of ourselves, our bodies and the space around us

• myths and experiences with other cultures (myths about the constellations and planets)

• designing and making instruments by hand
  o - for observing and recording celestial phenomena
  o - in order to have a clear, precise image stamped on our memories necessary because usual written texts with two-dimensional drawings and spoken lessons on complex phenomena have proven to be difficult to follow and to have little to do with our intuitions, emotions and personal effort.

In proposing this kind of teaching we remind people that the sky is everywhere and available to everyone, even poorly equipped schools in underdeveloped areas. This means that everyone can study astronomy simply by using the sky as a private research laboratory. And this kind of contact with the sky encourages children to be concerned about nature and the environment instead of wanting to manipulate and exploit it and science. It is our belief that the study of astronomy can contribute to re-establishing relationships between individuals, the environment and the planet.

In terms of method, it is important to keep in mind:
- the richness of mythology;

- the value of direct observation in the open,
- the value of long periods of systematic observation,
- the importance of addressing a subject from a cross-curricular point of view and not
    limiting it to traditional science;
- the attention paid to participants' initial concepts,
- the advantages of working together with our minds and hands to create instruments
    which can record three-dimensional celestial phenomena observed.

In both types of courses, local and outside of schools, we work on primary and fundamental elements.

3.1. *Primary and fundamental elements*

We work on primary cognitive elements which constitute the foundation of knowledge and that go back to the stage of learning in which understanding is thinking and doing, touching, moving, being thrilled.

Primary elements of the sky are, for example

The Local Horizon

The real local horizon (Lanciano et al 1996) can be seen everywhere out in the open: it is the dividing line between the sky and the Earth. If we stand in one spot and turn 360° we can trace the entire horizontal line with our gaze. It is generally not a circle and is usually not all level; in fact it is usually very uneven, and if we are in a city there will be buildings, light poles, antennas, maybe some hills, as well as trees. This is our real horizon which determines what time we can see the Sun in different periods of the year and what time it will set, i.e. disappear over the horizon. The astronomical, theoretical horizon is not useful when we are trying to decide what to observe in the sky and when, so we have to learn to consider our local horizon. This is why we observe it, draw it and use it to create our observatory.

The Globe

The same globe is used in all countries, but the inclination of its axis is not clear and significant to everybody. This is why we take it off its support when introducing it and set it in the position of a "parallel map"(Lanciano et al 1996), i.e. tilting its axis according to the latitude of a place, that is, directed toward the North Star. If Italy is at the top of the globe, with its tangential plane parallel to the table on which it sits, the Sun will do exactly what it does on Earth. We can see which pole is in darkness and which is illuminated, which countries are in daytime and where the Sun is at its zenith. This allows us to feel "central in relation to what we are observing", to become aware of our own particular vantage point and to connect the Earth we walk on with the globe we look at from outside.

This instrument stimulates us to think about what we feel to be our position on the Earth's sphere. We discover how much the fact that we think of North as high and South as low leads to unconscious conceptual and "geometrical"errors. We discover something new about the "sphericity"– physical, geometrical, astronomical and geographical – of our planet.

Meridians

A local meridian is every North-South line, and it is possible to find one at every point of the Earth's surface; meridians are not only those marked on the globe or map – theoretically it is possible to connect the two poles at any point on the globe. The problem involved in this kind of change in thinking is once again the problem of moving from the discrete to the continuum: from a discrete grid to great circles – to continuum circles which theoretically cover the entire surface of the sphere.

We work with fundamental elements because they involve the whole person. Some examples are painting, reciting and listening to myths, sounds, songs, becoming familiar with the night and celestial bodies (Lanciano 1993, Montinaro). In all these arts, these expressive languages, our hands, bodies, thoughts, emotions and hearts are directly involved. It is our conviction that it is impossible to follow a path of knowledge if we are resisting it in any way, if we are not ready to meet the unknown, something that could change our scheme of things, our basis of security. Learning always involves abandonment of the known, what we feel is ours which may be incorrect or incomplete but which reassures us. Imagine how difficult and frightening it must be for a small child to give up the comforting idea of a nice flat world for a spherical one in motion, suspended in the void like the moon and planets in the sky. Every discovery and conquest our minds achieve is accompanied by emotion, and the quality of the relationship we have with those who accompany and guide us and the objects we are studying determine whether our knowledge will be stable or uncertain, whether it will constitute a basis for new questioning and discoveries or will be a series of disconnected notions, easily forgotten. This is what psychology is telling us about the learning process, but it seems to be immediately forgotten or ignored when it comes to teaching individual subjects, when knowledge is broken up into separate subjects, when our main purpose in education becomes evaluating instead of teaching and offering opportunities for growth and discovery.

One of the most important factors in our work is that those who guide students be not only competent, but passionate about what they are offering to others and about education itself.

REFERENCES

LANCIANO, N., 1993, Il suono e le stelle, Cooperazione Educativa n. 6, La Nuova Italia ed, Florence.

LANCIANO, N. ET AL Dentro il cielo lecture notes, Quaderno n. 7 of the L.D.S., 1996.

MONTINARO, R. Le vie dei suoni, Il Crogiolo.

The Influences of the National Curriculum on Children's Misconceptions about Astronomy and the Use of these Misconceptions in the Development of Interactive Teaching Materials

By John H. Baxter

St. Luke's College, Exeter University, Exeter, England

It is now well established that children construct their own explanations for the easily observed astronomical events before they receive any formal education in astronomy (see Mali & Howe, 1985; Nussbaum & Novak, 1976; Vosnaidou, 1991. It is also generally accepted that childrens notions, or 'alternative frameworks' are tenacious and frequently pass into adulthood (Gunstone *et al*, 1981). Baxter's (1989) survey revealed a hierarchy of alternative frameworks about astronomy that became less naive as age increases, but also revealed that many pupils leaving school at the age of 16 years did not explain the easily observed astronomical events within a post-Copernican framework.

Until the introduction of a National Curriculum in 1989, astronomy rarely featured in English schools' science curricula (see Lintern-Ball, 1972; Baxter, 1991). Therefore, it is not surprising to discover that many children and adults (Durant, Evans and Thomas, 1989) have concepts about astronomy that bear a closer resemblance to those of the Dark Ages than the 20th-Century space age.

For over six years now astronomy has been an established part of English children's school science experience. The survey reported in this paper was carried out to discover if children's alternative frameworks have been affected by the more widespread teaching of astronomy.

1. Methods of Investigation

This study investigated children's ideas about the same four astronomical domains researched in the 1988 survey (see Baxter, 1989):

- Planet Earth in space.
- Day and night.
- Phases of the Moon.
- The seasons.

The study employed the same astronomy conceptual survey instrument developed for the 1988 survey (see Baxter, 1989, for full details of the survey method).

This sample comprised 120 children aged 9 to 10 years - 56 boys and 64 girls - taken from a number of primary schools in the south-west of England, including primary schools used in the original survey. According to the subjective opinions of the class teachers, the pupils covered the full range of abilities normally expected in state primary schools.

2. Results of the Survey

For clarity of presentation, the results obtained for a particular domain are followed by a comparison with those obtained during the survey of 1988. Prevalence trend diagrams are given for each of the domains investigated. The comparisons are made against the 10-16 year age range of the 1988 survey.

2.1. *Planet Earth and Gravity*

Pupils were asked to imagine that they had taken off in a space rocket and had been travelling away from Earth for one day. They look out of the window towards the Earth. They were then asked to draw how they thought the Earth would look. After completing their drawing, they were asked to draw in some people to show where they could live, then some clouds and then rain falling from the clouds.

The pupil' s drawings fell into four distinct notions and reflected those obtained in the 1988 survey. However, there appears to be a shift away from a naive flat Earth notion and a round Earth with 'UP' being directed towards the North, towards a view that represents - or closely represents - the accepted notion. See Fig. 1.

Only 3% of the sample in the 1994/5 survey produced flat Earth diagrams. This shows a reduction of 16% from the 19% recorded in 1988. The most notable difference between the diagrams of planet Earth drawn for the two surveys was the change in children's concept of the proportion between the size of the Earth and the height of the clouds. Fifty percent of pupils in the 1988 survey placed the clouds somewhere out in space. Ninety one percent of the diagrams drawn for the 1994 survey placed the clouds close to the Earth. See Fig. 2.

Although there has been a significant improvement in children's concept of planet Earth in space, a Newtonion concept of gravity does not appear to be the prevalent view. However, the consensus view has shifted from the naive notion that North is 'UP' and South 'DOWN' , to an intermediate notion that is represented by rain falling towards the centre of the Earth, but with people drawn with their heads pointing towards the North. No diagrams of this type were recorded during the 1988 survey.

2.2. *Day and Night*

Children's ideas about day and night have also shifted away from the more naive notions. None of the 1994 sample believed that it gets dark at night because the Sun goes behind a hill, and only 2% thought that it gets dark at night because the Moon covers the Sun. This compares with 4% believing in the influence of hills and 20% in the Moon making it dark at night as revealed by the 1988 survey. See Fig. 3.

The most significant change in children's explanations for day and night has been their rejection of the belief that either the Sun orbits the Earth, or the Earth orbits the Sun once a day. The 1988 survey revealed that 103%† of the sample explained the cause of day and night within the terms of these supposed orbits. By 1994 the percentage had reduced to 37%. 77% percent subscribed to the correct view for day and night. This compares with 34% recorded in the survey of 1988; an improvement of 43%.

2.3. *Phases of the Moon*

Children's explanations for the cause of the Moon's phases fell into three main types. This was a reduction of two from the five revealed by the 1988 survey. None of the sample considered that either the shadow of the Sun or a planet caused the phases. However, 19% thought that clouds produced the phases. This is an increase of 2% from the 17% revealed by the 1988 survey and it appears to be a notion that has persisted.

The most common explanation for the phases of the Moon recorded by the 1988 survey was that of the Earth's shadow falling on the Moon - 44% subscribed to this view. By 1994 this had dropped to17% and the number giving the correct explanation had risen from 28% to 48%; an improvement of 20%. See Fig. 4.

---

† It is interesting that the sum of the percentages subscribing to the various notions for the cause of day and night gives a total greater than 100%. This is caused by some pupils selecting more than one cause for this event and can be considered a measure of their uncertainty about this domain. In 1988 the total percentage subscribing to all possible causes was 161%. In 1994 this had fallen to 116%.

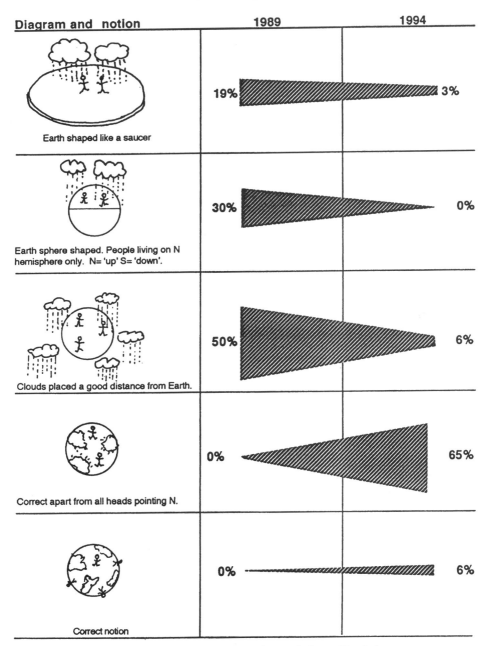

FIGURE 1. Children's changing notions of planet Earth in space.

2.4. *The Seasons*

The changes in children's notions about the cause of the seasons are almost identical to the conceptual changes that have taken place for the Moon's phases. None of the children from the 1994 sample believed that either a planet took heat from the Sun, or that the Sun moved to the other side of the Earth - 6% and 7% respectively subscribed to these notions in 1988. However, in the same way that children visualised clouds causing the Moon's phases, 9% of the 1994 sample considered that it gets cold in the winter because clouds take heat away from the Sun.

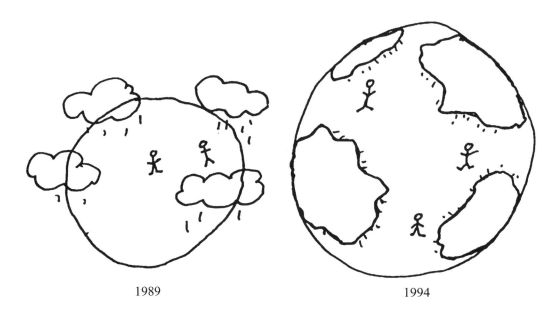

1989   1994

FIGURE 2. Placing of clouds in 1989 and in 1994.

This is an identical percentage to that obtained during the 1988 survey and this too appears to be a notion that has persisted.

The most common explanation for the cause of the seasons in 1988 was that the Sun moves closer to the Earth during the summer and further away during the winter. 59% subscribed to this notion in 1988. The number electing this as the cause of the seasons in the 1994 survey fell to 19%, while the correct notion was the most common view with 52% of the sample giving this reason for the seasons, an improvement of 32% on the original survey. See Fig. 5.

Although this survey suggests that there has been an improvement in pupils' understanding of the easily observed astronomical events, astronomy has still not taken its place in the mainstream of science. It is still uncommon for pupils to be given an opportunity to develop their science activity skills within the context of astronomy. This is primarily due to a self perpetuating triangle of factors.

The growth of computer generated interactive technology offers a unique opportunity to present pupils with activities that will enable them to develop their science skills within the area of astronomy. However, to date, the field of interactive technologies in education has had a history of being technology led rather than curriculum driven. Our knowledge of alternative notions in astronomy (and in other areas of science) gives a valuable framework for the development of interactive computer programmes because these alternative notions have been shown to be common responses, and to have a common progression pattern. Therefore, if the common responses are used in the design of interactive programmes, such programmes will move away from being the programmer's perception of pupils' knowledge and understanding, towards questions and possible pupil choices that are based on established research.

This means that those researching in the alternative framework movement are not just involved in the collection of pupil's wrong ideas. These misconceptions, and a knowledge of their progression, form a framework for the development of meaningful and well focused computer

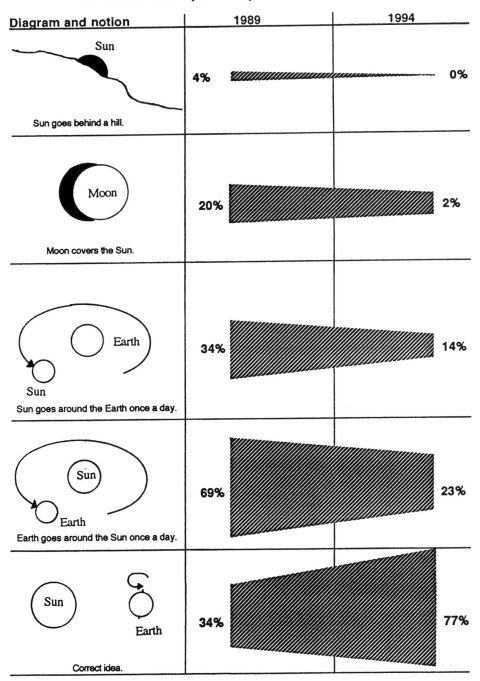

FIGURE 3. Children's changing notions about day and night.

generated interactive materials that are set to become a common feature of education into the next century.

In recognition of the above, a joint research programme has been set up between the University of Exeter and British Telecom (BT) that will develop focused materials for the teaching of astronomy (and other areas in the physical sciences) using our knowledge of pupils alternative

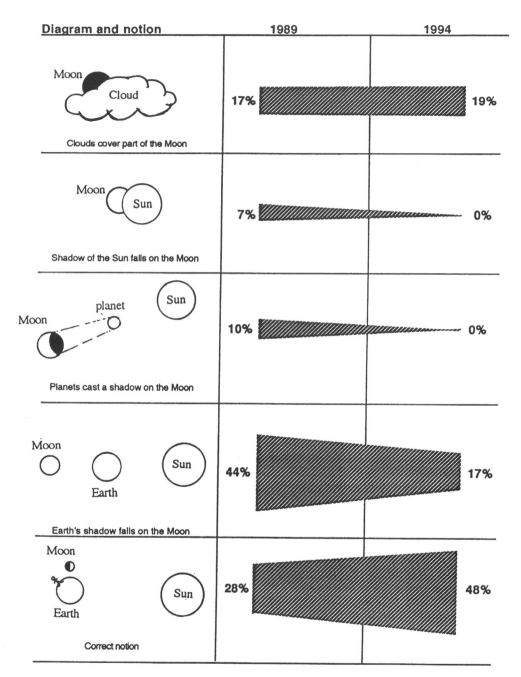

FIGURE 4. Children's changing notions on the phases of the Moon.

notions as a framework. The full range of on-line technologies will be used to distribute these materials, share good teaching practice and deliver online in service training to primary teachers, thereby making a substantial contribution to redressing the shortfall in primary teachers' knowledge (see Mant and Summers, 1993), and giving a unique opportunity to introduce teachers to interactive materials to which they would otherwise not have access, or be able to afford within

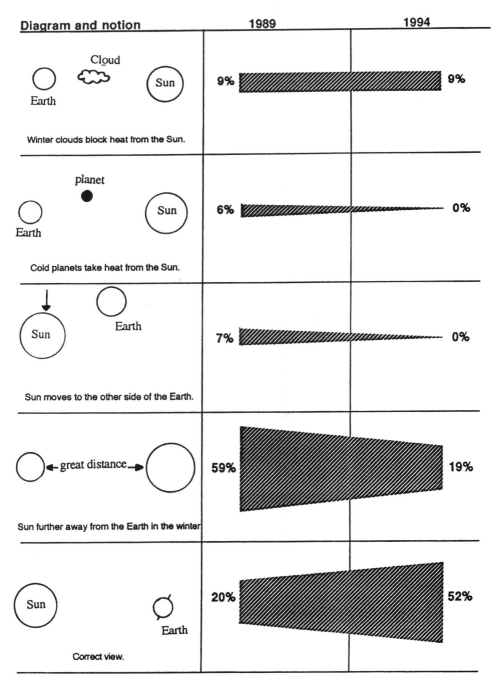

FIGURE 5. Children's changing notions about the seasons.

the ever tightening school budgets. Materials will be accessed on-line via the CampusWorld network and the CampusView server. The participating schools will form a network sharing their good practice via video-conferencing and videotaped lessons. Other schools will be able to access via B.T's CampusView server.

In this way teachers in one school can share their good practice in astronomy education with teachers in other distant schools. Pupils will be able to communicate with those in other schools via video-conferencing and E mail, thereby sharing their ideas and results. In this way well focused astronomy educational material will be produced, distributed and experiences shared, using the full range of on-line technologies now available to schools and colleges.

REFERENCES

BAXTER,J., 1989, *Children's understanding of familiar astronomical events*. Int.J.Sci.Educ.11, pp.502-513.

BAXTER,J.,
1991, *The National Curriculum: a challenge for astronomers*. Q.J.R.ast.Soc.32,pp.147-157.

DURANT,J.R., EVANS,G.A. & THOMAS,G.P., 1989, *The public understanding of science*. Nature. Vol.340 6th. July.

GUNSTONE, R.F., CHAMPAGNE,A.B.& KLOPFER,L.E., 1981, *Instruction for understanding: a case study*. Aust.Sci.Teach.J.27,(3),pp.27-32.

LINTERN-BALL,R.W., 1972, *England's astronomical education?* Q.J.R.ast.Soc., 13,p486.

MALI,J. & HOWE,A., 1979, *Development of Earth and gravity concepts among Nepali children*. Science Education,Vol.63,No.5,pp.685-691.

MANT,J. & SUMMERS,M., 1993, *Some primary-school teachers' understanding of the Earth's place in the Universe*. Research Papers in Education, Vol.8,No.1.

NUSSBAUM,J. & NOVAK,J.D., 1976, *An assessment of children's concepts of the Earth utilising structured interviews*. Science Education, Vol.60, No.4,pp.535-550.

SADLER,P., 1987, *Misconceptions in astronomy*. In: Proceedings of the 2nd. Int. Seminar on Misconceptions & Educational Strategies in Science & Maths, Vol.III,pp.422-425. Ithaca, New York: Cornell University Press.

VOSNAIDOU,S., 1991, *Designing curricula for conceptual restructuring: lessons from the study of knowledge acquisition in astronomy*. J. of Curriculum Studies,23,pp.219-221.

Role of Novel Scientific Results in Learning

By V. Vujnović

Institute of Physics of the University, P.O.Box 304, 10001 Zagreb, Croatia

1. Introduction

Introductory remarks on astronomy education in Croatia are given. Since the learning process is a complex intellectual and emotional process which should be supported during the interaction with the teacher, different approaches should be used. Tests could give useful insight into preconceptions. The following approaches should be balanced: historical approach, discovery approach (by the use of self-made tools and courtyard observations), and thorough inclusion of novel scientific results and views (to which a special precaution has been paid).

2. The Croatian Experience

This is a report about an experience in teaching astronomy to the students who will become teachers in physics or physics and mathematics. It should be stressed that astronomy in Croatia is not a standard subject in any schools, except as an elective course in some grammar and high schools; furthermore, astronomical concepts are partly exposed within physics.

The first step toward students should be mutual acquaintance. In order to test students' previous knowledge, I used 20-25 questions mainly of a general nature (starting in 1975). I had the opportunity to teach at all four Croatian universities: Zagreb, Osijek, Rijeka and Split. People in these towns may have different backgrounds. Zagreb is the capital of Croatia and cosmopolitan. Osijek is the center of Slavonia and belongs to an agricultural and Panonian environment. Split is heart of Dalmatia and situated on the Adriatic Sea - in the Mediterranean region. Without regard to differences in life attitudes, temperament and historical background of populations, the test showed a low level of general knowledge in natural sciencies, especially regarding comprehension of objects and scientific terms.

It is understandable that high school gives a sketchy knowledge of the universe but this knowledge is seldom useful; for example, pupils orient themselves with difficulty. They are unable to orient on the horizon (in space) and also in time by using celestial phenomena. Southern direction in a room? Local noon?

Common results of tests at the beginning of the study (knowledge obtained in the high school) can shortly be exemplified:

do not know:
tropical year
solar distance and (1975) its state of matter
meaning of zenith (confused with the superior culmination of the sun)

level of recognizability:
particular objects such as planetary nebulae

adopted concepts:
comets (object of significant interest which is restored after every apparition)

The learning process is a complex intellectual and emotional process which is influenced by a manifold of factors, including previous knowledge, self-experience, sensibility towards others' experience, and wish to comprehend the scientific views. New subject-matter can be accepted with adequate pre-knowledge, mathematical skill and basic knowledge of physical processes. Between opportunities which make the task of education easier one should look for available literature in the mother tongue, access to a planetarium, access to astronomical telescopes. Among important subjective motives there are: striving for a better mark (excellent, or very good); wish to get a diploma and to become an experienced teacher. All these factors and prerequisites have an influence on the students' learning process. And we have to help.

In fulfilling the aim, there is no unique method or educational technology of universal application. Various approaches (interactions between teacher and learner) should be balanced. In order to stimulate students in learning, as essential I consider:

- acquaintance with the development of astronomy, and repetition of the steps of great men
- attainment of practical skills in astronomical observations
- information on the latest news and results

One can find the experience of great astronomers very stimulating since students identify with them. The historical path gives the first motivation to the students who came from the high school where they are used to learn instead of doing research. For this purpose one can use the methods applied by the great to find the proper place of the Earth in the Universe (measuring the size of the Earth with Eratosthenes, relating the distance of the moon and sun with Aristarchus, determination of the sidereal periods of the planets with Copernicus, etc). After elaborating methods in historical order, space is opened to treat particular astronomical branches in detail.

The next important step is personal research practice and attainment of practical skills. Without personal experience all scholarship loses sense. Practical work is primarily designed for orientation on the horizon (elements of positional astronomy). For the purpose a self-made apparatus could be constructed consisting of gnomon, stellar and solar altimeter (in the form of a quadrant), cross- staff and star finder. The apparatus gives results the value of which depends on the attention paid to it. It can be used to determine meridian and cardinal points in the school-yard, moments of local noon, geographical latitudes and longitudes, length of the apparent solar day, obliquity of the ecliptic. Difficulties in comprehending the relation between the real and apparent motion of the Earth and celestial bodies are overcome by the practical work which also helps in developing and adapting the concepts.

As one of the most effective stimuli in the learning process I stress the role of new or newest scientific results. Its significance can be visualizued by an easy proof of the spherical Earth: instead of spending centuries on physical and other proofs, one snapshot taken by astronauts is sufficient to eliminate all doubts. A shortcut in perception reflects in a shortcut in cognition.

How to incorporate and use novel scientific results? I will illustrate this by two examples.

a)   Distance measurement of SN1987 From the tests and oral exercises I learned about poor students' understanding of the relation between the central angle and corresponding circumference. The relation between circular arc length, central angle and radius is the most simple mathematical relation - which pupils avoid since they have difficulties in comprehending the definition of the radian. By introducing the phenomenon of the recent supernova in the Large Magellanic Cloud, pointing out that it gave the first proof of neutrinos, and that it ejected a gaseous ring some 20000 years in advance of collapse – their attitude towards geometrical problems is changing. They follow as in a cartoon – paths of the light burst starting at the moment of the detected neutrino burst and directed towards the gaseous ring. They then solve simple arithmetic:

$$s = c \cdot t = r \cdot \vartheta,$$

where s is distance travelled, t time to illuminate the ring, r the supernova distance and ϑ the apparent ring radius.

Students were stimulated to solve the problem and attained experience in transferring the mathematics to other analogous problems. The angular method links many astronomical affairs and is far reaching; the formal problem is the same when measuring the size of the Earth by the method of Eratosthenes, or when measuring distances by the apparent galactic diameters.

b)   Planetary nebulae

For many years answers obtained during final examination about planetaries were not satisfying. Although some students could explain the evolutionary transition from red giant to white dwarf, others completely overlooked the existence of planetaries. The situation improved when an astronomy book was published in Croatian in 1989. However, qualitative change happened when students were really involved in the problem. We compared low-resolution images of planetaries which give an idyllic impression of simple geometry of a spherical shell leaving the rest of a star, with the higher resolution image of the Helix showing complex veils. Double bubble structure can be seen in another specimen. Deeper insight gives evidence of consecutive ejecta shown by Abell 30 (Borkowski et al,1995). This planetary shows a rounded shell of hydrogen expanding with a velocity of 40 km/s (typical for the planetaries), ejected some 9000 years ago, and more recent ejecta from 1000 years ago, which are devoid of hydrogen and rich in carbon – thus supporting the theory of nuclear evolution of the stellar interiors. Interaction of these ejecta with the radiation and wind of the central star reveals cometary structures.

Now, the phenomenon of planetaries can be compared with other instability phenomena, as with nova or pre-supernova ejecta. By contrasting these phenomena a new intellectual relation, "feeling" about phenomena arises and with higher understanding memorization is no longer mechanical.

What has happened in the rational process? New data, laborious problems combined with the hard task of incorporating them in the previous mental attitude, engaged much more intellectual forces than simple and easy problems. The seriousness of the problem invoked new processes and students were over-engaged. To get less, better to ask for more. During this engagement, the element of surprise is found very stimulative. Not surprise only in the bare sense of the sensation present in astronomical affairs (black holes or supernovas); equally sensational is the discovery that a little calculation with clear physical insight explains equally well a pendulum-clock and pulsating stars. Analogies should be exploited as much as possible.

3. Conclusion

In the complex learning process many factors and prerequisites may have an influence. Incorporation of new scientific results for which an extra intellectual engagement is needed, backed by curiosity and emotional capability, may be found stimulating. This raises the quality of the interactive process between teacher and learner.

For the application of temporary scientific results, a collection of exemplary problems with methodical hints would be very useful for lecturers at any educational level. Since science frontiers proceed at a quick pace, problems should be revisited at regular intervals; since lecturers are not in a position to be "science digestors" only, effort invested by an international group would be wellcome.

REFERENCES

BORKOWSKI, K.J., HARRINGTON, J.P. & TSVETANOV, Z.I. 1995, Astrophys. J., 449, L143.

The Jupiter-Comet Collision: some conceptual implications

By Dileep V. Sathe

Dadawala Jr. College, 1433 Kasba Peth, Pune – 411 011 India

1. INTRODUCTION

A typical science course at the high school level includes some information on planets and their moons. For example, it is well-known that Jupiter has 16 moons and Saturn has 18 moons. Add to this the enthusiasm of the public in the collision of comet Shoemaker-Levy 9 with Jupiter in July 1994. This immediately raises the possibility of a collision of a comet with a moon of Jupiter. Due to this possibility a strange fact about these moons comes into the picture, that is some of them are prograde in nature and some are retrograde. Can these two types of moons pose any problems in teaching ? The present situation in education leads us to believe that they can pose some problems. It is described below, in support of this answer.

Educators from many countries have observed that the Aristotelian ideas continue to persist among graduates, in spite of learning Newtonian mechanics in colleges also. This is evident, for example, in the fact that many students think that a tangential force acts on a body performing circular motion, instead of the centripetal force. So the greatest and *global* problem is how to get rid of the tangential force from the minds of students and how to impregnate the centripetal force instead.

According to the present author (Sathe, 1993) a majority of students will continue to ignore the centripetal force because:

i) the concept of work and the law of parallelogram of forces strengthen the idea that there has to be a force in the direction of motion and,

ii) our inability to decide the direction of motion on the basis of force, anticlockwise or clockwise, although we do assume that the gravitational force provides the centripetal force for the planetary motion. The second reason has been termed as the A/C paradox and it makes comprehension of Bohr's model of the hydrogen atom also very difficult where the electrostatic force provides the centripetal force for the electron's motion around the proton.

As prograde and retrograde moons of Jupiter are now prominently entering into discussions, due to the collision of 1994, the necessity of *accepting* the A/C paradox as one of the obstacles in the comprehension of basic concepts has increased considerably. The present communication considers the relevant findings from the past and suggests new directions for the future.

2. Why the JPR System is a Challenging one?

To understand the challenge posed by the system of Jupiter, one prograde moon and one retrograde moon – i.e. the JPR system – let us compare it with the system of Earth and Moon – i.e. the EM system.

In the EM system, one can easily get the equation of velocity of moon, assuming that the gravitational force provides the centripetal force. No problem of direction arises in this case. But in the JPR system, this assumption is *not* sufficient because it is necessary to explain *why* the prograde moon is orbiting anticlock-wise and the retrograde moon is orbiting clockwise, on the basis of gravitational or centripetal *force*. Therefore the JPR system is *much more challenging* than the EM system.

It is to be noted specifically that the use of initial conditions cannot solve this problem in the learning of basic concepts. The persistence of *global* misunderstanding, mentioned in the introduction, provides the validity of the foregoing statement. Hence the present author (Sathe, 1995) has appealed for the use of the JPR system instead of the EM system.

It should be noted categorically that the present treatment of uniform circular motion puts a lot of emphasis on the action of centripetal force. This is evident not only in textbooks but *even* in the *questionnaires* of educators. Educators always ask students to show some *force* acting on the body in circular motion. Therefore educators have to realize that students do expect that one should be able to decide the direction of motion anticlockwise or clockwise on the basis of force. Unfortunately, the present treatment can not fulfill this expectation.

3. Similarity between Kepler and Students

Lastly, it is to be noted that Kepler also has made an attempt to explain the planetary motion on the basis of sweeping action of lines of force, (Ebison, 1980). This is similar to the action of tangential force, usually thought of by even graduate students. Of course, in physics, Kepler's reasoning has taken the back seat, in view of Newton's monumental work. Nevertheless, *educators* of physics and astronomy also have to consider this similarity earnestly between the thinking of Kepler and students. Mathematical accuracy in physics cannot solve conceptual and educational problems.

REFERENCES

EBISON, M.G., 1980, Proceedings of the Intl. Conf, on Post Graduate Education of Physicists, Prague, Czechoslovakia, p. 160.

SATHE, D,Y., 1993, Bulletin 7 for Teaching Astronomy in the Asian-Pacific Region, p. 82.

SATHE, D.V., 1995, Physics Education, **30**,327.

Planetarium Education and Training

The Current Role of Planetariums in Astronomy Education

By William A.Gutsch, Jr.[1] & James. G. Manning[2]

[1] The American Museum-Hayden Planetarium, New York, USA
The Interntional Planetarium Society

[2] The Taylor Planetarium, Museum of the Rockies, Bozeman, Montana, USA

For decades, planetariums have been created to serve the cause of astronomical enlightenment – to offer people knowledge and understanding and a sense of place in a universe far bigger than themselves. It is an important role and one that we in planetariums continue to play – changing, we hope, as times, technology, educational philosophies, and our view of the universe change.

The first projection planetarium was demonstrated by the Zeiss Optical Company at the Deutsches Museum in Munich, Germany in 1923. By 1970, the height of the Apollo moon program, there were an estimated 700 to 800 planetariums in the world, half of them less than six years old. Today, 26 years later, that number has more than doubled to a little over 2,000.

The world organization of the planetarium profession is the International Planetarium Society with over 600 members in more than 30 countries. Based on figures compiled in the 1995 IPS Directory, we find that slightly more than half of the world's planetariums are located in North America, with large numbers also in Asia and Europe, but relatively few in other parts of the world. If we consider distribution by country, we find that half are in the United States, more than 300 are in Japan, and Germany ranks third with nearly 100. Nineteen countries have ten or more planetariums.

Some 33 percent of the worlds' planetariums are located in primary or secondary schools; 17 percent are at colleges and universities; 15 percent are part of museums and science centers; 7 percent are associated with observatories or other institutions. The settings of the remaining 27 percent are somewhat uncertain because of their location in parts of the world where communication is still conducted with some difficulty.

The heart of planetarium theaters is still the star projector. These devices range from the familiar dumbbell shape of the German made Zeiss, to inverse dumbbells like those fabricated by Tokyo-based Goto Optical Mafg., Co. to spherical or ellipsoidal models made by German, Japanese and American firms, to digital video projection systems known as Digistar and supplied by the Salt Lake City based firm Evans & Sutherland. Planetarium theaters vary from horizontal floored facilities to those with significant pitch. Some planetariums provide intimate settings while others have seating capacities of over 500. In addition, hundreds of portable planetariums such as Starlab operate around the world and provide astronomy education experiences in schools, auditoriums, shopping centers, and remote locations.

It is estimated that over 55 million people worldwide currently visit planetariums each year. Thus, planetariums represent one of the largest and most visible avenues for presenting astronomy and related subjects to the public. And this, in turn, gives planetariums an enormous potential and responsibility for supporting both formal and informal astronomy education.

Planetariums are well known for reproducing the naked eye sky as seen from any place on earth as well as demonstrating, in time-lapse fashion, many of the cycles of the heavens from simple diurnal rotation to the retrograde loops of the outer planets. But planetari-

153

ums also can create numerous other environments that encompass the audience, bringing them into the experience in a way that classrooms, books, television or the computer screen cannot. They can combine and effectively use audiovisual technology to help create these experiences. And they possess tremendous flexibility in how these audiovisual resources can be used. Indeed, many modern planetariums are utilizing exciting new equipment on the cutting edge of revolutions in multimedia storage, computer control and display to go well beyond the old lecture format of past planetarium experiences. Today, more than ever, the planetarium can truly reflect the exciting spectrum of astronomical discovery and take audiences on journey's from the turbulent atmospheres of other planets to the event horizons of black holes.

First and foremost, we strive to educate, in ways ranging from curriculum-based school lessons to popular-level programs. We also strive to enlighten for we want people not just to know but to understand and to incorporate this understanding in their lives. And yes, many of us also try to entertain—an endeavor based on the sound pedagogical principle that people simply learn and remember more when they are emotionally engaged in the subject matter. And not least, we strive to inspire. Our time with people is brief, and it is perhaps less important that someone remembers the diameter of Jupiter, than that he or she remembers Jupiter as an exciting and dynamic world that, in turn, can give us greater insight into our own. Such a person is also more likely to leave the planetarium to read more in a book, enroll in a class or come to an observatory open house.

In setting these goals, planetariums operate in all three realms of learning: in the thought-processing cognitive realm; increasingly in the psychomotor area, as we offer more interactive experiences involving physical action; and, as noted, we also operate in the affective domain, the realm of feelings, as we encourage greater appreciation and enjoyment of the universe around us and try to cultivate a sense of the adventure of science.

Public "Sky Shows" continues to be a major offering of many planetariums. They are offered in a variety of forms, from the traditional live-narrated current-sky program to automated, multimedia presentations on popular space-related topics. Most planetariums also present educational programs specifically designed for particular school grades to meet science curriculum objectives. Where possible and appropriate, we often create shows that relate to recent discoveries or current astronomy related news items. In this way, planetariums frequently serve as a respected and recognized source of scientific information in their areas and provide more in-depth coverage than can usually be given by the local or national news media. Most planetariums also present educational childrens programs specifically designed for different ages or grade levels. In places, cooperative efforts with popular television or motion picture companies have used character recognition to help make learning fun. Examples include works by one of the authors at New York's Hayden Planetarium in conjunction with the Children's Television Workshop and Lucasfilm Ltd.

Supporting our efforts are improving technologies. The new planetarium at the Forum Der Technik in Munich has a Zeiss Model VII projector, which uses fiber optics to create stars that look like true points of light. The Digistar computer graphics system allows the audience to travel through a variety of user generated and shared databases that can recreate the radio, infrared or X-ray sky as easily as the visible universe. Flights through the Yale Bright Star Catalog database, as well as galaxy data compiled by Geller, Huchra and others, are also easily accomplished.

Among the planetariums utilizing digital projection technology is the London facility which recently reopened after extensive renovations. Whenever possible, planetariums are updating themselves with new technology to better meet their goals and serve their public.

Computer systems are increasingly in use in planetariums today, both to control the planetarium projector and to automate auxiliary effects in programs. Video projection, pioneered by such planetariums as the one in Armagh, Northern Ireland is becoming an audiovisual staple. Some planetariums now have access to Silicon Graphics and comparable workstations for creating sophisticated video animation sequences for their programs and for distribution to other facilities. Most significantly, several companies in the U.S. and Japan are now developing computer controlled, interactive video systems that cover the entire planetarium dome with integrated, full color raster scanned images – a development which will completely revolutionize both the capabilities of the planetarium and the "look" of its shows. The line between what motion picture companies and planetariums can create and present is rapidly beginning to grey.

Advances in laser projection systems now allow for the creation of dynamic special effects from displays of the aurora, to solar flares to planetary magnetospheres. Jack Dunn at the Mueller Planetarium in Lincoln, Nebraska, has developed a program for people with visual impairments such as retinitis pigmentosa. Using the intense light of lasers to create star fields, he has given them back the night sky they thought they had lost forever.

Planetariums such as the Hansen Planetarium in Salt Lake City, the Buhl in Pittsburgh, and the Munich Planetarium have installed responder devices attached to each seat which allow the audience to vote on a choice of space destinations or topics within a program, respond to questions and even "fly" the planetarium theater as a kind of "spaceship of the mind".

Hands-on experiences are manifesting themselves in more traditional ways as well, especially in school programming. The Holt Planetarium at the Lawrence Hall of Science in Berkeley, California, has been a pioneer in interactive programming; its activity guide series called PASS (Planetarium Activities for Student Success), created by Alan Friedman, Alan Gould and others is currently being translated into Japanese for use in that country's many planetariums. Among its myriad activities are lessons involving the identification of features on the moon and the use of models to demonstrate the moon's phases.

Sheldon Schafer at the Lakewood Museum Planetarium in Peoria, Illinois, creates mysteries in which the time of the crime and the culprit can be determined by knowing when and where certain constellations appear in the sky. At the Suginami Science Education Center in Tokyo, Shoichi Itoh engages students in discovery through lessons in which they photograph the planetarium star field, create planetariums of their own, and find the constellations on their own using star maps.

Jeanne Bishop at the Westlake Schools Planetarium in Westlake, Ohio, reports that several Cleveland area planetariums are outfitting a Mobile Observatory with telescopes and computers for use by students; Jeanne plans to have some of her astronomy students prepare and conduct interactive lessons for elementary students using this equipment. And the staff at the U.S. Air Force Academy Planetarium uses its facility extensively for hands-on lessons of a special kind: Air Force cadet training in topics ranging from aeronautics to survival skills using a compass and the planetarium sky.

While our primary focus in most planetarium programs remains the physical aspects of modern astronomy, our fascination with the heavens in other ways can also be celebrated in the unique space that is the planetarium theater. Musical concerts, poetry under the stars, and live theatrical performance are periodic additions to many planetarium's schedules. At the Taylor Planetarium in Bozeman, Montana, one of the authors has hosted storyteller Lynn Moroney in a performance of Native American sky legends under the stars. And at New York's Hayden Planetarium, the other has produced live performances of African songs, dances, and sky stories.

But the efforts of planetariums extend to more than just the star theater. Astronomy classes, seminars, and workshops regularly combine classroom, planetarium, and outdoor learning at many of our facilities. Teacher workshops offer in-service training and resource materials to teachers of all grades.

Many planetariums have telescopes associated with their facilities even in the heart of major cities and offer regular observing programs. During the day, they show people the sun and/or project white light and hydrogen alpha images or the solar spectrum. At night, the planets or deep sky objects are introduced and special events such as eclipses or the occasional bright comet receive special attention.

At the Buhl Planetarium at the Carnegie Science Center in Pittsburgh, Martin Ratcliffe uses the Internet to link students with the Mount Wilson Observatory in California, letting them control one of the telescopes there and engage in research projects.

This past winter, planetariums in the U.S., Europe and Japan teamed with research scientists via television and the Internet to give students a real sense of the adventure of doing science. An American teacher and a high school student flew aboard the Kuiper Airbourne Observatory while students in schools and planetariums were linked to them and scientists during two observing runs. Students at Chicago's Adler Planetarium even got to control the KAO's infrared telescope via computer from the ground. In the late winter and spring, students in several countries were awarded three orbits on the Hubble Space Telescope. They got to choose their targets (in this case, Neptune and Pluto) and, assisted by Heidi Hammel of MIT and Marc Buie of the Lowell Observatory, got to see their images come down on live TV. They also got to process these images on the Internet and confer with Marc and Heidi on the results. A similar project, in conjunction with the Pathfinder and Global Surveyor missions to Mars, is now getting underway.

Planetarium efforts also extend to exhibits. For example, in the summer of 1994, The Museum of the Rockies curated an exhibit called "Pioneering Space", built around NASA scale models and chronicling the history of the U.S. manned space program. A few steps away, people could see a complementary program called "The Final Frontier" in the planetarium. Many planetariums have extensive exhibitions to solidify the astronomy experience–including those in London, Madrid, Los Angeles, Chicago and Hong Kong.

The Lakeview Museum Planetarium in Peoria has developed a scale model of the solar system that won it a place in the Guinness Book of World Records. The planetarium dome represents the sun, and scale models of the planets are placed at locations throughout the city and beyond–with Pluto 40 miles away! Each July, the museum sponsors a bicycle ride from the sun to the planets. Where else can you visit all of the planets in a single day and, in relative terms, even peddle faster than the speed of light!

In this paper, we have attempted to give a brief overview of some of the exciting and valuable contributions planetariums are making to astronomy education. Being limited by time, we have been able to chose only a few examples from the multitude of activities going on in planetariums around the world. In the future, our on-the-job limitations, as always, will be time and money. But we will continue to strive to maintain the bridge between the research community and the general public and use the tools available to tell the exciting saga of our efforts to understand this amazing universe in which we live.

The Use of the Planetarium in Nautical and Field Astronomy Education

By P.A.H. Seymour

Institute of Marine Studies, Plymouth University, Plymouth PL4 8AA, UK

1. Introduction

The universe of marine navigators and surveyors is basically a geocentric one. All calculations necessary for reducing celestial observations to obtain directional or positional information can be carried out within the pre- Copernican two sphere hypothesis. Some mature students on the degree courses have practical experience of navigation at sea but are not used to more abstract ways of thinking. However, most courses in navigation require students to understand the many corrections that have to be applied in astro-navigation. The planetarium can be used to illustrate the basic concepts of the two sphere hypothesis, although other methods are needed to understand the nautical almanac and the principles used in its calculation.

2. Coordinate Systems and the Planetarium

Obviously a planetarium is very useful to teach star identification, which all practical navigators should be able to do. Projected vertical circles, meridians, prime verticals, celestial equator and ecliptic, hour circles and projected protractors at zenith and pole are very helpful in teaching co-ordinate systems. These circles can also be used as an empirical introduction to the basic concepts of spherical trigonometry. Planetariums can also be used to demonstrate the effect of precession on right ascension and declination. However, since the planetarium emphasises the geocentric view point it cannot readily be used to explain the physics of precession.

3. Gyroscopes and Orreries

Many astronomy textbooks take the trouble to explain the dynamics of precession, using the spinning top analogy, but few explain the gyroscopic properties of a spinning body. Even physics students find rotary motion difficult at first, and most good physics textbooks go to great pains to explain and illustrate the concepts. On the other hand elementary astronomy texts, including those written for non-science majors, fail to explain the behaviour of a free spinning gyroscope in space. An orrery can be used to illustrate the parallel annual transport of Earths axis, but it is not obvious to some students, especially those not doing physics, what causes the axis to behave in this way. Marine navigation students have to understand the basic physics of the free-spinning gyroscope as an introduction to the dynamics of the gyroscopic compass, which is an important part of navigators training.

4. Geocentric Corrections

Some corrections which must be carried out by navigators and hydrographic surveys are geocentrically based. These are: dip of the sea horizon; refraction in the atmosphere; semi-diameter of the Sun; semi-diameter of the Moon (including augmentation of its semi-diameter); lunar parallax. Using slides and specially constructed teaching aids, it is possible to demonstrate some of these corrections in a planetarium, though in all cases the effect must be greatly exaggerated to be visible on the dome.

It is also possible to use the planetarium to illustrate the difference between the mean sun and the true sun, and the difference between solar and sidereal time. However, one has to step out of the pre-Copernican geocentric framework to explain the equation of time and the reasons for the difference between solar and sidereal time.

5. Heliocentric Corrections

Our courses on nautical astronomy go beyond the utilitarian aspects of celestial science and include some aspects of the history and philosophy of science. Since a great deal of navigation is now done via radio and satellite methods, the emphasis on astro-navigation has changed. Some students are committed sailors who want an understanding of more traditional methods of position fixing. Others are attracted by the educational value of classical positional astronomy and its applications.

The fact that the navigator can use an out-dated and limited model of the universe to obtain useful information comes as a surprise to many students. On the way to introducing Keplers Laws of Planetary Motion, it is possible to discuss the algebraic equivalence of the very different geometrical solar systems models of Copernicus and Tycho Brahe. In order to keep the advantages of the Copercican scheme, but still have a fixed Earth, Tycho had the Sun going around Earth, but all the other planets went around the Sun. The positions of the planets against the background stars can be obtained using either model.

Demonstration of the motion of the planets on the celestial sphere is of more than academic value to a navigator who wants to understand the nautical almanac. We can explain why the "v"correction has to be applied. Because the planets orbit the Sun they have a movement with respect to the Vernal Equinoxial Point which can be expressed as a change of sidereal hour angle. The SHA of the navigation planets lie between certain limits. This in turn gives rise to changes in the Greenwich Hour Angles. The *Nautical Almanac* adopts fixed values for the hourly differences of the various astronomical objects relevant to the marine navigator. These are: Sun and Planets 15°; Moon 14° 19'.0; Aries 15°02'.46.

Interpolation tables give the increment to be added to the GHA at the tabulated hour, for every second of each minute for all these bodies. To allow for the discrepany between the adopted value of hourly difference and its true value at the time, a quantity, denoted by "v", is tabulated in the daily pages of the *Nautical Almanac*. This is equal to the excess of the actual hourly difference over the adopted one, and it embodies information on the vagaries in the motions of these celestial objects.

6. History of Nautical Astronomy

Students on our BSc Marine Studies degree may also take modules in maritime history, including history of nautical astronomy. More specifically, we cover the beginnings of navigation based on the stars, the latitude problem and the longitude problem. The application of astronomy to position fixing at sea and the impact this had on progress in astronomy offers students an insight into important aspects of the history and philosophy of science.

6.1. *Finding latitude at sea*

A planetarium can demonstrate very clearly how to find latitude at sea, using Polaris and the Sun. Although Arab and Genoaese sailors could find latitude by Polaris by means of rather crude instruments, such as the Kamal, the Magic Kalabash and, possibly, the forestaff, the accuracy of their determinations was not very great. It was left to Portuguese captains, mathematicians and astronomers, who met at the court of Prince Henry the Navigator, to develop Pole Star and Solar Declination tables, which made a great improvement to finding latitude (Cotter, 1978). These improvements meant that the great voyages of discovery could be undertaken with more confidence.

6.2. *The longitude problem*

The outstanding problem in marine navigation just over three hundred years ago was that of finding longitude. Since accurate timekeepers did not yet exist, it was not possible to 'transport' the time of some reference meridian to sea. Two other possibilities presented themselves at this time. The first, due to Galileo, used the Moons of Jupiter, the other due to Sieur de St Pierre, involved using our own Moon.

6.3. *The Moons of Jupiter*

Paris Observatory, founded by Louis XIV in 1667, did much to develop the first method. Astronomers at Paris studied Jupiter's Galilean satellites to produce tables of their motions. Surveyors and geographers set up portable observatories along the coast and borders of France and by comparing their observations with the Paris tables could determine the longitude of places (Cotter, 1978). This procedure led to better maps of France, but the difficulties of using a telescope on board a ship made it useless at sea. Cassini's observations on Jupiter's moons led Roemer to discover the finite speed of light. We use a projection orrery, in our planetarium, to introduce students to the basic concepts of this method.

6.3.1. *The Lunar Distance Method*

The Royal Observatory at Greenwich was founded in 1675 by Charles II, specifically to solve the longitude problem by the method of lunar distances (Forbes, 1975). Essentially the method consisted of using the Moon as a celestial clock to tell Greenwich Time and comparing this with local time based on the stars. Successive Astronomer Royals at Greenwich made greatly improved maps of the sky and meticulously studied the motions of the Moon against the background stars. However, in order to produce a set of tables on the Moon's motion of use to a marine navigator, it was necessary to produce these tables several months, or even a year, in advance. This required a reasonably accurate theory of lunar motion. It is well known that the dynamics of our Moon is an example of the three body problem. Newton's geometrical work on lunar theory was cumbersome and was not accurate enough to calculate such tables. The planetarium is a powerful visual aid for teaching students some of the complexities of lunar motion and the basic principles of the lunar distance method.

7. Conclusion

The planetarium enables introduction of a wide variety of concepts relevant to maritime honours degree level education. It may be used not only to illustrate the more practical aspects of astro-navigation, but may also be used as a starting point for more advanced positional astronomy concepts. History of science and mathematical concepts may not easily capture the imagination of all students, but when linked to planetarium illustrations, even relatively difficult concepts can be made exciting. The planetarium has a unique role to play in maritime education and in training navigators and hydrographic surveyors.

REFERENCES

COTTER, C.H., 1978, *A History of Nautical Astronomy*, Hollis & Center.

FORBES, E.G., 1975, Greenwich Observatory, Vol. I, London, Taylor & Francis.

The Total Solar Eclipse of October 24 1995

By N. Raghavan

Nehru Planetarium, New Dehli 110011, INDIA

1. Introduction

The total solar eclipse of October 24, 1995, whose central line cut across the subcontinent of India, was only the second total solar eclipse visible from India in this century. The previous total eclipse visible from India occurred on February 16, 1980. At that time the print media filed widely varying reports on what the effect of seeing the eclipse would be, without much coordinated input from astronomers. With the new confused advice reinforcing old fears, almost the entire population literally hid indoors, fearing the worst. Many Indian astronomers silently resolved to themselves then, that public education must be taken up with the same level of seriousness as research programmes during the next eclipse.

2. The Background

The total solar eclipse of October 24, 1995 was visible along some of the most populated parts of India and took place during a season of generally clear skies. Elsewhere in the country the eclipse would be partial. So nation-wide, our class was a mere 900 million strong!

Even in the last decade of this century, astronomy education in India is very sparsely serviced below the post graduate level. Several new planetaria have been built around the country since 1980. Clearly they would play a role in public education. So our 900 million strong class could be apportioned between them as far as public education was concerned. But the school, undergraduate and amateur sectors continue to suffer from lack of focussed attention. Here, however, the numbers involved would be much smaller. The Nehru Planetarium decided to use this opportunity to design activities not only for the general public, but also for specific groups of school students, undergraduates and amateurs.

3. Public Education

Besides the staggering numbers, our planning for public education had to contend with two realities. Firstly, whatever material we produced had to be inexpensive and yet fully functional – so that a maximum number of people could benefit from it. Secondly, we had to chip away at deeply entrenched views, perhaps dating back several millennia! This was especially important as the eclipse coincided with the major festival of Deepavali in India and dire consequences due to the coincidence were predicted. We had to consciously devise ways of informing the public of the facts as they stood, without being pompous. For this we needed a starting point.

Early in 1995, we sought the help of various national dailies in locating people who had actually seen the total eclipse in 1980. They were specifically requested to outline what motivated them to see it, how they would describe it and if they suffered any short or long term ill effects. We received around 150 letters from all around the country. Most of the letters said that they or some member of their family had studied science and deliberately planned to see the celestial spectacle. They found it one of the most thrilling experiences of their lives. None had suffered any bad consequences. There were, however, two exceptions. One was written by some one who was only nine at that time and living in the partial eclipse region. He sneaked out and stared at the partially eclipsed sun wearing goggles and had eye problems to this day and another who had not seen the eclipse but sent me a cutting of the experience of a **doctor** who had lost his eyesight by watching an eclipse, excerpted from the Journal of the Royal College of Ophthalmologists. The 1995 eclipse was visible throughout the country. If through our urging people one million people saw it and 1% of them did so unsafely, it meant a thousand people with impaired eyesight. The publicity given to them would defeat the very purpose of the educationists. These two letters brought to centre-stage, the planetarium's public responsibility in clearly spelling out the danger of viewing the eclipse incorrectly, **without** scaring people off.

Another aspect was to examine in depth, the various negative practices that are associated with eclipses and guess how they may have originated. This could help in putting them in perspective.

Our strategy for public education, therefore rested on three pillars:
- Co-opt media to reach the largest number of people
- Emphasise the safety aspect
- Research ancient references to eclipses and attendant practices

Efforts in this direction led to a highly successful exhibition, a planetarium programme for the sky theatre and of course a press kit of past eclipse photographs and background material. Closer to the actual date of the eclipse, these served as the basis for many articles in the press. This early positioning of the Nehru Planetarium as the prime source of visual material brought many television crews to our door step, expanding our outreach literally a million fold.

4. School Education

To reach the school community, an audio visual show, a slide set and an information kit were produced. Copies were given to competent amateur astronomers, who visited the schools that requested help. Several voluntary organisations also took these to slum children who are normally out of the man made information loop but are well within nature's ! The highlight of the kit was a tiny mirror mounted on a cardboard altazimuth mount that could project the sun inside a room for safe viewing by many people - a perfectly adequate method for partial phases. The more enthusiastic among them were instructed to draw the phases of a projected image of the sun. Photography of the sun with a pin hole camera adequately protected by over exposed x-ray negatives was also taught by the members of the amateurs astronomers association to more keen students.

The nearest point on the central line of totality was 125 kilometres from the capital, Delhi. The planetarium itself arranged for a group of 400 school and college students to go to Ghata on the line of totality. All of them attended an orientation session at the planetarium where they were briefed on what to see, how to observe safety precautions and how to keep a record of the event. An activity book for school students was created and was used in manuscript form during a children's festival on eclipse day in interior Rajasthan, which experienced the earliest totality in the country.

5. Undergraduate Projects

The topics for the undergraduate projects emerged out of questions some of these students themselves raised during lectures that we gave in many colleges of Delhi. The first project was performed by students of Motilal Nehru College. It involved the construction of a three dimensional model to illustrate why there is no solar and lunar eclipse every, new and full moon day respectively. The cost of construction and the easy availability of the materials used were the two constraints imposed. After experimenting with cardboard, they came up with an acrylic version that worked quite well. Quite a few school teachers borrowed it for demonstration in their schools.

The second project was to make a demonstration video of the flash spectrum during the eclipse. Despite major problems in putting together equipment with no budget at all, a student from St. Stephen's college recorded successfully objective grating spectra using a replica holographic grating on video film. Some of the interesting frames have been digitised and hopefully emission gradients of the Balmer lines can be derived. But the primary value of the film lies in its use as a demonstration movie. It is like reliving the excitement felt by Charles Young in 1870!

The third project was to use a thermocouple to monitor the temperature variation during the eclipse. Although no comparison measurements were made before or after the eclipse the measured variation agreed fairly well with other results reported.

6. Amateur Experiments

The amateur astronomers association at the planetarium decided that it will organise a trip to the central line of totality. Site survey identified two locations, Ghata in Rajasthan and Bhind in Madhya Pradesh as being suitable from the point of view of morning seeing, availability of power and reasonable boarding and lodging facilities. The availability of the terrace of a degree college in Bhind marked it out as a preferred site.

The amateur astronomy movement in India has been at the nucleating stage for several decades. The high costs involved and the lack of access to books, magazines and equipment are only partly responsible for this. The tradition of doing something academic as a leisure time activity is not established in India. Especially the planning and practice so necessary to succeed in recording a phenomenon lasting less than a minute were alien concepts. So incentives were offered. Only those members would join the trip, who had a viable experiment to perform and had put together and tested their equipment in the full moon preceding the eclipse date! The trip thus became an observing expedition! Rudimentary but functional equipment including a heliostat was built and tested and nearly thirty members, from ages 5 to 55, each with an observing plan, were on board the bus that left Delhi for Bhind. Each member was rewarded not only with success but was transformed into a future eclipse chaser. In fact one sequence of calibrated photographs is being analysed for deriving electron densities out to two solar radii.

7. Conclusion

Reports showed that the public education effort had succeeded in a large measure. Interestingly a larger fraction of the city folk stayed indoors than people from towns and villages. What about school, undergraduate and amateur sectors? Although the number of people reached in the last three categories is minuscule compared to the first category, the success achieved is no less important. A twelfth century Arabian astronomer has said that when we place too much emphasis on mere popularisation of science, the quality of science itself falls. In present day India we are in some danger of this. It is for this reason that the Nehru Planetarium has addressed itself to the school, undergraduate and amateur sectors and achieved a modicum of success. In India the outreach of most planetaria to schools stops with concessional tickets to their public shows. It is necessary at the present juncture for large planetaria to shoulder the responsibility of school education as well as public education through well designed, sustained programmes especially those aimed at teachers. Again amateur activity in many countries is centred on the initiative of the individual, whereas in India it is necessary to deliberately foster it. I have no doubt that the IAU in general and Commission 46 in particular will devise programmes with this end in view for the benefit of star gazers in large parts of Asia, Africa and South America.

The Planetarium–a place to learn

By C. Iwaniszewska

Insitute of Astronomy, Nicolaus Copernicus University, Chopina 12/18, 87-100 Torun, Poland

I would like to dedicate this paper to the memory of professor Edith A. Müller, deceased at the age of 77 a year ago, on July 24, 1995, until her retirement working at Geneva Observatory. She had been at the very beginnings of our IAU Commission 46 in the late sixties, she had been its President in 1970 when we all met during the General Assembly in Brighton, she always took great interest in further educational developments. I am personally grateful to her for much helpful advice during Commission 46 meetings at the General Assembly of 1985 in New Delhi. Wonderful teacher and organizer, she was also an extremely kind lady.

While I am not aware of any connection of Edith Mueller with a special planetarium, yet I have chosen this short biographical note to introduce my first problem, not so very obvious when mentioning generally planetarium activities. Nearly every planetarium bears the name of a patron, who either made the existence of that institution possible through financing, or was well known in the town or country for his/her interests in astronomy, etc. Let me mention two examples: Luiz Erro Planetarium in Mexico City, and Jawaharlal Nehru Planetarium in New Delhi. While Erro had been very interested in astronomy – he once studied at Harvard and helped introduce modern astrophysics to his country – Nehru had been a national person: the Planetarium is next to the Nehru Memorial Museum; Prime Minister Indira Gandhi, Nehru's Daughter, attended in person the Planetarium opening. And now, I should like to ask all astronomers connected with a planetarium: Is it customary to tell people about the Patron of your Planetarium? Do you mention his/her existence to groups of visitors, tourists, students, coming there for the first time? Do you know that it would be an excellent occasion to bring forward a new topic, to give a living lesson in history, or geography, or even ... patriotism ?

In Torun, the city of Copernicus, we already have a University and a Museum bearing the name of that Astronomer, so the Planetarium opened here in 1994 was named after professor Wladyslaw Dziewulski (1878-1962), astronomer at Wilno and Torun Universities, first director of our Astronomical Observatory, whose name was also given to one of the lunar craters. Some photographs with a short biography of the Patron are displayed in the entrance hall of our Planetarium.

The Planetarium itself is built inside an old 19th century gasholder right in the middle of the Old City, which provides tourists with unique opportunities: between visiting an art exhibition in the old Town Hall, and looking at the immense interior height of the post-franciscan St. Mary's Church, one can go and relax sitting down and experience a wonderful excursion through the Universe!

The Planetarium building (dome diameter 15 m) contains a lecture room with multimedia facilities for 100 persons, the planetarium hall with 160 seats (recently arranged so that everyone will have a good view towards South) a small cafeteria with a sales stand for starmaps, astrophotographs, etc. staff rooms. The projector is a Zeiss (Jena, Germany) instrument.

The other nine Polish planetariums are either connected and administered by a school or Naval College, or they form parts of Museums and cultural institutions; in either case they get enough money for staff salaries, unrelated to the planetarium attendance. In Torun the situation is different, since the Planetarium is administered by a Fund connected with the Copernicus Museum, but must be self-sufficient financially; staff-salaries

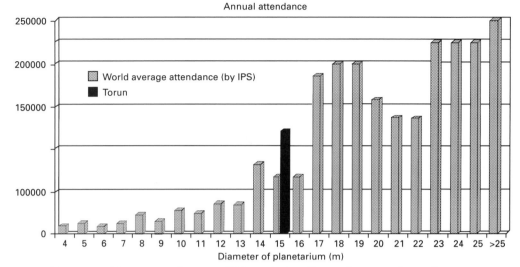

FIGURE 1. Torun attendance on IPS statistics of attendance vs dome diameter

as well as instrument repairs or new books are all paid from the entrance-fees of visitors. And the number of visitors may vary, it cannot be predicted accurately. For instance – it was unfortunate that an influenza epidemic last January considerably reduced the number of school groups visiting the Planetarium so that the staff did not get their New Year premiums! The number of performances was 1299 in 1994 with 102,837 visitors, while in 1995 the corresponding numbers were 1368 and 113,113, respectively. In 1996, during the height of the tourist season in May, the Planetarium had more than one thousand visitors daily. And according to the relation of annual attendance versus planetarium diameter prepared by the International Planetarium Society the Torun Planetarium shows a very high attendance for its medium size (Figure 1).

Large groups of tourists visiting our town from May till September attend one of the popular performances:

1. Copernicus and the Earth motion
2. The Blue Planet
3. The Great Encounter (of Comet Schoemaker-Levy with Jupiter)
4. Summer under the Stars.

Every programme is based not only upon the motions of the planetarium projector, but is enriched by well-adapted music and by slides, moving pictures from video-casettes, etc. Not only school children but also professional astronomers find, for instance, the flight over Venus surface is quite exciting "when the instructor's voice tells you what can be seen on your left, on your right, just like it would happen on a terrestrial excursion!"

Special attention has been paid from the very beginnings of the Planetarium existence to maintain good contacts with the provincial school authorities. As a result, regular conferences in teaching method for geography or physics teachers are being organised in the Planetarium lecture room: a lecture by one of the Torun University astronomers followed by a display of the possibilities of the Planetarium instruments and projectors. Six didactic programmes have been run during the past school year:

1. Wonderful Journey (beyond the Planetary System)
2. The Revolving Earth (consequences of diurnal motion)
3. Was Copernicus right? (consequences of Earth orbital motion)
4. The Solar Family (structure of the Solar System)

5. The Cosmic lights (birth, life and death of -stars)

6. The Star Islands (structure of galaxies).

Every programme could be adapted to the level of a particular class.

Nearly every Polish planetarium has still other types of specific activities, not necessarily related to the planetarium projector, but very important from the didactic point of view.

The largest Polish Planetarium in Chorzow (dome diameter 22 m) has been organizing Astronomical Olympiades for secondary schools for 39 years. They comprise three stages, the first at school, the second for neighbouring provinces, the final stage at the Planetarium. The jury is composed of ten professional astronomers, who not only allocate the final places, but are required to prepare the problems and exercises to be solved by students.

A small planetarium (dome diameter 8 m) built on top of a secondary school in Grudziadz (90 kms North of Torun) is the placewhere, for about 20 years, the final stages of an Astronomical Seminar take place. Secondary school students have to prepare their own scientific papers, based upon a chosen theme, or a computer programme, or own astronomical observations. Papers have to be presented at a meeting for every province, then the best go to the final presentation to Grudziadz. There are times when the jury members have to listen to about forty papers during the 3 days of the Seminar. This type of activity is prepared in collaboration with the Polish Amateur Astronomers Society whose members sit on the jury for this competition.

The Torun Planetarium is the place for the final meeting of the Astronomical Competition for Primary Schools (10-15 yrs). The Competition has been organised this year for the sixth time, jointly by the Polish Amateur Astronomers Society and the local school authorities. The first stage, some tests and simple astronomical observations, were prepared at home, and then twenty best students from 4 neighbouring provinces came to the Planetarium. I would like to tell here about two examples of didactic achievement. This year, the children have been shown video pictures of travel over the surface of the Earth, then Venus, and finally Mars. And after that: "You have looked at beautiful pictures, now give us some basic facts – write down the principal differences between the atmospheres of Mars and Venus"! And they did it; some children really presented samples of essential data. And they had to think quickly, having been given only five minutes time.

Another example. The group of seven best boys was given the last task: "Each of you must outline the most important information upon a given subject for 3 full minutes"! and the themes have been galaxies, or variable stars, or quasars, etc. And here also the children had to think quickly and to decide, which are the most important facts, how to present them to the audience, etc. I am thinking now about the answers of one of the youngest boys, Alexander, who took part in this competition for the third time. Two years ago, when he came here at the age of 9, he knew many things but he was too young, he did not know how to express himself, to present his knowledge. And this year, when he is already 11, he is more mature, he speaks more fluently and coherently about his favourite astronomical subjects. So I think young Alexander stands as a good example of the right effect of an astronomical competition upon the general abilities of children. Even if he will not become an astronomer in the future, Alexander has already gained one step in how to express himself, and that is something often not easily acquired. I see frequently my own university students finding great difficulties in expressing their knowledge coherently. So this is another unexpected result which will become useful for the general education of young people.

In more than one sense – the Planetarium is indeed a place to learn.

British Planetaria and the National Curriculum

By Steve Tidey

Association for Astronomy Education, Burlington House, Piccadilly, London, UK

1. Introduction

The UK is experiencing a relative Golden Age for planetaria, thanks in many ways to its national curriculum. In 1991 the British government finally bowed to many years of steady pressure by interest groups and introduced into a new and controversial general curriculum a requirement for pupils to attain knowledge about the Earth-Moon system, solar system objects and basic cosmology. Prior to this there had been no science curriculum for pupils aged under 11. Astronomy formed a small part of nature study. The science education of 11-16 year-olds depended on their GCSE syllabuses.

The purpose of this paper is to study what knowledge of the cosmos pupils are now required to attain, how the content changed when a revised curriculum was introduced in 1994, and how planetaria go about teaching the subject to schoolchildren. We will also look at how the curriculum differs in Scotland, and what 'A' level students have to learn about astronomy.

2. Background

From the late 1950s, when one of the first planetaria in Britain was built at Marylebone Road, London, up to 1991, some teachers had organised their school visits to these star theatres as an extra-curricula activity (except for those students studying astronomy at O-level) which required little or no preparation or class work afterwards. Generally speaking, however, most school parties turned up because they wanted to have a valuable learning experience about the Earth's place in the universe. Then, seemingly overnight, the government expected teachers to have detailed knowledge of the reasons for the seasons, tides, the Moon's phases, planetary motions, the Milky Way and many other difficult astronomical concepts. The radical changes in all subjects in the new curriculum caused problems in the education profession for a considerable time, and gave teachers a wide knowledge gap to fill.

Astronomy resources to help educators in their daunting task were available, scattered though they were over a number of publications not intended specifically for classwork, but unfortunately teachers did not know where to look for them and who to contact for more information on the necessary teaching methods. They desperately needed help, and so after having spent several decades on the fringe of school science education, British planetaria were ideally placed to lend a helping hand to a grateful teaching profession.

The planetarium scene in the UK is markedly different in some ways to that in the USA. The shock launch of Sputnik in 1957 sparked off a huge investment in American school science education. Within a few years hundreds of schools and colleges across the nation had built planetaria in their grounds, and to this day most domes are to be found there. In the UK, however, the trend up until the late 1980s had been for domes to spring up linked mainly to universities, museums and observatories such as those at Greenwich, Jodrell Bank, Liverpool, Armagh and Edinburgh. Schools had no option but to take coach parties of children to these places for their astronomy experience under the simulated stars, since barely a handful of domes were (and still are) directly linked to schools.

With the new curriculum, however, the scene was wide open for development by the cost-effective, American-designed Starlab mobile inflatable dome, which was destined to add a new dimension to astronomy education and greatly increase the number of pupils who could have the planetarium experience each year. Sure enough, the number of inflatables grew rapidly, satisfying tens of thousands of children and their relieved teachers. At the time of writing there are at least 23 Starlabs in the UK (several of which are owned by Local Education Authorities) and this figure represents roughly half of all British domes. Thanks largely to the curriculum, many of them are run as commercial concerns in different areas with arbitrary borders that do not infringe on other people's unofficial territory. Yet there are still large regions containing many schools not serviced by either a portable or fixed dome.

Traditionally, British planetarians have not been as technology-obsessed as their American cousins, but in the 1990s the equipment used at some of the larger domes has been brought up to date if only to address the expectations of audiences accustomed to sophisticated software in PC packages and video games, and the special effects seen in films. Several years ago technicians at the Armagh Planetarium, Northern Ireland, developed the world's first electronic interactive equipment which was later adopted by the dome at the National Galleries and Museums on Merseyside, in Liverpool. The design has since been copied by many other sites across the USA. Armagh now has a Digistar projector which creates computer-generated star images. The London Planetarium has also acquired one which was brought into service in 1995; this new technology has enhanced their school shows, opening new areas that could not be exploited with their old mechanical Zeiss star projector (which had long and faithful service from 1958 to 1994). Digistar's excellent graphical capabilities make topics such as planetary orbits and black holes easier to explain.

3. The 1991 national curriculum

The 1991 national curriculum was administered by the then Department of Education and Science and is taught in over 30,000 schools across England, Wales and Northern Ireland. Children aged 5-11 attend primary school, after which they move on to secondary school until the age of 16. Pupils aged 16 upwards have the option of studying subjects at 'A' level, a qualification which is needed if they hope to go on to university.

The curriculum divides the process of learning into four distinct Key Stages of increasing complexity. The astronomy content at every Key Stage in the old curriculum was grouped under the heading "The Earth's Place in the Universe". Key Stage 1, for pupils aged 5-7 still learning the basics of the world around them, required children to study a) the Sun's apparent daily motion across the sky, b) the Earth, Sun and Moon as separate spherical bodies, and c) the changing altitude and appearance of the Sun and Moon over time in a predictable manner. Children discovered that the Earth is a spinning planet which orbits the Sun, and these two factors cause night and day.

Key Stage 2, for ages 7-11, covered similar ground with the additional requirement of needing to learn that a) light travels faster than sound, b) the lengths of day and night and the year are determined by the Earth's motion round the Sun, c) the reflection of light enables objects to be seen in the sky, and d) the other planets in the solar system move in their own orbits.

Key Stage 3, for pupils aged 11-14, went deeper by looking at a) the Sun as the Earth's major source of energy, b) the solar system as a small part of a galaxy within a much larger Universe, and c) the properties of gravity and the fact that its force diminshes with distance.

Pupils aged from 14-16 in Key Stage 4 completed their basic astronomy education with a study of a) how energy is conducted by radiation, b) the quantative relationship between speed, frequency and wavelength of radio signals, c) the use of data about the planets and stars to speculate on conditions elsewhere in the Universe, d) the electromagnetic spectrum and how specific wavelengths are detected, e) gravitational theory as it relates to the motion of satellites, and f) theories about the origin and future of the Universe, and the observational evidence in support of them.

Once the dust had settled following the new curriculum's launch, and teachers had a grasp of the subject matter above with the help of the professionals, they were, broadly speaking, happy with the astronomy content.

The resulting new planetarium school shows were not put together in a vacuum. Local teachers were brought in at an early stage to be an active part of the programming. Sometimes they asked for certain topics to be omitted, as they were confident about covering those with relative ease in the classroom. Essentially, they wanted planetarians to use the star and planet projectors to explain the difficult concepts, such as lunar phases, planetary motion, constellations, time, seasons, daily changes in altitude and annual motion, that could not be done adequately in a classroom environment.

The shows were (and still are) almost exclusively live, and teachers often learned as much as their pupils, and they didn't mind if the presenter went off on a tangent (which was easily done, as so many topics in astronomy share common ground). They were surprised at the relative ease with which planetarians could explain complex concepts. It's amazing what you can do with an Earth globe, polystyrene balls and light bulbs! The experience sent them back to their classes feeling more confident about their ability to teach the subject.

The emergence of Starlabs has fortuitously coincided with a period in British education when schools are increasingly strapped for cash, and every pound spent has to be justified. The inflatables have therefore become popular with teachers for several reasons. To begin with, they do not have the time-consuming duty of organising a visit to a fixed dome, secondly the overall cost of inviting a Starlab operator to come on site for a few days is cheaper per pupil, and last of all the shows can be slotted into the school's schedule with relatively little disruption.

Schools were anxious for quality teachers notes to assist them with the new curriculum, and so the Association for Astronomy Education (AAE) a small but influential organisation, filled a gap in the market by co-publishing with the Association for Science Education (ASE) two sets of activities for primary and secondary children, under the title *Earth and Space*. With significant help from planetaria, approximately 16,000 copies have been sold to date, the majority to the teaching profession, and in 1996 updated editions will appear, incorporating the requirements of the revised 1994 curriculum.

Another aid for teachers has been the In-Service Training courses hosted by planetaria. They visit for either half a day or a whole day of practical demonstrations, which are designed to answer common questions about the practicalities of teaching astronomy in a classroom.

A similar idea organised by the Royal Astronomical Society (RAS) has been equally successful. Their regular Training the Trainers one-day courses continue to be well attended by teachers.

In recent years some British planetaria have turned to the acting profession for help with teaching the curriculum. Inter-Action, an educational charity which develops innovative learning and community projects and is based on the HMS Belfast (a decommissioned warship moored on the Thames at the *Victoria* Embankment) employs actors on short-term contracts to travel the country with several Starlabs. The actors use their drama training to present shows in a fanciful piece of drama and comedy, slipping in and out of the numerous mythological characters that fill the night sky. This highlights the more human face of astronomy for children, and the regular influx of fresh staff generates new ideas for shows.

In a similar vein, Peter Joyce, a professional actor, has become a familiar and welcome guest at planetaria, where he appears dressed as either Sir Isaac Newton or Galileo. As Newton, he regales his audience with tales of how history's greatest scientist developed his theories of planetary motion and gravity. To demonstrate the universal force, he picks a female from the audience and tells her he is attracted to her! Gravity then allows him to move to other areas, such as the Moon's motion round the Earth, tides and satellites in geostationary orbit. The appreciative audience are left in no doubt how much of what we take for granted today has its roots in his ground- breaking observations and insights.

As Galileo, Joyce relates the Aristotlean geocentric theory and talks of his own experiments dropping cannonballs from the leaning Tower of Pisa, and firing them from a cannon to study the force of gravity. Pupils also hear about the Copernican heliocentric revolution, and Kepler's studies which lead to the formation of his laws of planetary motion. Galileo reveals his delight with his first telescope and the wonder of discovering Jupiter's four largest moons, which his audience learn they can see for themselves with modest equipment.

3.1. *The 1994 curriculum*

In 1994, after several years of intensive lobbying from a teaching profession still having difficulties coming to terms with the new curriculum in every subject, the government finally scaled down its scope across the board to give teachers more time on each subject. There was a general feeling that, difficult though some aspects of the astronomy curriculum had been for teachers, overall it had been a success and popular with pupils. Organisations such as the AAE, ASE and RAS were consulted by the government for their advice on reshaping the astronomy content, and their feedback was largely adopted. The revised curriculum omits almost all mention of the subject in Key Stage 1, but as teachers had found that in the old curriculum 5-7 year-olds had difficulty with the concept of lunar phases, some of these changes were welcomed. The only astronomy topic Key Stage 1 now touches on is under the heading Light and Sound, which requires pupils to know that "light comes from a variety of sources, including the Sun", and that "darkness is the absence of light". Teachers who now want to talk astronomy at Key Stage 1, have to be creative and bring in the subject via areas such as Geography or History.

The amendments to the astronomy in Key Stages 2 and 3 (now found under "Earth and Beyond") were less radical, although the reasons for tides was the most notable omission.

3.2. *Teaching the curriculum*

Planetaria take a variety of approaches to teaching the curriculum. We shall study several of them.

Every year the London Planetarium caters for approximately 20,000 pupils who come through its doors to see curriculum-based shows. Teachers are given a comprehensive pack of support material which can be utilised in the classroom. It includes a quiz which tests pupils' astronomy knowledge before and after their visit, basic facts about constellations and deep sky objects and a solar system poster. For teachers there is a list of contacts for resource material, such as videos, books, telescopes, binoculars etc.

The exhibition area is divided into several distinct sections. In Planet Zone our nearest neighbours in space are the focus. Scale models are accompanied by video pictures and text with up-to-date information about each planet, whilst touch-screen monitors supply facts about comets and meteors etc. Other screens display information about tides, satellite orbits, the solar system's birth, the Sun's energy production, large-scale galaxy structure and what is visible to the naked eye in the sky at the current time. This area also has a three-dimensional model of the stars in the constellation Orion. Visitors see that the stars are at different distances from Earth, and it is a line-of-sight effect that makes them appear to us in the familiar pattern.

The other key area, Space Zone, also has touch-screens which cover more exotic subject matter such as star birth, stellar evolution, solar fusion, planet formation, cosmology, black holes etc.

The planetarium's Key Stage 1 show goes beyond the revised curriculum's limitations and, after introducing children to the concept of night and day and the motion of the Earth round the Sun, the Digistar II projector cleverly turns the planetarium into an imaginary spaceship that launches out of the dome of an imaginary observatory to explore the solar system. The audience land on the Moon and Mars, and travel to the outer reaches of our solar system before flying home.

The show is popular with pupils and teachers alike. So, too, are the other Key Stage shows, which give the audience more than the curriculum leads them to expect. The Key Stage 4 show climaxes with gasps of delight, as the audience is sent on an exciting trip down into a black hole.

In 1996 the planetarium successfully inaugurated what is hoped will become an annual nationwide Schools' Challenge Quiz, entitled "Rising Stars". It was a knockout competition featuring teams from 23 London schools competing for a Russian-made reflecting telescope. Some of the questions were based round constellations and other images on the dome.

At the Armagh Planetarium in Northern Ireland the pupils themselves are used as human models to explain seasons and lunar phases. They take advantage of a permanent scale model of the solar system situated in an Astropark, 25 acres of ground next to the planetarium. The dome houses a Digistar projector which displays a 3-D orrery, useful for explaining orbits and eclipses. The planetarium also has an outreach program, consisting of a purpose-built Astrovan which takes hands-on activities and a Starlab to schools. It caters for approximately 15,000 pupils each year.

Staff at the South Shields Planetarium in the north-east of England take a fresh angle on old and tried topics. For example, children are given worksheets that encourage them to count the number of feet in the zodiac constellation figures, and to discuss what happened to the Seventh Sister in the Pleiades! They also take part in light-hearted, fun astronomy quizzes and build fantasy rockets to explore the solar system.

Many planetaria give children information sheets containing basic facts about astronomy, encouraging them to look at the real sky and follow the Moon's phases, identify naked-eye planets, the larger constellations and do basic experiments with shadow sticks. Pupils also learn how to make simple inexpensive sundials, nocturnals, planispheres, build scale models of the solar system from cheap household items and design their own constellation shapes. At the South Shields dome simple rockets are launched with compressed water, and a Junior Astronomy Club meets regularly. A similar idea has proven successful at the Fort Victoria Planetarium, on the Isle of Wight, which hosts regular meetings of its Starlog Club for children. At Techniquest in Cardiff, pupils view sunspots on the projected solar surface.

The company Inter-Action believe that Starlab's adaptability is one of its greatest strengths, and that using the dome for nothing but astronomy shows will not utilise its full capabilities. In recent years they have found success developing new cross- curricula shows to cover other aspects of science in the national curriculum, some of which share common ground with astronomy. This innovation allows them to discuss topics which would not sit comfortably within a standard show about the night sky. It points the way forward for other Starlab operators who are thinking of expanding their scope, whilst staying true to their astronomy roots.

Before the revised curriculum was published, planetarians got wind of the changes and many of them, especially the sole-operators whose income derives largely from their Starlab curriculum-based shows, looked on it with foreboding. They foresaw their livelihood being seriously eroded overnight. However, the clear message from planetarians across the country since the changes were introduced could be summed up as, "Changes? What changes". Generally speaking, it appears that many teachers are happy to let them continue with more or less the same breadth of shows they were hosting under the old curriculum.

Extrapolating from a straw poll I conducted for the purpose of this paper, it appears that well over 500,000 schoolchildren visit British planetaria every year to attend curriculum-based shows. That figure will grow as more Starlabs and other fixed domes which are being planned today open for business. This is a tribute to both the effectiveness of those school presentations, and the enduring popularity with children of basic astronomy concepts which excite their imagination.

4. Scottish school astronomy

The national curriculum is not taught in Scotland. Their schools are instructed to cover specific broad subjects, such as Science, History, Music, etc., and it is left to teachers to decide what elements of those subjects they cover. Science gives them a convenient, broad scope, and fortunately many teachers plump for basic projects on Space and Astronomy simply because these have a proven track record with children at all levels.

Scottish planetaria tend to confine their schools programs largely to what pupils can see in the night sky, although some of the hands-on experiments offered at the Mills Observatory Planetarium in Dundee include such things as counting the numbers of galaxies and the lengths of cometary tails on large Schmidt plates, constructing simple planispheres and orrerys and studying small meteorite samples.

5. A-level Astronomy

Students hoping to get into university need to acquire qualifications at 'A' level. However, no examining board offers Astronomy at this level. Interested students can take a two-year 'A' level Physics course, one module of which is Astrophysics. The subject has been growing increasingly popular in recent years. Topics covered depend upon what examining body administers the course, but the common ones include: the microwave background, line spectra, the H-R diagram, the association between mass and energy, measuring distance by parallax, mass in binary stars, the fusion process, the transfer of energy to a star's surface, stellar evolution, black body radiation, red shift, special relativity, the Doppler Effect, the different means of classifying stars, the physics of optics and the Big Bang, etc.

6. Conclusion

British planetaria and the national curriculum have a lot for which to thank each other. Important ties have developed between planetarians and the teaching profession, ties which have given the former a respected position to comment upon and influence astronomy education in the UK. The increasing number of schoolchildren who see curriculum-based shows augers well for the planetarium profession, at a time when its finances are being squeezed from all sides. Teachers have gained much and will benefit still more in the years to come.

References

The following publications are of value.

Earth and Space: Workpack for Primary and Middle Schools 1990, Association for Science Education, Hatfield, UK.

Earth and Space: Workpack for Secondary Schools 1990, Association for Science Education, Hatfield, UK.

Science In The National Curriculum 1991, Her Majesty's Stationery Office, London, UK.

Science In The National Curriculum 1994, Her Majesty's Stationery Office, London, UK.

SECTION FIVE
Public Education in Astronomy

Public Education: the ultimatum for the profession

By M. Othman

Space Science Studies Division, 53 Jalan Perdana, 50480 Kuala Lumpur, Malaysia

1. Introduction

The impact of public education is without question in the 'public good' domain and hence there is really no need to justify the demand for it. However, some professionals and scientists remain unconvinced about the necessity for it. This paper will lay out the benefits it holds for the scientists, categorise the target groups and identify the methods of approach for each target group and finally outline some strategies that can be adopted to achieve the educational aims.

2. Benefits of public education for the professionals

Contrary to belief, the professionals have more to gain from public education than the public. There are several reasons for this.

The first of these is that public education calls attention to the scientist's work. The publicity generated through this will indirectly attract the attention of the relevant agencies or bodies that disburse grants, approve programmes or determine manpower requirements. In the light of budget cutbacks, downsizing demands and rationalisation exercises that are getting commonplace, the scientists will do well to create a public alertness to stave off these calamities. Public interest usually signifies a demand for the science or the field or the department and, therefore, the authorities might think twice before taking any negative action.

Secondly, it is obvious that through public education a scientist will be able to gain fame. This is not entirely without advantage – one day at a highway toll booth, the operator recognised me and waved me off. I saved 50c that day!

Thirdly, by getting involved in public education, the scientist gets a chance to reciprocate the tax payer's contribution to his science. This way he feels he is also making a contribution to society, a philosophical consideration that is increasingly getting important to many people.

3. The Professional's Apprehension About Public Education

But scientists and professionals continue to disparage and be apprehensive about the need for getting involved with the public. Why?

Firstly, professionals are generally suspicious of their colleagues who believe in public education because it is perceived as an attempt to get publicity. Any effort in this direction gains the jealousy, resentment and ridicule of others. This reaction usually stops him from getting actively involved. Of course this does not deter the genuinely interested person but it demands a bit more grit and perseverance.

It is also unfortunately true that professionals who have got involved in public education have sometimes had bad experiences. This refers to the situation where the education exercise is carried out through the media. Misquotations, deliberate or otherwise, misunderstanding of the subject and sometimes plain poor writing or reporting on the part of the journalist are sources of embarrassment and stress. Once bitten, the professionals shy away from all subsequent activities.

These problems are not insurmountable. Scientists and professionals in general can be made to understand the importance of public education after which they will be able to see that they should be encouraging their interested colleagues to do the battle for them rather than impeding them. The second conundrum can be overcome by training journalists in scientific journalism.

4. Categories of Target Groups

Target groups for public education can be divided into three main categorise: the mass media, sponsoring agencies and the public.

The mass media encompasses electronic and print and include television, magazines and newspapers. Sponsoring agencies is taken here to mean any organisation that promotes public education. They come in various forms: corporations, government agencies, universities, a board of Trustees etc. The last target group is the obvious cluster: the public themselves.

5. How to Make Contact

The approach one needs to achieve the goals of public education differs from group to group. A lot of innovative and artful methods can be utilised and in the end the approach one adopts will depend entirely on the educator. Broadly, we can identify the following peculiarities for each target group.

When we speak of the mass media, we talk about science journalists,most of them young and eager (at least they are in Malaysia), but not very experienced or learned. They want information and material fast and, most of all, they want these in easily readable and, if possible (alas), sensational form. To sustain their interest and assist them do their work we should try as much as possible to accommodate these requirements. In particular, they may want you to address the issue of how relevant your work is to man's needs. I think it is a good question, something all professionals should apply themselves to regularly. It is vital to note that these people are front-liners in the public education context and need to be nurtured (so that in the extreme they will not write rubbish) and cultivated to some extent (so that in the extreme they will not give you or your organisation bad publicity). It is important to make time for them because if you do not, it is very likely that they will get their information from somewhere else. As an example, I think research findings are best told by the researchers themselves and not by the Dean or the Head of Department.

Sponsoring agencies, like the mass media, make public education happen. Without them we will not have the funds or the means to carry out our educational programmes and hence their extreme importance. One significant aspect to note about sponsoring agencies is that some of the decisions they eventually make are often contingent on a few, or maybe even one individual, be they politicians, desk officers, secretaries, the chairman etc. How does one approach this and other problems?

To convince a sponsoring agency to support your public education programme, an interest must first be created through working papers and briefings which should be generously and beautifully illustrated. Remember the old adage: a picture speaks a thousand words. In my experience I find generally that selling the idea to a sponsor in the beginning a very formidable task. For instance, it took several years of hard work to convince the government that a planetarium for public education was essential and there was a demand for it. But the good thing to note is that once you have sold the idea, subsequent overtures are easier and it becomes a relatively simple matter to maintain their attention. At the National Planetarium today, some corporate sponsors seek us out to fund activities rather than the other way round. The prerequisite for this to happen is to make that first event you organise for and with them a successful one. Hence the key word is success: If you succeed in one public event, the rest is easy.

The public is the last but certainly the most important target group I want to address. How do you make the public desire to be educated? Let us focus on two different things: one is what we want them to acquire the knowledge of, the other is what they want to learn.

We want to communicate the following: the beauty, joy and essence of astronomy; facts and features; man's pioneering spirit in space; astronomy's connection with the other sciences as well as religion, the arts and human culture; and much more. Which are all very well, but what does the public want? Most of them want to be entertained. Sure, some of them want their intellect stimulated, but, by and large in my experience, if the atmosphere we create is boring and ordinary, we lose them very quickly. Communicating science through entertainment is vital philosophy professional faces: A 'gee-whiz-bang' kind of production versus a technical or scientific presentation. Of course the way out is to take a middle road. Scientists must adapt language and style while at the same time not lose the integrity of the science to be presented.

I found a paper which listed the following attributes for a public educator: clear and pleasant speaking voice, ability to explain things clearly, ability to write well, scientific aptitude, mechanical aptitude (in the planetarium), enthusiasm, dedication, interest, creativity, imagination, flexibility and humility. If anyone here found someone like that, I will be delighted to meet him or her. That aside, it is important to understand that these traits in an educator are a cornerstone of public education and that we should translate them into actual features of the education programme.

6. Strategies for Implementation

The strategies we can muster depend on our objectives. I believe a public education programme should have the following goals:
  (i) create and enhance public understanding of astronomy;
  (ii) strike strategic alliances to allow fulfilment of the mission;
  (iii) manage resources to preserve and expand the mission.
The following strategies can be adopted according to these goals:
Goal i)
  • provide up-to-date information for public through the Internet, pamphlets, brochures, bulletin boards etc;
  • organise seminars, exhibitions, observing session etc;
  • encourage astronomy clubs, societies etc;
  • provide formal learning programmes;
  • provide training for journalists and teachers
Goal ii)
  • make formal contacts with other organisations with similar interests for networking purposes;
  • develop joint public events with other organisations;
  • develop a marketing plan for sponsors
Goal iii)
  • manage funds, human resources etc. efficiently;

- conduct a study to identify what resources are required;
- create a volunteer programme.

These are only some of the strategies that can be adopted. A lot more can be listed depending on the programme's needs and resources.

7. Final Remarks

In today's setting, public education is a requisite if scientists and professionals wish to continue their research and associated activities unimpeded. Accountability to the public takes many forms but the easiest and most enjoyable is taking your activities and research findings to them. Scientists or professionals in general are not accustomed to this kind of enterprise, which is in stark contrast to their day to day routine, but it can be done and it is easy to become a convert.

I would like to add that in a country where there is no astronomy in the curriculum, no observatories and maybe even no professional astronomers, the single most effective initiative to establish astronomy in that country is a public education programme. Through the programme you can reach children, parents, teachers, decision makers, politicians, managers ad infinitum. When you have them in your embrace you can then reach for the stars.

The Role of Science Centres as Aids for Astronomical Education

By J. Fierro

Instituto de Astronomia, UNAM, Mexico

1. Introduction

Education is training, part of which is being able to handle information. At meetings such as this, one learns new ways to teach and to adapt ideas to one's culture in a way they can have a greater influence on the lay person (Pasachoff and Percy, 1990; Percy, 1996). A reason to promote science popularization is to give people a chance to experience *the pleasure of understanding*.

Traditionally written materials and planetariums were the ideal way to convey astronomical knowledge and to start an interest in science. Now the media, WWW and interactive exhibits are having a great influence on the lay person. Science centres are an important aid for education; they present astronomy in an attractive way, which is sometimes difficult to do at school. It is easier to teach something that pupils enjoy.

This paper will focus on science centres in Mexico; some of the ideas that we have used could help other developing nations with their projects. In order to grasp the differences between other countries and Mexico, I shall only mention that the average education is five years in large cities nad two in the country; 78 million, out of 95 million inhabitants, never buy a book, and only 1 million purchase more than 10 books per year; the introductory astronomy course that is taught to over 200,000 students per year in the USA is only taught to about 60 pupils per year at our National University.

We shall describe some of the activities that science centres can provide in order to aid public understanding of astronomy and the ways in which several very small museums have been installed in Mexico.

2. Science Centres

Science centres throughout the world have incorporated many of the resources used for public education: hands-on activities, formal exhibits (including texts, pictures and models), computers, workshops, theatre, publications and videos. Since every person is different, the more varied objects an exhibit presented, the more likely it will succeed in reaching a broader public.

Science centres have to adapt continuously to different audiences. For instance, in the USA elderly people are becoming an important part of the visitors, so labels have to have large type and provide interesting information for such a group (sometimes they may also have to include bilingual information).

I think every major city should have a good science centre with strong outreach programs, whereas small communities should focus more on small facilties, where emphasis is on workshops, teaching resources, and travelling exhibits, including major astronomical events such as comets or solar eclipses.

Every science centre should have ties with universities and draw from its local culture, otherwise it is very difficult for it to come up with new ideas and have a strong influence on the general public.

Mexico city has a large science centre: *Universum*, where two halls are dedicated to astronomy, with sixty interactive exhibits aimed at several age levels. Now a colonial ex-convent will house a science center whose theme will be electromagnetic radiation; it will also have an exhibit on the use of spectroscopy for astronomical purposes in its ex-chapel which has a domed ceiling. (Several displays have been tried out at book fairs where their designers receive feedback from the public.)

Universum has fostered displays in places as diverse as the Palace of Fine Arts accompanied by a performance of the National Symphonia Orchestra playing astronomical works, to lobbies of subway stations where people who would have difficulty going to a museum get a chance to see state of the art science exhibits (Fierro, 1993, 1994, 1995).

Universum has produced several low cost plays, with astronomical topics, that can easily travel to other communities; the latest is on evolution; it uses a darkened stage where actors, using flourescent garments and puppets, stage spiral galaxies, the solar system and the evolution from fish to frog, while a narrator describes main cosmological events.

2.1. *Hands-on exhibits*

Astronomical exhibits are some of the most popular, only comparable to the ones on our body and dinosaurs. Hands-on exhibits have proven extremely successful since there is direct correlation between the amount of activity and the quantity of information one remembers. They should appeal to all senses; for instance in Mexico we have exhibits on the way different worlds would smell, sound and feel if we could visit them.

There are splendid astronomical displays all over the world, many of which are adjacent to planetariums. I shall only mention a few example out of the hundreds of interesting displays that have been constructed all over the world that aid astronomy teaching. For instance, at the Deutsches Museum, the spectroscopy exhibit features a historical approach, challenging information, sturdy manipulating devices and topics for all age levels. As an example, it includes a sort of slide rule where the Doppler shift of a galaxy and a quasar can be measured by comparing certain absorption lines. In Spain there is an exhibit where the visitor has to elaborate a virtual message using a computer and has to convey it via a radio signal to some sort of unknown extraterrestrial being. This sort of activity motivates creativity and helps people understand the difficulty of trying to communicate with extraterrestrial intelligence. In Mexico, the display on seasons, first focuses on extreme conditions: Jupiter and Uranus, and then on intermediate ones: Earth and Mars. La Villete in France offeres materials for teachers and long term activities for school children.

Extreme care must be given when designing exhibits; the Ontario Science Centre has managed to create hundreds of interactive exhibits trying them out with visitors in a preliminary version, before constructing the final item.

In the developing countries hands-on apparatus are crucial in order to give students a chance to experiment. Many schoold have no laboratories and middle school teachers usually tend to give lectures.

2.2. *The Travelling Museum*

The Ontario Science Centre has a long tradition in travelling exhibits that have proven to be a most successful way of conveying scientific information and hands-on experiences to small communities. They can be placed at book fairs, cultural centres, schools and malls.

We have found that one of the most convenient places to place displays are large lobbies at subway stations; they can include simple workshops and public lectures. Hundreds of thousands of people use the subway and during school hours children can visit the exhibits at low cost (in Mexico the ticket for the underground is $ 0.15 U.S.).

Currently an exhibit on "the way Olympic Games would be held in other worlds of the solar system" is under design, sponsored by a sports-related National Lottery.

2.3. *Workshops*

Workshops are extremely important for education because students can create something manually. Apart from being good for conveying information and for the visitors' psychomotor development, students learn how to build objects: robots, models of extraterrestrial creatures, telescopes, sundials, three dimensional constellations or rockets, (Fraknoi, 1995; Jackson, 1994; VanCleave, 1991; Walusinski, 1989). Concluding a task is something education in Mexico strongly needs because many people have great projects but have difficulty turning them into reality.

2.4. *Public Lectures*

When science centres have strong links with universities they can provide high quality public lectures and panel discussions for a large variety of audiences if they use video conference facilties.

Listening carefully for the questions people address can be valuable to learn about their misconceptions and interests. One of the most successful books I have written is based on the answers to questions raised during my public lectures.

2.5. *Ushers*

Some of the science centres train ushers; they are usually undergraduate students. Many of them have such pleasant experiences at the centres that they later pursue scientific careers. Ushers can become the role models for small children. Amateur astronomers can be extremely good ushers specially for outreach programs.

2.6. *Plays*

Plays are attractive ways to convey scientific knowledge. Some of the scripts are simple enough so teachers can later encourage pupils to interpret their own plays. (One of the middle school geography books encourages students to make a play on the properties of the solar system.)

2.7. *Libraries*

Several science centres are equipped with libraries which hold collections that teachers and pupils can use; these may be the only existing specialized libraries. Several hold videos and private sections for small children. Science centres are producing audiovisual material that can be purchased by teachers. Some museums have visitor-oriented computers with on-line facilities where teachers can find information necessary to update their courses. Unfortunately most information on the WWW is in English, so developing nations must make an effort to include more material and point out the necessity of teaching a second language at school.

Most major museums and science centres have home pages on the WWW. The information varies greatly, including historical backgrounds, virtual visits to the major exhibits; a few include diagrams of interactive displays.

2.8. *Shops*

Science centres tend to have shops where instructors can purchase materials: science kits, information posters, videos, computer programs, books and magazines where they can find activities they can carry out before and after the visit, as well as workshops and updated information. In many cases such stores are the only places to buy educational items.

2.9. *Teaching materials*

Several science centres produce teaching materials. For instance the San Francisco *Exploratorium* and the *Ontario Science Center* have a series of books on experiments that do not require complicated equipment. Some edit newsletters, electronic and not, that include science information that can be very useful for elementary school teachers, especially in the case where local newspapers do not print science columns.

In Mexico a group of teachers edited a new magazine and asked scientists that collaborated with *Universum* to write small articles for them, including hands-on activities. Some of the astronomy articles have been inspired by the work done by the French group published in the *Cahiers Clairaut* and other CLEA publications (Walusinski, 1989-1996).

2.10. *Starlabs and small telescopes*

Starlabs, portable planetariums, are extemely useful teaching laboratories. They can be adapted to meet the requirements for every school level and should be included in the outreach programs of science centres. They can also adapt telescope viewing with their presentations.

In many schools in large cities a visit to a planetarium is compulsory, so it is important that they provide high quality shows. Here professional astronomers can provide invaluable aid if they referee the scripts.

2.11. *Amateurs*

Several science centres have planetariums and small telescopes; some shows can be conducted by amateurs, especially since they tend to be an enthusiastic group. They can contribue to outreach activities such as star parties and *Astronomical Extravanganza* (camping inside the dome). One must make sure that the amateurs involved not only convey enthusiasm but also have a sense of scientific knowledge.

Science centers can promote the creation of amateur groups especially if they have a planetarium or a telescope.

3. Very small science centers

3.1. *Science Wagons*

One of the most successful projects of promotion of science in Mexico has been called *A Wagon for Science*. It consists of discontinued train coaches, and more recently, trucks and airplanes, remodeled to hold laboratory facilities and libraries. Some of them are associated with theatres where public lectures, promoted by Mexico's National Academcy of Sciences, are given on Sunday mornings .

3.2. *Starting a Science Center*

Major cities in developing nations should have large science centers similar to those constructed in developed nations, and towns more workshop oriented facilities. Seven large and twenty-three small centers have been installed in Mexico; I shall outline some of their characteristics. One must be very careful when planning a new center: a museum at Aguascalientes is devoted to evolution; unfortunately its exhibits are rigid and difficult to update; it only serves 450,000 people and has no feedback by a scientific staff. A children's museum in Hermosillo is an almost exact copy of one in Mexico City, including the same mistakes on its labels!

An ideal way to start a new science center in a developing country is to purchase a

Starlab, a small telescope, install a Science Wagon and construct a few simple astronomical hands-on exhibits. These items should be tried out in small communities in order to get experience and obtain local support.

Developing nations must have as many interactive, travelling, exhibits as possible, not only because they have proven to be good for conveying scientific knowledge, but also because in many cases they are the only laboratories students will ever visit. It is necessary to have areas dedicated to workshops and other continually changing activities, otherwise the public will stop visiting and the museum will deteriorate rapidly from the lack of visitors. New science centers should use the vast experience developed all over the world but should use local skills.

All projects should have a strong link with teachers. They should provide follow up materials. Teachers, particularly in developing nations, need basic information that will help them cope with children's questions.

A good way to start getting a community's enthusiasm for a new science center is a "Market of Science" or a "Scientific Caravan". The market of science is a science fair where several interactive experiments and workshops are presented to the public. The scientific caravan is a museum on wheels where the fair travels to a large number of small communities. This kind of activity provides a good way to try out exhibits. Two universities in small cities in Mexico have begun this kind of activity with good results at Morelia and Sonora.

The final and most important suggestion is that any science center project must have strong links with local universities. This is fundamental in order to ensure conveying proper information and, especially, to constantly renew the exhibits.

4. Conclusion

Science centers can provide a wide range of aids for astronomical education. When new small science centers are opened in developing nations they should emphasize activities devoted to outreach programs, such as travelling exhibitis, workshops, teacher assistance, and promote self education. They should be associated with universities and draw from local cultures.

Even though science centers provide exciting expreiences they cannot substitute formal education; astronomers should collaborate with them in order to convey to a large population the pleasure of understanding.

REFERENCES

FIERRO, J. & DODDOLI, C., *Museum News*, May-June 1993, p.35.

FIERRO, J., *Asian-Pacific Teaching of Astronomy*, Vol. 8, Ed. S. Isobe, p1-15.

FIERRO, J.,, *Highlights of Astronomy*, **10**, Kluwer Academic Publishers.

FRAKNOI, A. Ed., 1995, *Project ASTRO*, Astronomical Society of the Pacific.

JACKSON, S., Ed., 1994, *How the Universe Works*, Reader's Digest Association.

PASACHOFF, J.M. & PERCY, J.R., Ed., 1990, *The Teaching of Astronomy*, Cambridge, Cambridge University Press, IAU Colloquium no. 105.

PERCY, J., Ed., 1996, *Astronomy Education: Current Developments, Future Coordination*, Astronomical Society of the Pacific Conference Series, Vol. 89.

VANCLEAVE, J., *Astronomy for Every Kid*, J. Wiley.

WALUSINSKI, G., Ed., 1989-1996, *Les Cahiers Clairaut*, CLEA.

The STAR CENTRE at Sheffield Hallam University

By John H Parkinson, D. Ashton, K. Atkin & B. Harrison

School of Science and Mathematics, Sheffield Hallam University, Pond Street, Sheffield S1 1WB

1. Introduction

The Star Centre is a national astronomy and space science base which
- facilitates public access to news and information
- promotes public awareness, interest, enjoyment and understanding.

The Star Centre meets these twin aims by providing an information service which can be accessed in a variety of ways and by offering a menu of public observing events.

The concept of a national astronomy base developed as part of the Centre for Science Educations growing portfolio of initiatives in both the formal education sector and the wider umbrella of the Public Understanding of Science. In December 1996 the Star Centre was launched with the aid of a Royal Society COPUS development grant and matching funding from Sheffield Hallam University.

This paper summarises the main activities of the Star Centre, gives some impression of the public response and outlines plans for future development.

2. The Star Centre in Context

The Star Centre reaches out directly to schools and the general public and is part of the growing network of long-term projects at the Centre for Science Education (CSE) within the School of Science and Mathematics as shown in Figure 1. The largest of these projects is the UK Research Council funded *Pupil Researcher Initiative* (PRI) in which school pupils in the 14–16 age range explore science topics through research briefs. The PRI provides resources, activities, strategies and support for science teachers and their pupils so that pupils will experience the excitement and relevance of science and engineering research and so develop a lasting interest and enthusiasm. All aspects of the research process are involved and there are opportunities for Science Fairs, Pupil Researcher Conferences and Roadshows. There is a national network for recruiting University Research/Graduate students who are trained appropriately before being installed as Researchers in Residence in schools where they have opportunities to discuss their work with pupils, further develop their communication skills and help to support the science learning of pupils. The Centre acts as the co-ordination focus for the PRI projects involving astronomy.

The Star Centre also has links to the Internet-based project *Schools On-Line* and to the *Science Alliance*, which connects primary schools with industry.

Stardome is a mobile Starlab planetarium supported in part by the University which reaches a total audience of around 12,000 people per year, mostly in schools but also at public events. Schoolchildren and adults often make their first contact with the Star Centre after a Stardome experience.

Sheffield Hallam University makes a major commitment to Public Understanding of Science activities which include the annual *Sheffield Science Festival*, the *National week of Science, Engineering and Technology*, various lectures particularly around Christmas, the *Yorkshire Science-a-thon* and, of course, *National Astronomy Week*.

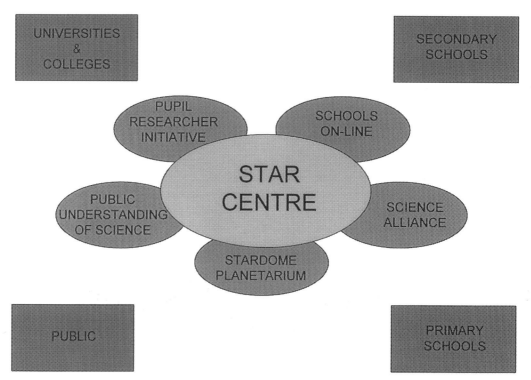

FIGURE 1. The network of projects within the Centre for Science Education at Sheffield Hallam University.

3. Activities

Star Centre activities are directed towards encouraging and enabling people to observe the night sky and to learn more about the Universe at large. The Centre aims to appeal particularly to interested sky watchers who would like to become recreational observers but we also cater for enthusiasts who wish to join an observing project. Some of the main activities are shown in Figure 2.

3.1. *The Information Centre*

Star Centre's information room is located at the University's Collegiate Crescent Campus. At the moment funds allow the base to be staffed by resident astronomers on only one day each week. Users write, phone, fax or e-mail to access information or pose questions on news, opinion, observation, equipment, events or background. The resident astronomers act as interpreters of information derived from the Internet, software, reference books and their own experience.

3.2. *Star Nights*

Public observing sessions from dark-sky sites in Derbyshire have been held on a regular basis. The two Star Nights held this year have attracted capacity audiences totalling 600 people. Unfortunately, both nights suffered typical British weather – intermittent clouds! The response, however, was very enthusiastic and visitors discovered that astronomy was more than just looking through telescopes and made use of a temporary information room, short talks, computer stations, video rooms and a cafe.

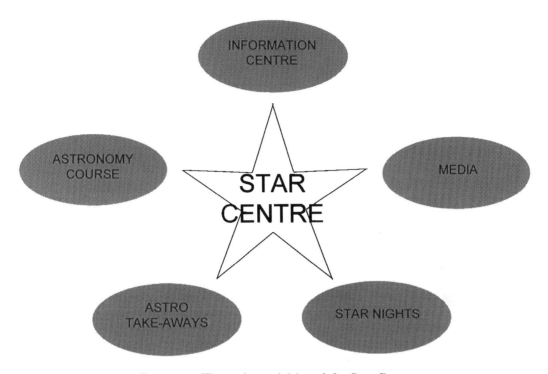

FIGURE 2. The major activities of the Star Centre.

3.3. *Astro Take-Aways*

These are smaller scale observing nights put on in response to requests from various interest groups. In the first half of 1996, 5 take–aways have been attended by a total of 200+ people.

3.4. *Observing Projects*

The Centre currently supports 2 projects:-

Starwatch UK — a light pollution survey established by the Royal Greenwich Observatory, the British Astronomical Association, the Federation of Astronomical Societies and others.

Meteor Watch — a meteor observation project with results fed to Star Centre for collation and analysis.

3.5. *Astronomy Course*

The Centre sponsors a 24 week course covering observing topics. This is over-subscribed with 20 members of the general public and a waiting list!

3.6. *Stardome*

Stardome has operated since 1993, giving 1,320 presentations to a total audience of 34,400 people in that time. The planetarium visits around 100 venues per year, 70% of which are schools. The remaining visits involve PUS activities such as events at *The Science Museum London, SET Week events, Association for Science Education, Towards 2000 Conference, Sheffield Festival, Association for Environmental Education, Star trek Conventions* etc.

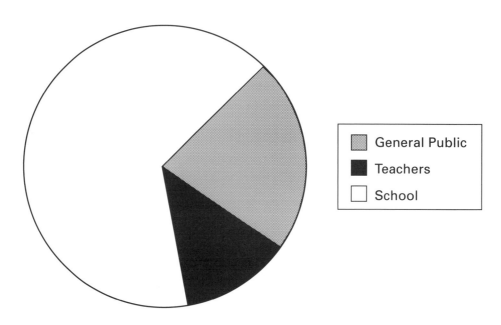

FIGURE 3. Analysis of the enquiries by type of enquirer.

3.7. *Media*

Star Centre provides a monthly 600 word astronomy column, for the Sheffield Telegraph, a regional newspaper, together with a regular slot on Radio Sheffield, where Star Centre astronomers explore a topic and then answer questions via a phone-in. The Centre also acts as a general enquiry point for the media at large.

3.8. *Starter Packs*

In response to user requests, Star Centre astronomers have produced a 24 page booklet for people who want to participate in recreational observation. The pack includes equipment guides, star maps and advice on observing planets and deep space objects.

4. Usage of the Centre

In the first half of 1996 the Star Centre received 360 enquiries. Figures 3 and 4 show how these were broken down.

Most schoolchildren request information and pictures about planets. Teachers often require classroom activities and the Star Centre supplies them using current Sheffield Hallam University educational project material.

Users will shortly be able to access Star Centre's own Internet site and the pages will reflect the aims of the Centre – to provide accessible information and encourage observation of the skies. The three themes of the site are:-

• Links with local and national organisations and events including the Bradford Robotic Telescope, societies, planetaria, observatories etc

• Observation - guides, maps, challenges, projects and equipment

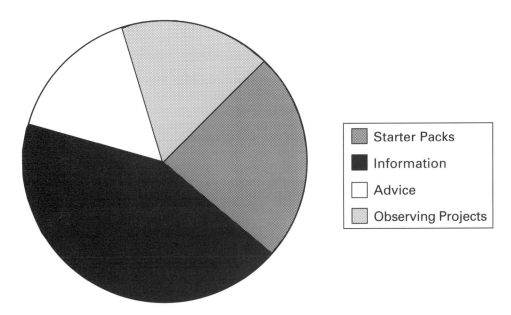

FIGURE 4. Analysis of enquiries by type of enquiry.

- Information and data on astronomical objects, including school activities.

The Star Centre URL is http://www.shu.ac.uk/schools/sci/star_centre/contents.htm.

5. Future Developments

Further funding in the form of grants and sponsorship is being sought to develop the activities of Star Centre as we continue to enhance the national profile of the information centre, develop activities for use in schools and extend our public observation programme. The Centre will seek to raise awareness in astronomical establishments of its ability to deal with enquiries from the public. Visitors to planetaria and observatories could use the Star Centre as an ongoing and accessible follow-up to their visit.

Star Centre will gain a more substantial international profile when its Internet site comes on-line during the summer of 1996. As well as acting as an information source, the interactive nature of the contents will help the Centre establish itself as a co-ordination base and the hub of a network of observers, particularly those new to the enjoyment of recreational observation.

5.1. *Schools*

Star Centre staff are primarily astronomy educators. Utilising active learning principles which underpin many of Sheffield Hallam University's school's projects, the Centre aims to produce school-based resource packages targeted at the National Curriculum science unit *Earth and Beyond* at both primary and secondary levels.

Through the *Science Alliance* project, the Star Centre is to contribute to an astronomy day for primary schools in the spring of 1997. This will be a pilot for further one-day conferences for schoolchildren at different levels, including those choosing the A-level Physics astronomy option.

Young astronomers will be targeted in the *Astronomy Suitcase* which the Star Centre has been asked to produce for Scout and Guide groups. As well as helping youngsters gain their astronomy badge, the suitcases will provide a model for practical-based resource packages for schools.

5.2. *Public Observation*

Star Nights have proved very popular and more public sessions are planned. These will be themed nights featuring Comet Hale-Bopp (we hope!) and the Geminid meteor shower. In National Astronomy Week, the Star Centre will work with other local groups to offer five observing nights at an observatory. On 1996 October 12, the Star Centre astronomers will take to the streets, offering Sheffield shoppers a chance to view the partial solar eclipse from a city centre square.

5.3. *The 1999 Total Solar Eclipse*

The Star Centre will play a major role in increasing awareness and safe enjoyment of the 1999 August 11 total solar eclipse. This will be the first such eclipse visible from the UK mainland since the 1926 event which was very badly affected by weather. Totality only occurs in Cornwall but the whole of the British Isles will experience a major partial eclipse. It is hoped to extend the PRI framework to place suitable material in every school in the country enabling students of all ages to share the excitement of carrying out scientific investigations of this cosmic event. Expeditions from schools around the country will be organised to observe totality from Cornwall.

6. Summary

The Star Centre has established a firm local presence and has become an integral part of the continuing educational and Public Understanding projects initiated at Sheffield Hallam University. In the future, an increased profile will involve public participation through the Internet and contact with the information centre, together with educational projects to enhance interest and learning in schools.

Underlying all the developments is the philosophy that the fascination of astronomy - new discoveries, different worlds, bizarre objects, cosmic evolution, enjoyment of the night sky - should be made easily accessible to people of all ages.

How to Suceed in Convincing Municipalities to build Astronomy Centres: the experience of Campinas Region

By A.L.K. Bretones[1] & P.S. Bretones[2]

[1]Centro Educacional - SESI 403, Campinas, Brasil

[2]Faculdade de Educação, Universidade Estadual de Campinas, Campinas, Brasil

1. Introduction

The objective of this work is to make known the astronomical activities in the region of Campinas, the process of developing municipal cooperation and the general conclusions that reflect this process.

This research has been done by means of interviews with people related to the creation of astronomical centers in the region of the city of Campinas that is located in the state of São Paulo in Brasil (Fig. 1 and 2).

The conditions studied are related with this region but many ideas could be used in developing countries or others.

Nowadays there are works in the areas of research, teaching and popularization in many institutions besides the individual efforts of teachers and amateur astronomers in general. At two local universities (Unicamp and Puccamp) the discipline of astronomy at the graduate level is still optional. There are no groups or departments of astronomy but various students have obtained the MSc degree in areas such as black holes, active galactic nuclei, cosmology and teaching of astronomy. There is also a small planetarium, a naked eye observatory, some astronomy clubs as well as various amateur astronomers who contribute with systematic observations of the sky in many fields such as Sun, planets, occultations, comets, astrophotography, CCD astronomy and even with the use of spectroheliograph, coronograph and a Schmidt camera. Some teachers conduct astronomical activities in many schools of the region within disciplines like sciences or physics as well as lectures, exhibits and observations of the sky. There are six municipal observatories (Campinas, Americana, Itapira, Piracicaba, Diadema and Amparo), with continuing activities of observational programs, public observation, exhibits, and other activities directed to students. These are not observatories that put an emphasis on research. They spend more time on educational, cultural and touristic activities.

With such a quantity of resources and assets we can ask: how can we succeed in convincing municipalities to build astronomy centers?

2. The Process

Various factors contribute to this. The first factor is the local conditions that helped this process. The city of Campinas has about one million people and a population of high cultural and economic level. Besides the existence of the universities, there is the proximity of São Paulo city, the great financial and population center of Brazil and also the existence of neighbouring cities of great agricultural, industrial and commercial production.

The observatory of Campinas was the first municipal observatory in the country and has influenced the opening of others. There have been astronomical activities in the city

FIGURE 1. Geographical positions of Brazil in South America and the state of São Paulo in Brasil

of Campinas since the beginning of the century. There was an amateur astronomy group and teachers that besides activities in the university, lectured on astronomy.

In 1976 the mayor of Campinas, being also an amateur astronomer, looked to the Capricórnio Observatory staff to build a municipal observatory. That institution was established in another city and had already an academic background and high level amateurs. The mayor then summoned these people together with other astronomy teachers to staff the new observatory. The founders, J. Nicolini and N. Travnick and other teachers of the region had a great influence on the cultural development of the students and promoted it with their work and assistance of the media.

For a city to have an astronomy center it's necessary to be provided with a minimum of financial conditions and a population with good quality of life. It's even better if the city has some touristic potential and good geographical location. It's also important to have people interested in astronomy living in the city, an astronomy club or society where parallel events occur. Besides, it is important to have support from the media that increases the interest in astronomy matters and in reading of astronomy articles.

The last return of comet Halley influenced public opinion creating a strong lobby within the municipal administrations. Nowadays even without comet Halley, we have celestial phenomena that encourage these initiatives. Astronomy is always present in our lives. We can also take advantage of the passage of a comet, an eclipse, etc.

Its also true that people of good intellectual level, technical expertise and equipment are available for the city administration provided by the project's authors. The municipality does not spend much money, can use cheaper land in the outskirts, which is convenient for astronomical activities to build a small building. On the other hand, the municipality

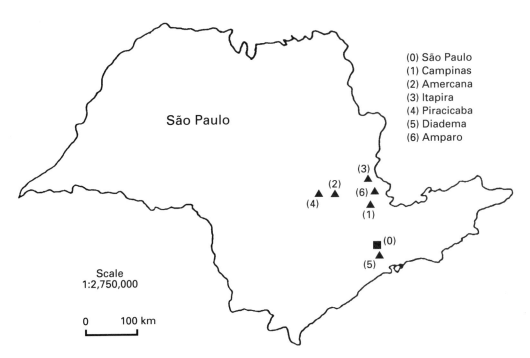

(0) São Paulo
(1) Campinas
(2) Amercana
(3) Itapira
(4) Piracicaba
(5) Diadema
(6) Amparo

FIGURE 2. Geographical positions of the cities with minicipal astronomical centres: Campinas, Americana, Itapira, Piracicaba, Diadema and Amparo

also offers services to the community because it receives public and students in these centers, which is politically convenient. The municipality will therefore offer something accessible to the population and not merely an obsevatory dedicated just for research and a few interested people.

If the project's authors make available instruments, library, pedagogical material, and if possible, the dome, this also will be important to convince a municipality and show that it will not spend much money on the project.

It's important for the project's authors to know that it will be necessary to spend some of their own money in the beginning and in the maintenance of the undertaking.

This process takes time. First and foremost it is necessary to convince the secretary of culture or education that he must approve the project. Literature about the importance of the project must be well organized and as clear as possible. It is important to stress that news on astronomy happens on a daily basis, that people are interested and it has educational importance.

In general, the best occasion for a preliminary contact is after the first two years when the administration has paid the debts left by the former government and has taken care of other important projects. To initiate the preliminary contacts in the last year of administration is not advisable because it is carrying out projects already approved. But it is difficult to say which is the best occasion. It depends on many factors related to the administration's priorities. Another phenomenon typical of Brazil, is that there is some easiness for an astronomer to persuade the municipalities to build astronomy centers. This happens because we have a country where the government is formed by a learned elite that makes decisions without consulting the people — the vast majority.

The most important aspect is to look for the right people for the first contact. This can be the secretary of culture or education and even one of the mayor's auxiliaries or someone who has influence over the decisions and can see the importance of an astronomy center. Many projects have not been established because the wrong persons were chosen for the first contacts. Sometimes its difficult to find the right person. One may have to try many times before someone happens to appear who is interested in the project.

The next step is to go to the mayor and to explain the idea and the political dividends. The project depends fundamentally on the mayor. If he does not like the idea the project will not be established. But this will not be enough; it's also important that the mayor's auxiliaries like the project. The administration's office staff, the cultural and educational people and the engineers should always be consulted. There have been cases when the mayor was accessible and wanted to carry out the project but it was blocked by his auxiliaries who alleged financial difficulties among others.

But when the mayor and his auxiliaries have will and vision of the importance, the process is faster. The city would not need to have an astronomy club or society or even previous support of the media. The mayor must issue a decree approving the project and later the city council should approve it. It is important to have the acceptance of the city councillors by political maneuvering without preference for any political party. If the community accepts the project it will be difficult for the councillors to block it for it now belongs to the city. It may be necessary to speak about the project in a city council session. It is necessary, that besides other important works, to establish that an observatory is also important. If we had only to think about the populations' problems there would not be education, culture and leisure.

Something that can make things easier is to take advantages of ongoing projects like an ecological park, a science museum etc. There you need to be flexible for the place must be of easily accessible to the public and visitors. Choosing a remote place, where the administration would have to install the basic infrastructure like running water, light, telephone, roads, etc, would make the project more expensive and make it more difficult to convince the municipality. Afterwards a contract of services and equipment lease are made. The contract is valid for a year and renewable. If the project's authors want to become employees of the municipality this would add additional problems. If the astronomical center is established by law, and there is community support, there will be employment stability.

As far as the planetarium of Campinas is concerned, it has been established as a result of the presence of the observatory. The federal government needed to install two telescopes and since the municipality decided to house them the city was presented with a planetarium. The federal government does not incur the risk of installing a planetarium in a city which does not have astronomical activities and potential to develop this field. The planetarium is a project accepted more easily by the municipality because it is profitable. Its sessions have a guaranteed public attendance since it does not have to rely on good weather.

If the observatory is situated in a big city and is having problems and if the municipality does not solve them, the commuinity and media demand the solution of such problems. But this is soon forgotten because the city has so many other problems to take care of and other attractions as well. The smaller the city the more attention an observatory gets because generally it is one of a few attractions in town. When there are problems they are taken care of by the appropriate channels without any further pressure from the media or the community. These problems are then internally solved. In a small city, people know the mayor, and demand the solution of several problems directly with him.

In the case of Brazil and other countries as well, these observatories can be used

THE PROCESS

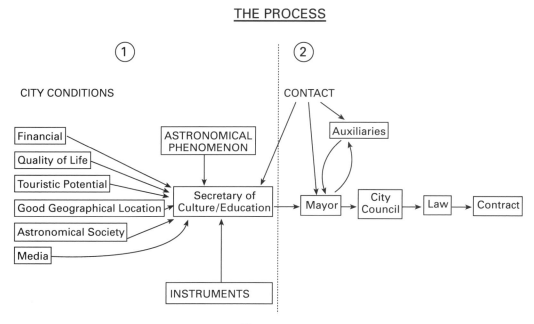

FIGURE 3.

as a means of employment of astronomers. As a municipal employee the astronomer does not have an adequate career plan. A municipality has no structure and means of assessment of the professional progression of an astronomer as it would in the university. The published papers, for instance, generally do not mean anything for the astronomer's career plan inside the municipality. In the case of an astronomer having a contract of services, his or her scientific production would be taken into account at the time of the contract's signature only. Later it would be of little value at the contract's renewal time. In a small city it much more difficult for the municipality's employees have low wages.

As a means of assessment the municipalities, in some cases, request an annual report. But the most important thing is to keep many activities frequently reported by the media. It must be well done work, so this way, the municipality recognizes that the astronomy center has been useful for the community. This way, the contracts are renewed every year. These projects can sometimes benefit from political opportunism. The mayor has a four year term and the observatory is usually built in the last two years and begins its operation in the final days of the term.

There have been instances where projects were cancelled after being approved. Further-more, there have been cases where a municipal administration installed an observatory and the next administration abandoned the project. The establishment of an observatory in a city is a curious fact. Sometimes it has all the elements to be a successful project but it is not. Even after it has been approved by the city council the mayor may still have more interest in other works. By law the mayor has the power to veto such a project at any time and even against the community's interest. This would hardly happen in a region where an astronomy center has been established with the community's support.

Maintenance has been the great problem of these centers. It happens that people do not understand the importance of these projects. It is difficult to find two consecutive mayors who are well informed about the project. The fact that the astronomy center is established by law is no guarantee that it will thereafter function well.

After a mayor has established the astronomy center others usually continue to support the project but not so enthusiastically. They have not given all the necessary support but have not failed to recognize the importance of the project. In one case, the mayor almost extinguished the observatory, because of political rivalry with the former mayor. In smaller cities, with around 100,000 inhabitants, there are unique political characteristics. Once a political group is installed it is difficult to have any significant change in the status quo. In some cases this has been an advantage. A mayor may not support an established astronomy center but would not extinguish it because he is a friend of the former mayor. Every mayor gives more attention to their own projects as a way to have political gains. The problem is that the decisions can be imposed by a mayor who sometimes does not understand much about astronomy. He sometimes wishes to add inadequate or incompatible improvements that can be deleterious to the project, e.g. street lights. It has been necessary to work this way because the mayors have only a political vision of the matter.

3. Conclusions

The process described has its problems because it can create instituitions directed by people without scientific background. There are no criteria for choosing the people who run the project. This process can be just political.

From our point of view we believe that the importance is not only to convince the municipalities but to persuade the people about the significance of these astronomical centers and by this way to influence the local government.

If the project is rooted in the people, children and parents, a new mentality can be created. This result has been better in smaller cities because its people are proud to have such an institution. In this case the people can be considered an ally and democracy is done.

An observatory should have more activities related to teaching in the schools. It would also be interesting to popularize astronomy in places out of the observatory such as city suburbs and even shopping centers or squares for it to be accessible for everyone.

3.1. *Acknowledgements*

We would like to express our deepest thanks to C. A. Argüello for the orientation of the project and A. C. Negreiros for the critical reading of the text and IAU for a Travel Grant. We wold like to thank Dr J. Percy for the suggestion of the subject of this research.

4. References

Special inquiry for this report.
Personal records of interviews with N. Travnick, C. Corsini and C. E. A. Mariano.
Personal information from several institutions.

Public Information Project of the Total Solar Eclipse of November 3, 1994 in Paraná State, Brazil

By R.H. Trevisan

University of Londrina, Campus Universitário-Londrina - Paraná -CEP 86051-970, Brazil

This project had two principal objectives: to communicate safe methods to observe the Sun, so as to prevent ophthalmological accidents to people during the total solar eclipse of 3rd November 1994, and to collaborate with the primary school teachers in the science classroom, illustrating the classes, motivating the students to observe sky phenomena.

1. Introduction

In January 1993, a commission called "ECLIPSE 94" *Executive Commission*, of the *Brazilian Astronomical Society* was created to coordinate assistance with arrangements for observing the total solar eclipse of 3rd November 1994, that in Brazil was total in the western part of Paraná State, in Santa Catarina State and in a Rio Grande do Sul zone. Professional astronomers from Brazil and from several parts of the world were mobilized to observe this eclipse. The biggest interest in this phenomenon was because the next one of this type, in Brazil, will only occur in the year 2046, and will be visible in Paraíba State. The general coordination was done by Prof. Dr. Oscar Matsuura, from the Astronomical and Geophysical Institute of University of São Paulo.

Following the suggestion of the *Working Group on Eclipses* of the *International Astronomical Union*, this commission decided to amplify their action, assuming the articulation of a large publicity campaign about eclipses, close to the common people. Such a campaign was aimed at giving technical and astronomical information and at preventing ophthalmological accidents to people during the total solar eclipse of November 3, 1994. Utilising this fact, we decided to use this campaign to collaborate with the teachers, principally in high school, illustrating science classes, and motivating the students to observe sky phenomena. This commission was helped by people linked to official institutions that, because of vocation, count as an adequate substructure to articulate this campaign in their own regions. These people were the state coordinators (people in universities, observatories, planetariums, museums and research institutes).

2. The General Directive of the Campaign

The general directive of the national campaign (Matsuura, 1994) was as follows. The states of the region where the eclipse would be total (in the south of Brazil), would have a *state coordinator*, to coordinate the information in the region. This *coordinator*, would designate some *Regional Supervisors*, and they would designate the special monitors, to work with them in their regions. The attendance should include all the people, in town and country. The distribution of information such as the distribution of materials to observe the Sun safely, should be promoted by the special monitors, previously trained and instructed. These monitors were university students and teachers, amateur astronomers, teachers of high school, etc. Four months before the eclipse, in the winter holidays, the special monitor group would organise together a training course of three days, attended by the regional supervisors, about eclipse theory and observation using cheap and safe materials. The Executive Commission would give basic texts and original images to official or private organisations, to make folders, posters, booklets, etc. The coordination of the campaign at a local level, involved universities, planetariums, amateur associations, official organisations of the state and of the cities. The local coordinator organised courses to train new monitors.

3. The Communication Campaign in Paraná State

The beginning of the campaign in Paraná, was in March 1993, when the regional supervisor was named. The campaign followed the program in Figure 1.

Six *special monitors* chosen among physics students of the University worked with the state coordinator, in the general coordination. They worked 20 hours a week, and were paid by official support.

The regional supervisors chosen were: teachers from four universities of the state, educational professionals from Government Education Department of State, city hall, cultural center, high schools and physics students. There were twelve elements in Paraná State. Their function was to coordinate communication of the event in their region, forming monitors of the second and third generation, promoting courses, talks, fairs, interviews and events in general, with the collaboration of the city halls of the cities and of the Government Education Department of State, coordinated by the state coordinator of the project.

The training of the Paraná *regional supervisors* was with all Brazilian supervisors (92 in total), and was done within the general coordination (Oscar Matsuura, and Roberto Boczko from IAG/USP). The training of these people, allowed them to achieve a "multiplier effect" in communication. The course was of 24 hours, in three days, in Chapec City (Santa Catarina), in the region of totality. All the costs were paid by the local city hall and by national research funding agencies. Each one of the 92 people, instructed about 400 new monitors in their regions.

Most of the *financial support* (US$ 5,760) of the campaign in Paraná, was given by the Ministry of Education. This subsidy was for daily expenses, air tickets, consumables for propaganda materials and to pay the students. Besides this, we counted on the support of the Department of Physics of University of Londrina – Paraná, the telephone company, city newspaper, the Agronomic Institute and some contractors.

The Government Education Department of State worked with the eclipse state coordination through their net of 25 "Regional Centres of Teaching", localised in 25 different cities of the state. Their participation was in several ways: paying for their representatives to go to the capital of the state (Curitiba) to attend a course of about 20 hours, with the state coordinators, where they could learn all about eclipses; administering courses to primary and high school science teachers in all public schools of the state; sending notes and folders to all these teachers and schools, teaching them how to observe the eclipse safely and stimulating them to observe the eclipse with their children. The program of these courses was:

- The history of eclipses
- The phenomena of eclipses and their occurrence – local circumstances of the eclipse
- The Sun: a general view; Solar atmosphere and wind; Archaeoastronomy; Observational solar eclipse safety methods
- Didactic experiments for the solar eclipse; Observation with binoculars and small telescope
- Solar eclipse photographs
- Debate: strategy of the campaign; Interaction with the community.

FIGURE 1. Program of the principal events

Date	Event
June 1993	State coordinator designation
July 1993	Regional supervisor designation (twelve elements in Paraná, State)
August 1993	Poster presentation in a meeting of astronomers, with the strategy of the campaign
October 1993	Sending submission of the eclipse project to financial organisations to ask for financial support
February 1994	First contact of the state coordinator with the Government Education Department of State: definition of the aims of the project
February 1994	Special monitors and monitors of the second generation designated
March 1994	Contact of the regional supervisors with the city halls of the principal cities of the State
April 1994	Folders and posters development
April 1994	Starts the contact with Press
May 1994	Course for regional supervisors (92 people), coming from seven Brazilian states
August 1994	Start the folder distribution
August 1994	Training course for special monitors (25 people), coming from 25 different cities of the State
September 1994	Training courses for science teachers in 25 cities of the state, coordinated by the special monitors
Sept/Oct. 1994	Fairs, expositions; poster competition by the students, and talks in many cities of the State, coordinated by the regional supervisors and special monitors
October 1994	Start the poster distribution
October 1994	Journalism course specialising in Science and Technology course to train journalists, with emphasis on eclipses
October 1994	Photos, videos and kits show
November 1994	i) "See the Eclipse in the Squares" extension project in Londrina City ii) "Seeing the Eclipse with the University of Londrina" extension project in Foz do Iguaçu and Criciúma Cities
December 1994	Educational video "Eclipse 94" to be distributed to schools

The objective of these courses was: to communicate the solar eclipse of November 3rd 1994; the precautions to take to observe the eclipse and the Sun; to motivate the children and the general public to observe astronomical phenomena, utilising the Universe as a laboratory. Two experiments were designed for children and ten teaching instrument kits to observe the Sun by projection. They were given to teachers who instructed their students how to use them. The kits were distributed in the schools science fairs.

The contact with the press was fundamental to the success of this project. To do this, we started the contact about one year before the eclipse. Near the date of the event, we gave a Journalism Course Specialising in Science and Technology with emphasis on eclipses to train journalists of the region, with the objective of giving accurate information to the people in a simple and didactic way, always with the orientation of a monitor, technical information about safe solar observation to prevent ophthalmological accidents.

The principal propaganda materials were: folders and posters. There were 15,000 folders inviting people to look at the eclipse, and to look safely. They were distributed to the regional supervisors in different regions, and to the Government Education Department of State to be reproduced freely. It was distributed to all professional Brazilian astronomers; universities, private schools, city halls, shopping malls, churches, associations and syndicates, business offices, etc. The posters (1,500) were distributed by the monitors to the Municipal Bus Company, banks, schools, inviting people to see the "Eclipse in the Square" by projection. The "Eclipse in the Square" was a project operated in Londrina City. Big paintings were used in the squares. The number of people in the squares of Londrina City, where the totality of the eclipse was 90%, was estimated to be 6,000 people.

4. Results

Target Population reached by the Project Eclipse 94 in Paraná		
Group		**Number**
Primary school teachers from official state schools		2,040
First/second grade teachers from independent schools		*
Students of primary/high/independent schools		*
Students of primary/high public schools		1,207,832
University students/teachers of Paran State		*
Students of first grade from Londrina municipal schools		28,000
First grade teachers from Londrina municipal schools	(directly)	75
	(indirectly)	*
Journalists and journalism students		60
Londrina City population	(directly)	6,000
	(indirectly)	*
Education technicians from funding agencies		412
General population		*
* = Impossible to assess.		

We estimate that 80% of population from south of Brazil was reached by the national project *Eclipse 94*.

Many students and teachers developed their own interest in astronomy.

Information Papers Produced

Scientific education articles in authorised reviews	02
Interviews (newspapers, TV, radio) (registered)	45
Scientific events (posters, talks, round table)	05
Talks to general public	13

We can say that the objectives of the campaign were totally reached. The two principal factors that permit us to say that are: i) the large number of articles in press and ii) **no eclipse related accidents in people were reported**, which is rare in a total solar eclipse.

Acknowledgement

Our thanks to the physics student Sheila Aparecida Faraco, who worked together with the general coordination of the project.

REFERENCES

MATSUURA, O. *The Divulgation Norms of the Eclipse 94 Brazilian Campaign*, personal communication.

Solar Eclipses and Public Education

By J. M. Pasachoff

Williams College-Hopkins Observatory, Williamstown, MA 01267, USA

Solar eclipses draw the attention of the general public to celestial events in the countries from which they are visible, and broad public education programs are necessary to promote safe observations. Most recently, a subcommittee of IAU Commission 46 composed of Julieta Fierro (from the National University of Mexico), the Canadian professor of optometry Ralph Chou (from the University of Waterloo) and me provided information about safe observations of the 24 October 1995 eclipse to people in Pakistan, India, Cambodia, Vietnam, and Guam. An important point is that there are advantages to seeing eclipses, including inspiration to students, and that people must always be given correct information. If scare techniques are used to warn people off eclipses, when it is later found out that the eclipse was not dangerous and, indeed, was spectacular, these students and other individuals will not trust warnings for truly hazardous activities like smoking, drugs, and behavior that puts one at risk for AIDS.

A total eclipse of the Sun is the most spectacular sight that can be seen, in my view, both from its physical and from its emotional impact, with the otherwise powerful Sun disappearing in the middle of the day. Though public interest in eclipses may be intense for only the immediate days preceding them, we can nonetheless take advantage of this interest to carry across important scientific ideas. The notion that the Universe is understandable and, in important ways, predictable, is a powerful idea that acts against the ideas of superstition and pseudoscience that are so rampant. Though I certainly think the diamond ring effect and totality are spectacular, I am most moved at first contact, when the moon barely kisses the Sun, because then it becomes clear that the predictions of astronomers were correct, and that I have not been led by a hoax into a wild-goose chase.

It is interesting to note that no fewer than three papers at this IAU Colloquium are devoted to public education through solar eclipses, those of Nirupama Raghavan from India and Rute Helena Trevisan from Brazil, as well as my own. Indeed, Dr. Raghavan reported that over 150 million people had the results of her educational intervention. The 1991 eclipse went over Mexico City; the 1995 eclipse went over Calcutta; and the 1999 eclipse will go over several major cities in Europe, so it is common for large numbers of people to become interested in what is going on in the sky, and how to watch it safely. Television is not an adequate substitute for viewing an eclipse directly, but we must work to make sure that people appreciate that truth.

We astronomers must be out in front in providing useful and correct information. All too often, the impression is spread early on, months before the eclipse, that eclipses are hazardous and that it is easy to go blind from watching it. The kernel of truth in that report is often overwhelmed by overbearing scare techniques. That is one reason that when I formed a subcommittee of Commission 46 on the Teaching of Astronomy of the IAU, I was glad to include Prof. Ralph Chou, professor of optometry at the University of Waterloo in Canada, and an expert on eclipse filters and eye safety. Prof. Julieta Fierro of the National University in Mexico City joined us on this subcommittee, bringing her experience in public education at the favorable total solar eclipse that graced her country in 1991.

In fact, it is not as easy to damage your eyes by looking at the Sun as many people assume. On a normal day, the eye blink reflex keeps you from staring at the Sun, while in the minutes before totality the total flux of sunlight may not activate this reflex though the specific intensity of the solar photosphere remains only slightly diminished (by limb darkening) from that at the center of the solar disk. Still, a brief glance at the Sun does not harm, just as people's field of view often crosses the Sun by accident on a normal day at the beach. What I like to emphasize is the following idea: never stare at the Sun, not at an eclipse and not on a normal day. Even when I view the partial phases of an eclipse with a suitable filter, I never look for more than a couple of seconds, and then I look away for a while. And one must always be careful when looking at the Sun with any optical aid, such as a telescope or binoculars, because the focused Sun is much more harmful than the unfocused Sun.

The number of cases of eye damage one hears of is very small, especially given the huge numbers of people who watch eclipses. In fact, if one were to survey a large section of the population, one would find damage in many eyes, and one wonders if most of the reports of eye damage were really caused by the eclipse. After the annular eclipse in 1984 in the United States, after which newspapers reported eye damage to a girl in Racine, Wisconsin, who had stared at the Sun for 45 minutes–something unusual and difficult to do even during an eclipse–we checked up months later and found that she had almost entirely recovered. So I ask readers, when they hear of eclipse eye damage in their own country, to try to check up on the actual circumstances and on the eventual results.

Public education programs can be valuable, and the opportunity of an eclipse leads local newspapers to run many eclipse-related articles. Since the 1980 eclipse in India, enough representations have been made by Indian scientists and educators that large numbers of Indian citizens were willing to be outdoors, to watch, to eat food, and to carry out other activities during the 1995 eclipse, a change commented on by local newspapers.

Public information programs about eye safety must be carried out for the partial solar eclipse of October 12, 1996, and the total solar eclipse of August 11, 1999, for both of which most sites are in Europe. Though no totality will occur in Europe in 1996, most people won't travel to the zone of totality in 1999 so would find the 60% 1996 eclipse of equal interest. Thus this year's partial eclipse can act not only as a bonding between European citizens and the sky but also as a trial run for the 1999 eclipse that will receive so much more publicity.

Ignorance of the true nature and value of eclipses has led to unfortunate barring of students from seeing eclipses. I have seen this problem often in the two dozen eclipses of my experience, including such varied sites as Australia in 1974; Manitoba, Canada in 1979; and Virginia, United States, in 1984. What happens when students who were locked in basement classrooms on the opposite side of the building from the Sun get out of school? They might meet a friend who says, "Didn't you see the eclipse, it was beautiful." "No," they say, "the teacher said we would go blind." When they learn from the friend about the spectacular sight they missed, how will they respond to information from the same teacher when they are told, "don't smoke," or "don't drink while driving," or "don't take drugs," or, most recently, "don't put yourself at risk for AIDS." Thus the net result of the excessive warnings about eclipse eye safety may be negative, with people put at increased risk over the risk they would have had by merely being properly advised about safe ways to observe the eclipse.

Solar eclipses provide a tremendous opportunity for us astronomers and astronomical educators to reach millions of people to show them how interesting the heavens are, and what real science as opposed to pseudoscience can do. Through planning and proper diffusion of interesting and precise materials to the public, we can bring hundreds of millions of people into contact not only with astronomy but also with the capabilities of science in general.

Acknowledgments

My recent work at solar eclipses has been supported by the U.S. National Science Foundation through its Atmospheric Sciences Division (grant ATM- 9207110), its Education Division (grant DUE-9351279), and its Astronomy Division through its Instrumentation Program (grant AST-9512216); the National Geographic Society through its Committee on Research and Exploration (grant 5190-94); and the Keck Northeast Astronomy Consortium. I also thank the Bronfman Science Center and the Safford Fund at Williams College for their support of the efforts of my students and me.

REFERENCES

PASACHOFF, J.M., and COVINGTON, M., 1993, *The Cambridge Eclipse Photography Guide*, Cambridge University Press;

WILLIAMS, S., 1996, *UK Solar Eclipses from Year 1*, Clock Tower Press, Leighton Buzzard,UK.

The Role of Amateur Astronomers in Astronomy Education

By J.R. Percy

Erindale College, University of Toronto, Mississauga ON, Canada L5L 1C6

1. What Is An Amateur Astronomer?

Let us begin by defining "amateur astronomer". According to a dictionary, an amateur astronomer is "someone who loves astronomy, and cultivates it as a hobby". At IAU colloquium 98 (The Contributions of Amateurs to Astronomy), Williams (1988) discussed this issue at length. He proposed that, to be an amateur astronomer, one must be an astronomer - able to do astronomy with some degree of skill; he then defined an amateur astronomer as "someone who carries out astronomy with a high degree of skill, but not for pay".

Unfortunately, the word "amateur" has negative connotations to many people. This is partly because of the unfortunate choice of the word; "volunteer astronomers" might be a better choice. It is partly because there are indeed a few amateurs whose ideas and attitudes might be judged rather bizarre - but the same is true for some professionals. There might even be a hint of snobbery, especially in cultures in which qualifications (as opposed to ability) are paramount. Professionals certainly respect the contributions of the "superstars" of amateur astronomy: Frank Bateson, Robert Evans, Patrick Moore and the like. We tend to hold these people as examples, though very few amateurs are willing or able to contribute at this level. There are thousands of "rank-and-file amateurs" worldwide. They can and do contribute significantly to the advancement of astronomy.

I prefer to define amateur astronomer extremely broadly. In this case, their education, knowledge, skills at instrumentation, computing, observing, teaching and other astronomical activities could be anything from zero to PhD level in astronomy or a related field. Many amateur astronomers are professionals in other scientific or technical fields. When you think of it, there is a wide range of interests and abilities among professionals, or even in one professional at different points in his/her career. Some become amateurs when they retire. In some countries now, many undergraduate and graduate students in astronomy are taking jobs in other fields. We should strongly encourage all these people to continue to contribute to astronomy, as volunteers. It is also possible that a very capable and persistent amateur astronomer – such as Patrick Moore in the UK, and Terence Dickinson in Canada – can eventually establish a career in astronomy and become a professional! Amateur astronomers are united by one characteristic - their interest and enthusiasm for astronomy. Considering that there are at least ten times more amateurs than professionals, they are an ally which we should not ignore.

Almost every child is interested in astronomy, and many of them aspire to become an astronomer. They grow out of this phase of course, especially when they realize that the number of jobs in professional astronomy is limited. By recognising and valuing amateur astronomy, we can offer our students another option – choosing a different career, but continuing their interest and involvement in astronomy as an amateur. This may be an especially useful and realistic option in the developing countries, and in countries such as China and Russia where business careers are now much more lucrative than academic ones.

2. What Is Astronomy Education?

Earlier in this meeting, I defined astronomy education by quoting Fraknoi (1996): he pointed out that astronomy education takes place, not only in the classroom, but in many informal environments as well. His point was addressed to professional astronomers and educators; for the purpose of the present paper, I will reverse Fraknoi's point. Amateur astronomers are accustomed to contributing to informal or public education but, in this paper, I want to point out that amateurs can contribute to formal or classroom education as well.

3. The Contributions of Amateur Astronomers to Astronomical Research

The contributions of amateur astronomers to research are well known, and were discussed at length at IAU Colloquium 98 (Dunlop and Gerbaldi (1988)):

- Comets: discovery, imaging, astrometry, photometry
- Variable Stars: discovery and photometry
- Novae and Supernovae: discovery and photometry
- Solar Astronomy: sunspots, solar flares, sudden ionospheric disturbances
- Meteors and Aurorae: systematic observation
- Instrumentation, Software and Computation: development and use
- Occultations: total and grazing; lunar, planetary and asteroidal
- Asteroids: discovery, astrometry and photometry
- Double Stars: measurement
- Eclipses: solar and lunar

Some fields are, of course, more fertile for significant contributions than others. These contributions have a significant impact on education. The discovery of comets such as Shoemaker-Levy 9, Hyakutake, and Hale-Bopp, have a profound effect on public interest in astronomy.

Students can also use amateur astronomers' data and techniques for a variety of projects. At this colloquium, I displayed a poster, with Mattei and Saladyga as co-authors, describing "Hands-On Astrophysics" – an education project of the American Association of Variable Star Observers. This project uses variable star data and observation to develop and integrate a wide range of skills in math, science and computing. Presently, it is targeted at high school students, but it can be used, with minor modification, at university level. It uses visual observations from the AAVSO's International Database, made primarily by amateur astronomers from around the world. It also uses the charts and techniques which have been refined by AAVSO observers, to enable the students to make measurements from the real sky.

The AAVSO hopes that this project will not only contribute to science education, but that it will also attract young people to amateur astronomy. The AAVSO and other such organisations have noted a decline in interest in serious amateur astronomy – at least in North America. The average age of AAVSO observers is increasing, and this does not bode well for the future.

My students use AAVSO data for projects (Percy et al. 1996). They also use data from an Automatic Photometric Telescope developed by amateurs Lou Boyd and Russ Genet (Percy and Attard 1992). Software written by amateurs is widely used by students. The techniques developed by amateur astronomers for the observation of the sun, moon, planets and other objects are the same ones used by students in their practical lab sessions, so that books such as the three-volume *Compendium of Practical Astronomy* (Roth 1994) are useful references for astronomers and teachers.

The majority of amateur astronomers are content to read about astronomy, or learn about it through other channels, or to be recreational sky observers. Some belong to clubs or larger associations; others pursue their hobby in a more independent way. Individuals, clubs, and associations of amateur astronomers can support astronomy and astronomy education in many other ways.

4. Amateur Astronomers As Supporters Of Astronomy

Amateur astronomers of all kinds provide grass-roots support for both research and education in astronomy. Many observatories and planetariums – including those in my city of Toronto – came about, in part, because of the efforts of amateurs. In many cases, the facilities were funded by generous contributions from amateur astronomers. The AAVSO, for instance, owes its existence to the generosity of Clinton B. Ford. This may be especially important in countries such as the developing countries, and the former Soviet Union, where there is little government funding for astronomy, but where former astronomers or astronomy students are now business people. At the risk of sounding mercenary: I think we should become aware of rich and influential people who have a special interest in astronomy. Their interest was usually sparked by some interaction with a professional or amateur astronomer, at a public lecture, star party, or other such event. This is a strong argument for public outreach, by both amateurs and professionals.

Another area of common concern to both astronomers and educators is light pollution. Amateurs, with their broad range of interest, training and experience, are often seen as "ordinary citizens" rather than as lobbyists, and carry considerable weight with local politicians

5. Amateur Astronomers And Public Education

When most people think of amateur astronomers' contributions to astronomy education, they probably think first of public education. This was certainly the main topic of discussion at IAU Colloquium 98. The list of possible contributions is a long one:

• TV and Radio Programs: Patrick Moore's monthly TV program is legendary. In my city of Toronto, a long-running series of programs on community cable TV was produced – very professionally – by local amateur astronomers.

• Newspaper and Magazine Articles: The largest-circulation newspaper in Canada has published a weekly astronomy column since 1981 by former amateur astronomer Terence Dickinson.

• Books: Patrick Moore and Terence Dickinson are prolific authors of books; one of Dickinson's books has recently been turned into a movie concept for director Steven Spielberg – an achievement which very few professionals have shared. In both the industrialised countries, and in countries with very few professional astronomers, amateurs often fill the need for public information about astronomy. Their writing skills are often as good as, or better than those of professionals, and they find it easier to write in non-technical language.

• Courses and Lectures: In most cities, the lectures organized by the local astronomy club are a rich resource for both the public, and for students with a special interest in astronomy. Many colleges, school boards, planetariums and museums, offer non-credit astronomy courses which are given by amateurs. The Astronomical League, the "umbrella organisation" of astronomy clubs in the US, has produced a useful booklet on how to organise a public lecture or course. (Astronomical League: 2112 Kingfisher Lane East, Rolling Meadows, IL 60008, USA; 73357.1572@CompuServe.com).

• Planetariums and Science Centres: Almost all of these facilities use amateur astronomers to organise star parties, and many use amateurs to present shows. In my city of Toronto, most of the original staff of the planetarium were local amateur astronomers.

• Public Observatories: These facilities are much more common in Europe than in North America. There are exceptions, and many North American public observatories came about through the efforts of the local astronomy club. In many public observatories, amateurs play an important role in scientific work, as well as in educational programs for students and the general public.

• Star Parties: Star parties in the parks are a popular and important activity of amateur astronomers; they enable people – both young and old – to get an "eyes-on" view of the universe. Amateurs assist at the weekly public programs of my university's David Dunlap Observatory, north of Toronto, as well as at the many events which they themselves organize.

• Displays: There are many venues where groups of amateur astronomers can set up demontrations and displays, and talk informally and enthusiastically with the general public. My local astronomy club organizes such events in libraries, in shopping malls, in museums and science centres, and at the annual Canadian National Exhibition which attracts almost two million visitors during a three-week interval each year.

• Electronic Resources: Amateurs, who are often skilled at computer science and related fields, have been quick to set up Web sites of interest to both students and the public.

I want to make special mention of Astronomy Day (or Week), a program of the Astronomical League. The Astronomical League sets and promotes Astronomy Day in North America, and have produced a useful handbook for organising a successful Astronomy Day program. This handbook is obtainable, at modest cost, from Gary Tomlinson, Public Museum of Grand Rapids, 272 Pearl St. NW, Grand Rapids, MI 49504, USA.

In Toronto, Astronomy Day is a collaborative program of two universities, a planetarium, a science centre, and the local astronomy club – with the local astronomy club taking the lead. This year, we added a teachers' workshop to the program, with Professor Michele Gerbaldi as the very special guest. This workshop was co-sponsored by the local science teachers, science co-ordinators, and teachers' union, so it brought additional partners into the astronomy education sphere.

Astronomy Day (or its equivalent) is celebrated in many countries around the world, though not necessarily on the same date. I would like to see the IAU, through Commission 46, officially support the concept of Astronomy Day, with the detailed circumstances of the celebration to be determined by the local organisers.

6. Amateur Astronomers and Formal Education

Amateur astronomers support astronomy in the schools and universities, but usually "at a distance". In the US, the Astronomical League sponsors the National Outstanding Young Astronomer award. Astronomy clubs in Canada offer scholarships and awards to outstanding students; they provide judges for science fairs, and speakers for special programs at schools. They may even do class visits, or arrange star parties for students. But their relationship with the teacher and students is not an ongoing partnership. To achieve that, amateur astronomers must understand the nature of teaching and learning, and the realities of the classroom environment. The teacher must understand the background and motivation of the amateur astronomer, and not feel intimidated by their own lack of knowledge of astronomy.

7. Amateur Astronomers and Elementary School Education

Two of the main problems of elementary school astronomy education worldwide are (i) the lack of teacher training and confidence; and (ii) the problem of arranging observing sessions, especially at night. Both these problems can be overcome through partnerships between teachers and amateur astronomers. Bennett, elsewhere in this book, describes an exemplary program: the Astronomical Society of the Pacific's Project ASTRO. The keys to success in this and other such programs are: (i) effective orientation and training of both the teacher and the amateur, and (ii) development of resources, such as books and slides, to support the program. The external evaluations of Project ASTRO were very positive. They showed, among other things, that amateur astronomers were even more successful than professional astronomers in forming successful partnerships with teachers.

8. Amateur Astronomers and Secondary School Education

At the secondary level, the problems of teacher training and confidence are less, but the amateur astronomer (or his/her club) can still play an important role by providing access and instruction in the use of telescopes, either in the school or in some other location. Another interesting connection is the fact that, according to the paper of Sadler and Luzader (1990) at IAU Colloquium 105 in Williamstown, the majority of teachers of astronomy in US secondary schools considered themselves to be amateur astronomers! Clearly, then, by supporting amateur astronomy, we are supporting astronomy education!

9. Amateur Astronomers and University Education

University astronomy education is normally seen as the preserve of professional astronomers, and that is certainly the case in major research universities. But at smaller colleges in North America, and at universities in the developing countries, amateur astronomers provide access and instruction in the use of telescopes. I think of the contributions of Eduardo Parini in Paraguay, who made his private observatory available to the IAU Visiting Lecturer Program. In the UK, some advanced amateur astronomers serve as tutors at The Open University and, in at least one college in the US, they are brought in as visiting lecturers in introductory astronomy courses. If they are interested and capable, they may even serve as instructors, especially if there are no professional astronomers available. After all, the "hands-on" approach is a very effective one in university, and amateur astronomers may be best suited to this approach.

10. Summary

I have tried to make several points, most of which are probably obvious. The most important one is that amateurs are a large, diverse, enthusiastic and skilled group of people who can relate very well to students, teachers, and the general public. We should respect them, cultivate them, train them, learn from them and work with them in our common goal to increase awareness, appreciation and understanding of the universe. I have worked with amateur astronomers for 35 years, through the Royal Astronomical

Society of Canada, the American Association of Variable Star Observers, the Astronomical Society of the Pacific, and other organisations, and I count amateurs among my best colleagues and friends.

In order for this to happen, there have to be links between professional and amateur astronomers, and other educators – particularly at the local level. In June 1995, a major conference on astronomy education was held at the University of Maryland (Percy 1996). It identified, as a high priority in astronomy education, the formation of coalitions and networks of astronomy educators, who could learn about the best available ideas, resources and materials for astronomy education. In my city of Toronto, such a coalition has existed because of the organisation of Astronomy Day each year. In 1995, the coalition was mobilized to deal with a sad event: the closing of the McLaughlin Planetarium – one of the world's major planetariums. We hope that the quick and strong response of professional and amateur astronomers, teachers, and people from every walk of life, will lead to the refurbishment and reopening of the planetarium. It was certainly a sign that interest in astronomy is shared by a wide segment of our population.

REFERENCES

DUNLOP, S. & GERBALDI, M., 1988 *Stargazers: The Contributions of Amateurs to Astronomy*, Berlin, Springer-Verlag

FRAKNOI, A., 1996, *Astronomy Education: Current Developments, Future Coordination* ASP Conf. Series, Vol. 89, 9-25.

MATTEI, J.A. ET AL., 1996, *Astronomy Education: Current Developments, Future Coordination*, ASP Conf. Series, Vol. 89, 247-248.

PASACHOFF, J.M. & PERCY, J.R., 1990, *The Teaching of Astronomy* IAU Colloqium 105, Cambridge, Cambridge University Press.

PERCY, J.R., ed. 1996, *Astronomy Education: Current Developmen, Future Coordination*, ASP Conf. Series, Vol. 89.

PERCY, J.R. & ATTARD, A., 1992, PASP 104, 1160-1163.

PERCY, J.R., DESJARDINS, A., YU, L. & LANDIS, H.J., 1996, PASP 108, 139-145.

ROTH, G., ed. 1992, *Compendium of Practical Astronomy*, Berlin, Springer-Verlag.

SADLER, P.M. & LUZADER, W., 1990, *The Teaching of Astronomy*, IAU Colloquium 105, Cambridge, Cambridge University Press, 257-276.

WILLIAMS, T.R., 1988, *Stargazers: The Contributions of Amateurs to Astronomy*, Berlin, Springer-Verlag, 24-25.

Astronomy to understand a human environment

By S. Isobe

National Astronomical Observatory, 2-21-1-Osawa, Mitaka, Tokyo 181, Japan

Astronomy is an important science in understanding a human environment. However, it is thought by most politicians, economists, and members of the public that astronomy is a pure science having no contribution to daily human activities except a few matters relating to time. The Japanese government is studying a reorganisation of our school system to have 5 school days per week, instead of 6 days per week, and this July its committee made a recommendation to reduce school hours for science and set up new courses for practical computers and environmental science. I currently made a proposal. It is very difficult for most of the school pupils, who will have non-scientific jobs, to understand science courses currently taught in school, because each science is taught independently from the other sciences. Therefore, their knowledge of sciences obtained during their school period does not greatly help their understanding of global environmental problems. We should present several stories to connect all the related sciences in order to give those pupils ideas in the understanding of global environmental problems. I believe that astronomy is able to play an important role in this context.

1. Expansion of scientific items to be taught.

Items which should be taught at school increase depending on time. Although items in language courses, mathematics courses, art courses and gymnastic courses increase little, those of social science courses increase gradually, but those in science courses do so drastically in recent decades. Therefore, it becomes much more difficult to teach all the necessary pupils. Pupils at the lower level of an elementary school have an interest in science, especially in astronomy. However, their interest is not in motions of the sun, moon, planets and starts on the celestial sphere, but in black holes, the big bang, star formation, comets and so on. Since in an elementary school and a junior high school, they are taught only about objects inside the solar system, they are gradually losing their interest in astronomy and also in science. In Japan, it has been said in this decade that most pupils do not now like science, because they are taught rather difficult scientific items. Figure 1 shows percentages of children, getting into a senior high school, and into a college, and a university, and these equivalences depending on time. Before 1960, only 10% of children went into a university level, and therefore, lecturers in a class at the lower level were not targeted for the pupils going to the university level. Teachers had the possibility to make an attractive lecture.

2. New recommendations of governmental consulting committee for education.

In 1947 after the second world-war, a completely new education system was introduced under the US occupation. Therefore, this system imitated the US education system. Since then, there have been 6 minor revisions. In all the cases, the number of school hours for science courses was gradually reduced. At the first stage, the science course was divided into 4 parts, namely physics, chemistry, biology and earth science. To set up earth science, there were deep discussions, and a conclusion was reached whereby

211

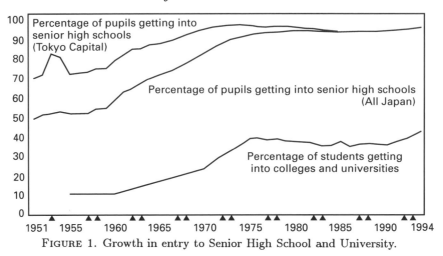

FIGURE 1. Growth in entry to Senior High School and University.

astronomy was included in it instead of in physics. In 1969, a new part was set up, fundamental science, which was changed to science I and II in 1988, and to integrated science in 1994. All efforts to combine all the science parts into one part were not successful. Form 1980 our government has requested companies to introduce 5 workings days per week. The number of companies having accepted this system is now over 80%. To follow this up, the Japanese government set up a committee in the spring of 1995 to study a re-organization of our school system to have 5 rather than 6 school days per week, and the committee made a final recommendation to reduce school hours for science and introduce new areas such as practical computers (internet etc.) and environmental science this July. Although a final decision on school hours for each course will be made by a committee for school hour allocation, it will certainly start with school hours for science courses being reduced because of two effects, reduction of the total school hours and an addition of new courses. It will not be possible to teach the science being carried out as the present 4 parts of the science course, such as physics, chemistry, biology and earth science, with such reduced school hours. We need a drastic change of science course to meet this recommendation and to take into account matters in section 1.

3. Proposed change for science course.

It is very difficult for most school pupils, who will have non-scientific jobs, to understand science courses currently taught in school. As shown in Fig. 1, only 10% of all the pupils in 1960 went to university and about 5% of them became natural scientists and engineers. This ratio is probably not different between 1960 and the present. When one considers our scientific curriculum, school pupils are able to develop very high ability in science if they absorb all the scientific matter taught in a school class. Furthermore, physical, chemical, biological, and earth science phenomena are taught independently. However, since science phenomena which pupils see in their daily life are combinations of those phenomena, and there is practically no case of purely physical, chemical, biological and earth science phenomena, school pupils feel that their knowledge of sciences obtained during their school period does not much help their understanding of daily life sciences. For example, since an electric microwave oven is used to heat up food, it may sometimes happen that one tried to dry the wet hair of cat using this oven. There is no ability to combine the two pieces of knowledge of microwave and heating despite being taught both items at school. These years, there are many discussions and also activities on environmental problems by so-called environmental groups. However, because of their shortage of ability to combine different areas of scientific knowledge, it happens many times that they take out garbage from their area (e.g. their own houses) but move it to another (e.g. their neighbor's house). That is, we are not able, these days, to escape from global environmental problems: one should always consider whether an action to solve an environmental problem does or does not affect the global environmental problems. To solve these matters, I will propose a way in which 20 to 50 stories to connect all the related sciences should be prepared in order to give those pupils idea in understanding global environmental problems. As examples, we can prepare the following stories. 1) The cycle of water, 2) Solar light, 3) Rise and decline of the dinosaurs and 4) Shape of the Earth.

Each story should contain physical, chemical, biological and earth scientific items, mixing items from each part. Astronomy is able to play an important role in this context. For the example 1), water was collected on the Earth when it was formed from asteroid and comet collisions during its early history and those collisions are seen today as meteors, fireballs, and asteroid collisions forming craters. For the other examples, we are also able to include astronomical items. It is not necessary for pupils to learn all the stories. They can choose several stories for each grade. One can increase slightly difficult items for each story depending on the grade and also include new topical items such as the event of the Shoemaker-Levy 9 comet. Then, pupils can enjoy interesting scientific items and also learn the inter-relation of 4 parts of sciences. We should also prepare a curriculum for pupils who have good scientific ability and a high possibility to be scientists or engineers. This type of pupil should learn fundamental items and methods of 4 parts of sciences. Amounts of fundamental items increase depending on grade during a period of elementary school, but at a grade of junior and senior high school we should prepare a different curriculum for those two types of pupils.

4. Conclusion.

Pupils who have the ability to be scientists and engineers can enjoy the current curriculum. Therefore, we should prepare this fundamental scientific curriculum. Pupils who will have non-scientific jobs can also enjoy a new curriculum with stories as described above, and get an understanding of the importance of science, both in everyday life and as applied to global environmental problems.

Selling Our Southern Skies: recent public astronomy developments at the Carter Observatory, New Zealand

By W. Orchiston, B. Carter, R. Dodd & R. Hall

Carter Observatory, PO Box 2909, Wellington, New Zealand (Wayne.Orchiston@vuw.ac.nz)

1. Introduction

Carter Observatory is the gazetted National Observatory of New Zealand, and opened in 1941 December. From the start, the main function of the Observatory was to provide for the astronomical needs of the citizens of, and visitors to, the Wellington region, and today this remains one of its four recognised functions (Orchiston and Dodd, 1995). The other three are to conduct astronomical research of international significance; provide a national astronomy education service for school students, teachers, and trainee teachers; and assist in the preservation of New Zealand's astronomical heritage.

Since 1992 the Carter Observatory has undergone major restructuring as a result of acquiring an aging Zeiss planetarium and an accompanying visitor centre (Van Dijk, 1992), and in response to major changes in Government funding policy. As a result, there has been a wholesale revamp of the education and public astronomy functions (see Orchiston, 1995b; Orchiston and Dodd, 1996). This paper focuses on the latter area, with emphasis on development of the Visitor Centre and the publications program.

2. The Visitor Center

The focal point of the Observatory's in-house public astronomy programs is the Visitor Centre, comprising a foyer area with "Space Shop", the planetarium chamber, an audio-visual theatre (that also doubles as a meeting room), a small video room which also houses a public-access PC (featuring the "Orbits"program), and the "Dome Room"where the Observatory's historically-significant 23cm Cook photovisual refractor (see Andrews and Budding, 1992; Orchiston et al., 1995) holds pride of place. There are also the mandatory wheelchair-access toilets, and adjacent to the theatre is a small kitchen.

Access to the Visitor Centre is not a problem for most visitors using public transport, since the Carter Observatory is only a short distance from downtown Wellington and can be reached by bus or cable car. Those coming by private car face more of a challenge, given the lack of public car parking nearby. Adequate external signage is vital for any public facility, and in this regard the Carter Observatory is now well provided for. At nearby entrances to the Wellington Botanic Garden there are large colourful maps and conspicuous signposting indicating the location of the Observatory, and this information is also readily available in brochure form.

In order to create an attractive "astronomical environment"in the Visitor Centre the walls and ceiling of the foyer have been painted black, and external lighting has been minimised (except for the large "shop front"window near the entrance doorway) so that beamed spotlights can have maximal effect. The astronomical mood is further enhanced by the attractive astronomical murals that have been painted on the foyer walls and ceiling by one of the authors (R.H.), while panel displays feature in the audio-visual theatre.

In addition to providing an accessible, warm, friendly astronomical environment, we aim whenever possible to supply our visitors with background information which will help them optimise the benefits of a visit. This is achieved by giving every visitor a single A4 *Visitor Orientation Sheet*, which contains a list of what there is to see and do in the Visitor Centre; a map of the Centre; and background information about the Carter Observatory and the Observatory's Golden Bay Planetarium. We have made full use of the monthly statistics on overseas visitors to New Zealand supplied by the New Zealand Tourism Board, and now have *Visitor Orientation Sheets* in the following languages: English, Chinese (two different dialects), German, Japanese, and Korean. In the 1995 calendar year, native-speakers using these languages accounted for over 90% of the 1.4 million overseas visitors to New Zealand.

When the Carter Observatory was presented with the Zeiss planetarium facility by the Wellington Planetarium Society in 1992 it also inherited a number of thematic planetarium programs made during the 1970s and early 80s. A special challenge for Observatory staff has been to produce new up-to-date programs, and since the end of 1993 four have been completed. Brief accounts of the first three appear in Orchiston and Dodd, 1996. With previous experience to draw on and extensive feedback from the public, in 1996 the "Project Team" decided to embark on a major refurbishment of the planetarium chamber prior to the launch of the most recent program, "Journey to the Centre of the Galaxy". As a result, the projection system for the seven slide projectors was altered to make use of just half of the dome, and lap-dissolved images from two IMAX-type lenses were beamed onto the same dome space (instead of opposite one another, as previously). The seating configuration in the planetarium chamber was also extensively modified to accommodate this new unidirectional projection system; the speakers were relocated in order to provide an improved sound system; the interior surface of the 6m dome was resealed and repainted; and a light-free "tunnel" was constructed leading from the darkened planetarium chamber to the illuminated visitor centre foyer (thereby, if necessary, allowing people to leave during a screening without disrupting other patrons).

A major facility in the foyer is the "Space Shop" which has grown since 1992 to encompass a wide range of stock for the following specific "target audiences": general public, school students, teachers and trainee teachers, amateur astronomers, and overseas tourists (but particularly those from northern hemisphere countries). In addition to books, booklets and pamphlets, the shop has a good selection of posters, cards, post cards (both astronomical and non-astronomical), astronomical jewellery, holographic products, and educational products (e.g. kits, computer programs, sky globes). There are also tektites for sale, and the following special Carter Observatory stock lines: ball point pens, "astrofrogs" (novelty gifts), and T-shirts and sweat shirts. Also available, on compact disk, is the original background music composed especially for our latest planetarium program by an amateur astronomer who works professionally in the music industry.

Shop product is sourced both from within New Zealand and overseas, and staff are always on the lookout for suitable new stock lines. Perceived popularity, quality, and calculated retail price are key selection criteria. Considerable shop revenue is generated through the Observatory's Summer and Winter Mail Order Catalogues which we generate in-house. These now run to 14 A5 pages, and are distributed to schools, libraries, astronomical societies and individual amateur and professional astronomers.

3. Carter Observatory Publications

One of the most notable developments in public astronomy at the Carter Observatory since 1993 has been the in-house publications strategy. The aim has been to provide different types of publications for different perceived target audiences. This involved modifying the existing publications, introducing new types of publications, and (for purposes of product branding and marketing) adopting a "house style". A black cover with white print was selected as "astronomically- appropriate", but this is relieved by a feature coloured image on the front cover and a coloured photograph of the Observatory itself on the back cover.

The annual *Astronomical Handbook*, which has been in production since 1972, was one of the first publications to undergo wholesale change, in 1994. A new cover, with an astrophotograph of the Great Orion Nebula by award-winning New Zealand amateur astronomer, John Drummond, replaced the former monochrome cover, while the range of astronomical data inserted was drastically changed. A further innovation was the introduction of an "Articles Section" along the lines of that in Patrick Moore's annual *Yearbook of Astronomy*. The first "new look" *Handbook* ran to 72 pages (up from the previous 48 pages), while additional astronomical information and an enlarged Articles Section extended the 1996 issue to 132 pages.

The aim with the Articles Section has been to provide a range of topics, but with some emphasis on New Zealand astronomy, and to include at least one article per issue by a noted overseas astronomer. The 1995 *Handbook* contained six articles ranging in length from 4 to 10 pages, while the seven articles in the 1996 issue were between 6 and 11 pages long. Patrick Moore and Duncan Steel (Anglo-Australian Observatory) featured in the 1995 edition, while Heather Couper and Andrew Taylor (University of Adelaide) have articles in the 1996 Handbook. Although now resident in Australia, both Steel and Taylor completed their Ph.D.'s in New Zealand. At just $NZ9:95 (plus postage, if necessary), we believe that the 1996 *Handbook* represents exceptional value for money.

Another existing publication to undergo a drastic revamp was the *Annual Report*, starting with the 1993-94 issue. Again, the "house style" was adopted, and the contents were restructured. More line drawings and half-tones were introduced, making this publication more "user friendly" for amateur astronomers, funding bodies, potential corporate sponsors, and Government officials. The 1993-94 *Annual Report* ran to 38 pages, while the 1994-95 issue was 48 pages long.

In order to provide cheap, readily-available and up-to-date information for the general public, overseas visitors, school students and New Zealand amateur astronomers, the Carter Observatory introduced its *Information Sheet* series in 1993. To date, 16 different *Information Sheets* have been generated (Table 1), and these A4 photocopied in-house publications range from 2 to 10 pages in length. Given the paucity of authoritative books and research papers on New Zealand astronomy, the historically-oriented *Information Sheets* have proved particularly valuable for they provide the most detailed accounts available – indeed, sometimes the only accounts available – on particular aspects of our astronomical history and heritage. Meanwhile, second editions are currently being prepared of *Information Sheets* 6 and 11, where the changing state of our knowledge has rendered the original versions obsolete.

A new publication pioneered in 1996 May was a monthly *Carter Observatory Newsletter*. This A4 publication runs to just 8 pages, and contains information on the night sky, the Observatory's current public astronomy and education programs, general Carter Observatory news, and summaries of selected material from recent *IAU Circulars* and astronomical magazines and journals. By pooling this information, the *Newsletter* brings together material that three different staff members were assembling monthly for the media, the Wellington region's five visitor information centres, and for two different astronomical society newsletters. The *Carter Observatory Newsletter* is now supplied free of charge to New Zealand astronomical societies, and they are encouraged to reproduce any material of interest in their own newsletters (with due acknowledgement). Meanwhile, individuals, libraries and schools who wish to subscribe to the *Carter Observatory Newsletter* pay an annual fee of $NZ36.

Other types of Carter Observatory publications are planned, and the first monograph in the Carter Observatory Occasional Papers series is in active preparation.

Table 1: Carter Observatory Information Sheets

No.	Year	Title	Pages
1	1995	A Brief History of the Carter Observatory	10
2	1993	Coming to Terms with Size, Time and Distance in Astronomy	4
3	1993	John Grigg: New Zealand's Pioneer Cometary Astronomer	4
4	1993	A Guide to Purchasing Telescopes and Binoculars	8
5	1993	Careers in Astronomy in New Zealand	2
6	1994	The Quest for the Elusive Tenth Planet	4
7	1994	The Enigma of the Tektites	2
8	1994	The Southern Cross: Our Very Own Constellation	4
9	1994	James Cook and the 1769 Transit of Mercury	4
10	1994	Ronald McIntosh: A Remarkable New Zealand Astronomer	4
11	1994	Worlds in Collision: Comet Shoemaker-Levy 9 and Jupiter	2
12	1995	The Historic Astronomical Observatories in the Wellington Botanic Garden: A Brief Introduction	8
13	1995	A History of the 23cm Refracting Telescope at the Carter Observatory	6
14	1995	Charles Rooking Carter: A Brief Biography	4
15	1995	Is There Water in Space?	2
16	1996	The Enigmatic "Uncle Charlie": Algernon Charles Gifford	4

4. Other Public Program Initiatives

In addition to the foregoing developments, in 1994 Carter Observatory staff began producing 30-minute thematic audio-visual shows, using two slide projectors and a tape with commentary, sound effects and background theme music. Two shows have been prepared to date: "Galileo: A Space Odyssey", and "Ra: God of Light". Wide-angle lenses on the slide projectors throw 4m by 2.4m images onto the end wall of the audio-visual theatre. But despite their obvious visual appeal, on twice-weekly "Public Nights" the video screenings, audio-visual show and two different planetarium programs do not usurp the rightful place of sky-viewing through the 23-cm telescope, for we believe that visitors if possible should also see "the real thing" (c.f. Bell, 1993 and Mullaney, 1993).

Another significant development since 1993 has been the establishment of three different monthly radio programs. Two of these, of 15 minutes and 30 minutes duration, are beamed locally, while the third program (also of 15 minutes duration) goes out nationally. On all three programs we provide astronomical news (often with a New Zealand focus) and "Sky for the Month". Fortunately, we have not had to find sponsorship for any of these programs. As a spin-off from these high-profile presentations, Observatory staff are increasingly appearing on local and national television.

Another high-profile innovation is the annual Carter Memorial Lecture series. Launched in 1993 with New Zealand's most distinguished space scientist, Sir Ian Axford as speaker (see Orchiston, 1995a), this series aims to provide New Zealanders with up-to-date information by talented speakers of international repute. The 1995 Lecturer was Professor Heather Couper, and she presented Carter Memorial Lectures at five different New Zealand centres. Meanwhile, former Astronomer Royal, Sir Arnold Wolfendale, delivered the 1996 Carter Memorial Lecture in Wellington, and as with both of his predecessors attracted a capacity audience of around 300.

Finally, the Observatory has sought to take its message to a diverse national and international audience through the World Wide Web. In late-1995, Dr Timothy Banks (Research Associate) and one of the authors (W.O.) established a Home Page containing wide-ranging information about the Observatory, its staff and its activities. During the first six months interest increased dramatically, and as at 1996 June the number of accesses was >3,500 per month from around the world. Our Home Page address is:

http://www.vuw.ac.nz/~bankst/carter.html

5. Discussion

A major problem, and one which is seen to typify many much larger institutions (e.g. see Griffen, 1993), is the totally inadequate space available in our Visitor Centre, and the current functional mix is a necessary compromise. In the three years from 1992, visitor numbers have trebled from 11,495 to 33,038, and the visitor "saturation point" has been reached. In the long run, it is vital that the Visitor Centre is expanded to provide discrete spaces for public astronomy and education, and increase the range of options available to visitors. Specifically, we require an interactive display gallery, space for a program of changing mini-displays, and a microcomputer gallery with a range of different software packages. Because of funding restrictions, it is unlikely that this program of expansion will occur in the foreseeable future, and so the challenge now is to make the best possible use of the minimal facilities and the limited staff and funding at our disposal. Our focus, therefore, will be on continuing to produce captivating, visually-appealing, instructive new planetarium programs and audio-visual shows, and increasing the number and range of in-house publications. We also recognise the didactic value of well-prepared displays (see Fierro, 1994), and another short-term objective is to develop a major display on astronomy in New Zealand. When funding becomes available, we would also like to begin making our own videotapes with a specifically New Zealand or southern sky slant.

6. Conclusion

Carter Observatory has undergone a major revitalisation since 1992, and now offers a wide-ranging public astronomy program which targets a number of discrete audience sectors. Further developments are planned as additional funding comes to hand.

Acknowledgements

We are grateful to the British Council and the International Astronomical Union for grants which made it possible for the one of us (W.O.) to visit London and present this paper at IAU Colloquium 162.

REFERENCES

ANDREWS,F.P., & BUDDING,E.,1992. Carter Observatory's 9-inch refractor: the Crossley connection. Southern Stars **34**:358-366.

BELL,J.U.,1993. Live vs. video. *The Planetarian*, **22**(3); 24-25,32.

FIERRO,J.,1994. Getting started: an astronomy hall for a science museum, *Teaching of Astronomy in Asian-Pacific Region*, **8**, 1-15.

GRIFFEN,I.,1993. Armagh goes for gold in its silver year. *Astronomy Now*, **7**(9), 39-42.

MALLEY,M.,1993. The Carter Observatory Library. *Southern Stars*, **35**, 110-111.

MULLANEY,J.,1993. The role of the telescope in the planetarium. *The Planetarian* **22**(3), 6-7.

ORCHISTON,W.,1995a. DR WILLIAM IAN AXFORD 1995 "New Zealander of the Year". *Australian Journal of Astronomy*, **6**, 75-79.

—,1995b, Public astronomy and education at the Carter Observatory: some recent developments, In Orchiston,W., Carter,B., and Dodd,R.eds. *Astronomical Handbook for 1996*, Wellington, Carter Observatory, 59-69.

—, and Dodd,R.,1995. *A Brief History of the Carter Observatory, Carter Observatory Information Sheet No. 1*, Wellington.

—, and —,1996, Education and public astronomy programs at the Carter Observatory: an overview, *Publications of the Astronomical Society of Australia,* **13**, 165-172.

—, Andrews,F. & Budding,E.,1995. A History of 23cm Refracting Telescope at the Carter Observatory, *Carter Observatory Information Sheet No. 13*.

VAN DIJK,G.,1992.The New Zealand Astronomy Centre. *Southern Stars*, **34**, 403-406.

Astronomy Education in Latvia - problems and development

By I. Vilks

Astronomical Observatory, University of Latvia

1. Primary school

School education in Latvia, as in many other countries, is divided into two stages: primary and secondary education. Primary education is compulsory. Every year 30 000 new school children start attending primary school. This is a potential audience that can study astronomy fundamentals. During first grade studies school children learn the basics of natural science which include some elements of astronomy. These lessons are given once a week. At this stage children's interest in the Universe is great; therefore the most active teachers use some out of curriculum activities to give the schoolchildren an idea about the stars, planets and other celestial bodies. The science curriculum itself contains very few elements of astronomy (Karule, 1995). Even more many teachers have problems teaching science at the elementary school, because they are afraid that it is too sophisticated. This situation should be corrected, but at the moment no teacher training in science is planned.

In higher grades of primary school some astronomy elements are taught in different disciplines. In geography there are some topics about the Earth, Seasons and Tides (Klavins, 1992). In physics there are some topics about Eclipses of the Sun and the Moon (Kokare, 1992). And that is all. It leads to the situation that a young person, graduated from primary school, has heard nothing about constellations, Moon phases, comets and many other astronomy questions.

The expected positive changes are the following: new textbooks of the basics of natural science, where more attention is paid to astronomy, are being developed. The textbook for 1st grade has already been published (Vaivode et al., 1995), the others will follow. During this year the education programs of primary school will be revised. Just now teachers have a good opportunity to prepare proposals on how to introduce more science in the curriculum for primary schools.

2. Secondary school

The situation with astronomy education in secondary school is better. Latvia is one of the few countries where astronomy is taught as a separate discipline. Astronomy is a 70 hours course by choice. The main aim of this course is to give the students the notion about the basic structures of Universe and to explain the most important astronomy facts and laws (Ros, 1995). Special attention should be paid to the spatial configuration and the evolution of celestial bodies. For better understanding, the description of astronomical objects is connected with their visibility in the sky (Vilks, 1995).

Two new documents: *Curriculum of Astronomy* (Vilks, 1993) and *Standard of Knowledge* (Vilks, 1995) have been prepared. These documents define the aims and the content of the astronomy course and the level of knowledge which should be achieved. A new astronomy textbook for Latvian schools (Vilks, 1996) has been published this year. It has replaced the old astronomy textbook (Vorontsov- Velyaminov, 1987) that was used for 50 years.

More teaching materials are to come soon. The first book is *Sky Guide* (Vilks, 1996)

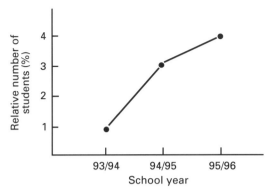

FIGURE 1. Relative number of students, studying astronomy at secondary school during last three school years.

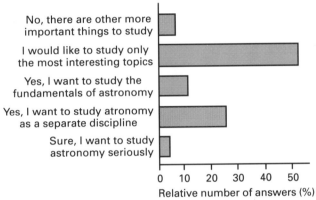

FIGURE 2. Distribution of 400 students answers to the question "Do you want to study astronomy at the secondary school?"

which will help the students to find the constellations and make the observations of different celestial bodies. For teachers *Didactic Materials in Astronomy* (Vilks et al., 1996) will be issued, too. The latter will contain some exercises and lesson planning patterns. The most advanced teachers can also use the materials from *Project STAR* (Project STAR, 1991), *The Universe at Your Fingertips* (Fraknoi et al., 1995), and others. But in most schools a lot of necessary teaching aids are still missing. They are star charts, sky globes, slides, videos, astronomical software and other things. Of course, a lot of astronomical information can be found on *Internet*, but at the moment only few schools in Latvia are connected to this computer network.

During the last three school years the number of students studying astronomy has gradually increased, but nevertheless, it is very low: only 4% of students are choosing astronomy to study (Fig. 1). At the same time, judging by the results of the questionnaire which has been made this year by *The Association of Astronomy Teachers* and covered approximately 400 students, more than half of them would like to learn the most important astronomy topics at the secondary school (Fig. 2). This means that more astronomy fundamentals should be introduced in other disciplines.

The possible solution is to teach some astronomy problems (18 hours) in the basics of natural science. This is a new discipline for Latvian secondary schools, supposed to be introduced in the secondary school curriculum starting from the next school year. A textbook for this new discipline is under development now. More astronomy fundamentals (35 hours) could be introduced in physics during the next few years.

3. Teachers' activities

Last year *Latvian Association of Astronomy Teachers* was organized. The purpose of this organization is to promote astronomy education in all levels — school, university, public education, correspondance education and mass media. The Association includes not only teachers, but also other persons interested in teaching astronomy. During a one year period several teachers' meetings concerning educational methods have been held, and some new ideas have been generated. The most active teachers have already started preparing some new teaching aids: exercise books, overhead materials and others. Teachers help organize *The Open Olympiad (Competition) of Astronomy* which is held every year in Riga, the capital of the state. The most gifted astronomy students from all over the country take part in this Olympiad. The Association has established some contacts and the information exchange with Lithuanian and Russian colleagues. Now astronomy teachers from Latvia are looking for mutual cooperation contacts with *The European Association of Astronomy Education*.

4. Public education

Since 1947 *The Latvian Astronomical Society* has been functioning. This Society unites professional and amateur astronomers, astronomy teachers and the people interested in astronomy. Regular meetings concerning actual astronomy and space exploration questions are held and expeditions to eclipse sites are organized.

In Riga *The Museum of Space Exploration* was opened in 1987. It is located in the building, where Fridrich Tsander, one of the rocket constructing pioneers lived at the beginning of the XXth century. The exposition of the museum is devoted to the history and some recent achievements in space exploration.

Unfortunately there is no planetarium in Latvia. In the former Soviet Union a *Zeiss planetarium* existed but it was situated in an old orthodox church. Some years ago this building was given back to the church and planetarium was closed. Now the authorities of the Riga School Management Board are looking for the possibility to buy a new planetarium in Sweden.

In the *Astronomical Tower* of the University of Latvia sky observations with a 22 cm reflecting telescope are available for students of schools and universities and for the general public. The Moon, planets, double stars, galaxies, star clusters and nebulae are being demonstrated. During the *Hyakutake* comet visibility period the Tower was visited by more than 1 000 people.

Every year in August during the *Perseids* meteor stream maximum the *Star Party "Aquila"* is held in the country area where the darkness of the sky is appropriate. During the Star Party students and amateur astronomers observe the sky with some home-made telescopes, and they establish some new contacts and share their experience.

Two astronomical periodicals are issued in Latvian. They are – the popular science magazine *"The Starry Sky"* and the *Astronomical Calendar*. "The Starry Sky" is the only popular science magazine now published in Latvia.

This year the home page of **Astronomical Observatory of University of Latvia** has been created by the University students. It contains the most comprehensive information about astronomy activities in Latvia. The texts are given in English and Latvian. The home page address is: http://www.lanet.lv/members/LU/astro/.

Acknowledgments

The author wants to express his gratitude to the International Astronomy Union and the Scientific Organizing Committee of the IAU Colloquium No. 162, "New Trends in Astronomy Teaching", which has awarded him the Travel Grant thus enabling him to take part at this Colloquium and the possibility to present this paper.

REFERENCES

KARULE, L., 1995, "Recommendations for Teaching the Basics of Nature Science. Grade I - IV", Riga, Latvian Ministry of Education and Science, in Latvian.

KLAVINS, J., 1992, "Standard of Knowledge in Geography for the Primary School", Riga, Latvian Ministry of Education, in Latvian.

KOKARE, V., 1992, "Standard of Knowledge in Physics for the Primary School", Riga, Latvian Ministry of Education, in Latvian.

PROJECT STAR, 1991, "Science Teaching through its Astronomical Roots", Cambridge, Harvard-Smithsonian Center for Astrophysics.

ROS, M. R., Ed., 1995, Proceedings of 5th International Conference Teaching Astronomy, Vilanova i la Geltru, Universitat Politecnica de Catalunya, 278–280.

FRAKNOI, A., Ed., "The Universe at Your Fingertips. An Astronomy Activity and Resource Notebook", San Francisco, Astronomical Society of the Pacific.

VAIVODE, E., CISA, L., SKUTE, A. & SOMS, J., 1995, "You and I and All Around Us. Basics of the Nature Science for 1st Grade", Lielvarde, Lielvards, in Latvian.

VILKS, I., MURANE, I., ISAKOVS, M.& DUDAREVA, I., 1996, "The Didactic Materials in Astronomy"(in preparation), Riga, Latvian Ministry of Education and Science, in Latvian.

VILKS, I., 1993, "The Experimental Curriculum of Astronomy", Riga, Latvian Ministry of Education, in Latvian.

VILKS, I., 1995, Acta Universitatis Latviensis, Volume 600, Astronomy 20, Riga, University of Latvia, 77 - 82, in English.

VILKS, I., 1995, "Standard of Knowledge in Astronomy for the High School", Riga, Latvian Ministry of Education, in Latvian.

VILKS, I., 1996, "Astronomy Textbook for the Secondary School", Riga, Zvaigzne ABC, in Latvian.

VILKS, I., 1996, "Sky Guide"(in preparation), Riga, Macibu gramata, in Latvian.

VORONTSOV-VELYAMINOV, B., 1987, "Astronomy Textbook", Riga, Zvaigzne, in Latvian (translated from Russian).

Teaching Astronomy at Sydney Observatory

By N.R. Lomb & J. Kitchener

Sydney Observatory, PO Box K346, Haymarket 2000, Australia

1. Introduction

Sydney Observatory is a museum of astronomy and a public observatory. It is Australia's oldest existing observatory and is now a branch of the Powerhouse Museum, the largest museum in the southern hemisphere. With 65,000 visitors each year, the observatory is popular with the public. Visitors can come during the day to see exhibits and audiovisuals and in the evenings on telescope viewing sessions. They can also take part in school holiday workshops, adult education courses or a telescope-making course. In addition, many school groups come along during the school terms to extend the astronomical knowledge of their students. Other professional services provided by the observatory include an annual guidebook with up-to-date information for the sky as seen from Sydney and an astronomical information service for the public and the media.

In this paper we will mainly discuss selected aspects of our educational activities, exhibitions and equipment, highlighting recent developments in the 1990s.

2. Recent Innovations in Education

2.1. *Open Nights*

A maximum of only 45 people can be accommodated at any one time in one of our regular evening sessions. During school holidays this is nowhere near enough to meet the demand. When there is a major astronomical event, we like to give more people a chance to look through our telescopes. At these times we organise open nights at which up to 1 000 people can attend.

Special open nights have been held to view lunar eclipses, a favourable opposition of Mars and the ring plane passage of Saturn. During the collision of Comet Shoemaker-Levy 9 with Jupiter we held six open nights, each attracting over 1 000 people. A recent open night targeted Comet Hyakutake. For this event we chose a Saturday night just before the comet disappeared from view in the southern hemisphere. Although viewing could only start at midnight, about 500 visitors attended. As these events attract much media interest, the open nights receive much free publicity.

Open nights are also held on less rare events. In the school holiday time when the observatory is very busy, a particular planet or constellation is targeted. For example, the target could be Jupiter, Saturn, or the Southern Cross.

On an open night the two telescopes in the domes are used as well as telescopes in the grounds. These include a 14 in (35 cm) Celestron Schmidt- Cassegrain which has a fixed pier in the grounds of the observatory and three smaller portable telescopes. In addition, we encourage people who have just completed a telescope-making course at the observatory to bring along their finished telescopes. Talks with up-to-date images are held at regular intervals, continuous videos are shown and the exhibitions are open. Refreshments are provided by school students raising money for their school. Numerous guide lecturers are available to assist the visitors.

Open nights give a wonderful learning opportunity to many people who otherwise may not visit the observatory. By not having to book there is no prior commitment and the participants can make a decision to attend on the night.

FIGURE 1. Sydney Observatory in the 1860s.

2.2. *Camp-ins for Children*

These are held for secondary school students (approximately 12 -17 years of age). Generally, there are two held in each year, one in the winter school holidays and one in the summer holidays. They are very popular and there is usually a waiting list for them. Twenty students attend each camp-in at a time.

If the night is clear, there are several viewing sessions held throughout the night interspersed with other activities. The format of the camp-in can vary but ordinarily it includes a planisphere-making workshop, a demonstration of our computer programs and finding astronomical information using the Internet. On a cloudy night an astronomical quiz and a session in our tiny planetarium are included. Our use of computer programs and the Internet is discussed in more detail in Lomb, 1996.

The students at the observatory are well provided with nourishment: dinner, supper and breakfast are served and extra food and drink are available on demand. The students are asked to bring warm clothes and sleeping bags, though understandably many do not get around to sleeping during the night. In the early hours of the morning the computerised Meade telescope (section 3.2) is available for their use. The students can choose objects that they want to view from a catalogue and then find them using the telescope. They can also rest or watch pseudo-science videos.

The camp-ins are a wonderful learning experience as they provide enthusiastic youngsters with an opportunity to feel at home in the impressive environment of an observatory. They are constantly being stimulated and the activities are varied. There is also a lot of informal time when the students and our specialist staff can interact.

2.3. *Camp-ins for Adults*

The observatory has had many inquiries from adults who felt disappointed that we only offered specialised and fun activities for children. Last summer we gave adults the opportunity to stay overnight. In spite of exceptionally bad weather at the start of the night, the response was tremendous. Fortunately, the clouds eventually cleared away. We started later than with the youngsters and offered supper and breakfast. The program included a planisphere-making workshop, computer programs and accessing the Internet, as well as viewing through to daylight.

The adult camp-in was a great success with most participants staying up all night. There was plenty of opportunity for informal interaction with the staff and many interesting discussions ensued.

2.4. *Midnight Suppers*

The best viewing hours are after the normal closing time of the observatory. To make the observatory more accessible and to give the viewing a more sophisticated feel in summer we trialed midnight viewings with a champagne supper. We targeted the last quarter moon which can never be seen at our regular sessions. Again, despite initial bad weather, there was an excellent response from the public. Later the sky cleared enough to observe the moon.

These midnight suppers have been repeated, sometimes with a special theme such as the Rebirth of the Sun held on our winter solstice in June. This particular event attracted 65 participants including a group of new age people who specialise in reciting chants to nature.

Midnight viewing attracts people who want exciting experiences with a difference. It gives an excellent exposure to astronomy for an audience who would not normally come to a regular viewing session.

3. Viewing the sky

3.1. *Light pollution*

For members of the public the highlight of an evening visit to Sydney Observatory is the chance to look through a large telescope. Although the observatory is in the centre of a large, light polluted city, the planets and many of the signposts of the southern sky are still visible. These include the double star Alpha Centauri, the globular cluster Omega Centauri and the famous open cluster Kappa Crucis.

Recently there have been some threats of further increase in sky brightness. One of these was from the floodlighting installed to illuminate a nearby office building. Each of its four 125 metre high concrete columns was lit from below by six powerful lights aimed almost directly upwards. The high reflectivity of the concrete meant that beams of light, extending high above the building, were reflected off the concrete columns.

We protested to the local city council who had authorised the installation and we were supported by an amateur astronomy group who has a long association with Sydney Observatory. The amateur astronomers organised a petition of over one thousand signatures. As well, we have tried to encourage the discussion of light pollution in the media. A January 1996 front page article in the *Sydney Morning Herald* concentrated on the slow blotting out of the stars of that icon of the southern sky, the Southern Cross. This article caused considerable interest and led to many radio interviews. As a result of these protests, the illuminations of the building has been reduced by a slight reaiming

FIGURE 2. The historic 29 cm refracting telescope.

of the lights and council has imposed a curfew of 10 pm. Possibly, what is even more important is that the city council is now aware of the problem of light pollution.

3.2. *Telescopes*

There are two main telescopes at the observatory. In the south dome there is a historic 29 cm refractor dating from 1874. In the other dome there is a computer controlled Meade LX200 16" (40 cm) Schmidt-Cassegrain telescope, which was installed in early 1995. The historic telescope is useful and provides an excellent way of demonstrating 19th century astronomy. However, the Meade has now become the preferred instrument for the public and the staff at the observatory.

The Meade is much easier to use as it is operated by push buttons instead of the physical manipulation necessary with the historic telescope. Its main advantage is its high pointing accuracy of a few minutes of arc. As a result any object aimed is within the field of view of the eyepiece. This overcomes the greatest difficulty at a bright sky site, which is to find the objects to view. If a telescope can be pointed in precisely the right direction, even quite faint objects are visible through it.

With the Meade telescope we are trialing the use of a CCD camera. A 500 mm focal length telescope has been installed on the main telescope and to this a CCD camera with small pixels has been attached. The field of view of the CCD camera approximately matches that through the eyepiece. In this way we can offer the public the direct look that they demand and yet show an image of the same object on a monitor. The image allows features of interest to be pointed out to visitors so that they know what to look for when they peer through the eyepiece.

4. Exhibitions

4.1. *Picturing the Sky: the Photographs of David Malin*

This is a small exhibition celebrating the work of the famous Australian astrophotographer, Dr David Malin. It comprises

• six large (1.0 × 1.2 metre) colour photographic prints. Each print has a descriptive label and a sky map indicating the position of the object.

• a display case with the lens and a photographic plate from a pioneering study of the southern sky made from Sydney Observatory over a century ago. There are modern prints made by David Malin from the original negatives.

• The highlight of the exhibition is a touch screen computer interactive which allows visitors to take a journey in a spaceship to a variety of spectacular southern sky objects. At each object the recorded voice of Dr Malin provides expert guidance.

4.2. *By the Light of the Southern Stars: Planned for 1997*

By the Light of the Southern Stars is a major new exhibition that will utilise the whole observatory building. It is being planned for the second half of 1997. The exhibition is being developed by a team comprising of ourselves, designers and other colleagues from the museum. Our initial plans for the exhibition have just been accepted by the museum management.

Preparation of these initial plans has been a long and arduous process. The difficulty was trying to fit in many conflicting requirements into a building that has small rooms, little display space and that was designed as a working observatory and residence, not as a museum.

The current plan provides a satisfactory balance between historical and modern material. It begins with the astronomy of the Australian Aboriginal people whose stories about the Sun, the Moon and the stars reflect their belief in the spiritual unity of Nature and human beings (Haynes et al, 1996). It continues with an overview of Australian astronomy from James Cook, whose trip to observe the 1769 Transit of Venus led to European settlement, to the major instruments of the present day such as the Australia Telescope. Other historical sections include one on the main research achievements of Sydney Observatory and another discussing the relationship between time and navigation that led to the establishment of Sydney Observatory.

One of two largest rooms in the building will be targeted mainly at school groups. Through a mix of computer-based interactives, objects and working models it will cover aspects of the school syllabus such as the planets and the solar system, stars and constellations and the daily and yearly cycles of the sky. The other large room will be dedicated to modern astronomy and will emphasise the work and results from some of the main Australian research establishments.

5. Discussion

At Sydney Observatory we aim to interest and educate the public in astronomy. Currently we largely rely on our educational programs to do this as there are only a limited number of exhibits on display. They include the new *Picturing the Sky* exhibition (section 4.1) and remnants from the *Hands on Astronomy* exhibition discussed in Lomb and Wilson, 1990. The new educational initiatives such as the open nights and the camp-ins have been highly successful with the participation of many members of the public, both adults and children.

The most important activity at a public observatory is looking through a telescope. Thus it is essential to maintain and if possible to improve the facilities for viewing the sky. A major improvement has come with the addition of a large modern telescope and the provision of the capability of showing a CCD image. Our experience with the camera so far indicates that visitors do get more out of their look through the telescope when explanations are first made with a CCD image of the same object. However, it seems to make little difference whether the image is live or recorded from a previous occasion.

We are attempting to prevent further degradation of our night sky by fighting irresponsible attempts at light pollution. This is not only crucial for Sydney Observatory itself and amateur astronomers living in the city but in the long term for research astronomy as well. Sky glow from Sydney is already visible 400 km away at Siding Spring, Australia's premier site for optical astronomy (McNally, 1996).

Much work still has to be done in preparing our new exhibition for 1997. We have to make sure that it is of interest and of relevance to our current and to potential visitors. It has to integrate with our educational and other activities. We are confident that *By the Light of the Southern Stars* will enhance the profile of the observatory and introduce many Australians and overseas visitors to astronomy and the southern sky.

REFERENCES

HAYNES, R., HAYNES, R., MALIN, D. & McGEE, R., 1996, *Explorers of the Southern Sky*, Cambridge, Cambridge University Press.

LOMB, N.R., 1996, Publ. Astron. Soc. Aust. **13**, 173-178.

LOMB, N.R. & WILSON, T., 1990, *The Teaching of Astronomy*, Cambridge, Cambridge University Press, IAU Colloquium no. 105, 357-359.

McNALLY, D., 1996, Q.J.R. astr. Soc. **37**, 129-151.

Developing Science Education and Outreach Partnerships at Research Institutions

By K. L. Dow

Harvard-Smithsonian Center for Astrophysics, Cambridge, MA, USA

Harvard Graduate School of Education, Cambridge, MA , USA

1. Introduction

Like many research institutions, the Harvard-Smithsonian Center for Astrophysics† (CfA), has been actively engaged in education and public outreach activities for many years. The Harvard University Department of Astronomy, the formal higher education arm of the CfA, offers an undergraduate concentration and a doctoral program. In our Science Education Department, educational researchers manage ten programs that address the needs of teachers and students (K-12 and college), through advanced technology, teacher enhancement programs, and the development of curriculum materials. The Editorial and Public Affairs Department offers several public lecture series, recorded sky information, children's nights, and runs the Whipple Observatory Visitors Center in Amado, AZ. In this environment of successful programs, the High Energy Astrophysics (HEA) division, one of seven research divisions at the CfA, has initiated, or partnered with other institutions, development of several new education and outreach programs. Some of these programs involve partnerships with the education community, but all of them have been initiated by and involve scientists.

Astronomical research in the HEA division is mainly focused on x-ray astronomy and the development of advanced x-ray instrumentation. Historically, involvement in education and outreach programs, including presenting public talks, school talks, the development of slides sets, and the publication of popular articles, has been informal. With management support, however, this long-lived tradition of informal education and outreach recently has coalesced into several formal programs that target three specific groups: college students, K-12 students and teachers, and the general public. This paper describes the projects which have been developed in the last three years, including the *SAO Summer Intern Program, Astronomy in Motion, Live from the Smithsonian, Science Information Infrastructure, An International Exploration of the Universe, Everyday Classroom Tools*, as well as some of the lessons we have learned about developing education and outreach programs in a research setting.

All of these formal programs complement the ongoing activities of individual researchers and are but a part of an overall effort in the HEA division to make our expertise, our facilities, and our excitement about science more accessible to teachers, students, and the general public. More information about these and other programs is available from the author or from the HEA division education and outreach homepage: http://hea-www.harvard.edu/scied/ .

---

† The Smithsonian Astrophysical Observatory (SAO) is a research bureau of the Smithsonian Institution. Research at the Harvard University Department of Astronomy is carried out at the Harvard College Observatory (HCO). Together, the two institutions comprise the Harvard-Smithsonian Center for Astrophysics.

2. Outreach Programs

2.1. *Programs for College Students and Teachers*

Smithsonian Astrophysical Observatory Summer Intern Program

The National Science Foundation (NSF) sponsored Research Experiences for Undergraduates (REU) program at the Smithsonian Astrophysical Observatory (SAO) in Cambridge, MA, provides talented undergraduates with an opportunity to actively participate in an astronomical research project. The primary goals of the *SAO Summer Intern Program* are to expose college students to an actual research environment in order to help them focus their career goals, to encourage their development as physical scientists, and to empower students to become full participants in a scientific and technological society. Students spend 10 weeks at the SAO beginning the second week of June and are recruited from US colleges and universities that offer programs in the physical sciences.

While this program is run by the HEA division, students may work with a senior scientist from any research division. Students also are assigned to a graduate student or post-doctoral fellow "mentor" during their appointment. We believe that the experience of assisting a senior scientist in the supervision of a participant in the program may encourage these "mentors" to work with undergraduates later in their careers.

Students engage in a research program in a variety of areas, including observational and theoretical astronomy, extragalactic and galactic astronomy, interstellar medium and star formation, laboratory astrophysics, supernovae and supernova remnants, and planetary science. While they spend the majority of their time working on a research project, regular lectures and meetings also are held throughout the summer. These gatherings are designed to further expose students to the wide breadth of modern astrophysical research. Other program activities include local field trips, a colloquia series, discussion evenings, and the opportunity to present research results at a professional meeting. The program concludes with an "Intern Symposium" where the participants present the results of their work. The overall structure of the program is designed to provide students with challenges of increasing difficulty. We feel that as students master more difficult tasks, they achieve greater self-reliance as researchers. The *SAO Summer Intern Program* is one of several similar programs offered at research institutions across the US.

Project JOVE

As a mentor institution for NASA's *Project JOVE*, SAO offers research opportunities for faculty members that cover a wide range of possible astrophysics topics. The primary goal of Project JOVE is to establish three year joint venture partnerships between NASA and institutions of higher education which have had little or no involvement in the Nation's space program. Four faculty members from three colleges have recently participated in *Project JOVE* at SAO in the HEA division. The experience includes a 10 week research internship during the first summer. Current areas of interest for some HEA staff members particularly interested in Project JOVE are outlined at the following url: http://hea-www.harvard.edu/scied/jove/jove_sao.html.

2.2. *Programs for K-12 Students and Teachers*

Science Information Infrastructure

The *Science Information Infrastructure (SII)* is a consortium of science centers and museums linked together to bring current research topics to the public. The HEA division at SAO, in partnership with the Center for EUV Astrophysics, is working to create a network of on-line resources created by teams of teachers, science museum personnel, scientists and technical staff. Thirteen curriculum modules have been created so far and can be accessed from World Wide Web sites at the Exploratorium, the Lawrence Hall of Science, the National Air and Space Museum, and the Science Museum of Virginia. All of the modules can be found on the Center for EUV Astrophysics *SII* homepage (http://www.cea.berkeley.edu/Education/sii/sii_sii.html). The resources being created are intended for use by anyone, but especially in the classroom setting as well as for exploratory learning.

In the *SII* model, the small, local focus teams of teachers and individuals from the science museums and partner research institutions create on-line curriculum for science and math classes based on NASA data and related information and networked science research services. The life cycle of the materials and on-line tools involves testing in the classroom, evaluation, modification and revision and further testing in different environments. The *SII* is intended to build a sustainable framework for this process and accommodate new and experimental activities that use networked technology in innovative ways. *SII* is a three year program funded by NASA's Public Use of Remote Sensing Data Program.

An International Exploration of the Universe

Two Russian and two Massachusetts High Schools are working with an SAO scientist in Cambridge, MA and an Institute for Space Research scientist in Moscow to study the multi- wavelength aspects of astronomy. As part of the theme of multi-wavelength astronomy, the scientists introduce the "missing mass"question through the analysis of clusters of galaxies using NASA data, study galaxy redshifts and the expanding universe, and utilize public-access robotic telescopes. Astronomical observing using small telescopes complements the computer- based activities to view "ideal"multi-wavelength objects such as the Orion Nebula, the optical counterpart to the probable black hole Cygnus X-1, globular clusters, open clusters, and M31. Observing nights are also offered to help forge links between the communities and the schools. This project fosters a wide range of communications between different school systems and promotes international understanding. *An International Exploration of the Universe* is funded by a grant from the NASA Initiative to Develop Education through Astronomy program.

City Stars: A Teacher Workshop on Astronomy

In collaboration with Charles Hayden Planetarium staff, SAO scientists developed and offer two annual astronomy workshops, *City Stars*, for Boston area middle and high school teachers at the Boston Museum of Science. The workshops cover basic astronomy concepts and demonstrations of inexpensive hands-on astronomy activities. Participants receive a teacher's packet consisting of a workbook with descriptive guidesheets for each activity, a cassette tape with astronomy songs, solar system slide set, spectroscope, disposable outdoor camera, sundial template, star finder template, Styrofoam ball and pencil, Saturn poster, and Museum of Science brochures. Workshop participants complete a formal evaluation of the in-class experience. *City Stars* is funded by a grant from the NASA Initiative to Develop Education through Astronomy program.

Everyday Classroom Tools

Everyday Classroom Tools is a three year program funded by NASA's Public Use of Remote Sensing Data Program. HEA division scientists and programmers are working with two Massachusetts elementary schools to develop an inquiry-based astronomy curriculum for elementary school students. The use of technology to learn about the world is another major goal of the project. Macintosh computers with CD-ROM capabilities and direct access to the Internet connect all the participating teachers and permits communication among the teachers and the program personnel. *Everyday Classroom Tools* staff also have developed a series of hands-on astronomy activities called "Eyes on the Sky, Feet on the Ground,"partially funded by a Smithsonian Outreach grant. These materials are available on-line at http://hea-www.harvard.edu/ECT/the_book/index.html.

2.3. *Programs for the General Public*

Astronomy in Motion

English and Spanish posters on "The Sun"and "Jupiter"were displayed on the Massachusetts Bay Transit Authority (MBTA) subway in late 1995 and early 1996. These posters use eye-catching graphics to heighten interest and appreciation of astronomy among the general public. The posters were distributed to Boston area schools, along with introductory material on the Sun and Jupiter, a list of related books, World Wide Web sites, and classroom activities. We are presently working to expand the project to also include transit systems in New York City, San Francisco, and Washington, DC. The *Astronomy in Motion* pilot project was funded by the National Aeronautics and Space Administration.

Live from the Smithsonian

Science museums around the country often enlist astronomers to occupy an "Ask-an-Astronomer"booth on Astronomy Day. While these booths are tremendously popular with museum visitors of all ages, they are short lived. Recent developments in video conferencing hardware and software now provide the opportunity to enhance this activity, and to involve greater numbers of astronomers.

In the past, teleconferencing with scientists through satellite downlinks has proven immensely popular with the museum-going public. At the Boston Museum of Science, for example, students and teachers followed oceanographer Bob Ballard and colleagues on a marine archaeological expedition, and chatted with Russian cosmonauts aboard Mir. *Live from the Smithsonian* will expand the capabilities of teleconferencing by encouraging the kind of in-depth, iterative discussion that leads to deeper understanding.

As a test of the application of video conferencing in a typical museum environment, we will place a video conference kiosk on the floor of the "Stars"Gallery at the National Air and Space Museum (NASM) that will provide Museum visitors with the opportunity to "Ask-a- Smithsonian-Astrophysical-Observatory-Astronomer"questions during regularly scheduled sessions. To prepare NASM visitors to talk with astronomers, the kiosk will also be set up as a touch-screen to provide the visitor with an introduction to the astronomical research presently conducted at the Smithsonian Institution. Screens will be available to describe the Smithsonian Astrophysical Observatory programs, where the Observatory is located, who will be available that week for discussions, and what their specialties are. The kiosk will also describe representative research areas – the mysteries – along with "frequently asked questions"from the public.

While a few other science centers and museums (Exploratorium, Oregon Museum of Science and Technology, Ithaca Science Center), currently have video conferencing capabilities, this will be the first time that this new technology will be incorporated into an exhibit. It will be a test for considering the use of video conferencing and remote maintenance of updatable information in Smithsonian galleries that deal with contemporary issues in science.

The main educational goals of this project are: encourage general public and K-12 student interest in astronomy, involve at least 60 astronomers in a three month public outreach activity, show that astronomers come from many backgrounds, promote greater awareness of the research side of the Smithsonian Institution and its activities beyond "The Mall,"and to form a partnership (between astronomers, educational researchers, exhibit designers, curators, teachers, students, and the general public) that may lead to future collaborations. The main museum-based goals of this project are: explore efficient and easily updatable means of presenting contemporary information about astronomy and space research to the public in a major and very demanding museum environment, and to explore ways to promote greater contact between the research side of the Smithsonian and the Smithsonian's patrons. The *Live from the Smithsonian* project will open in the Stars Gallery in the Spring of 1997.

3. Conclusion

The marriage of science and education at the Harvard-Smithsonian Center for Astrophysics is similar to all such unions. Each partner is distinct and brings strengths, weaknesses, and idiosyncrasies to the relationship. Further, the partners need not lose their individuality by working together and may even produce some interesting offspring.

Recall the Fred Astaire and Ginger Rogers film *Shall We Dance*. In this film, think of Fred as science and Ginger as education. Fred is a famous ballet dancer who wishes he had the freedom to be a tap dancer. Ginger is a famous tap dancer who wishes she had the respect that ballet dancers command. Fred and Ginger only know each other by reputation, but Fred vows to marry Ginger before he even meets her. Then through a series of events, familiar to any fan of 1930's movies, it is incorrectly announced in the paper that Fred and Ginger are already married. In the end, however, love conquers all and they do marry and live happily ever after.

It is my hope that science and education, like partners in any marriage, will come to see that the longer they are together, the more they realize how much they have in common. The marriage of science and education will undoubtedly lead to the birth of new ideas, new approaches to developing and disseminating curriculum, and new methods of conveying the excitement of science to students and the general public, in ways that we cannot foresee today. Finally, those who think that Fred was a genius, and that this somehow diminishes Ginger's contribution to their films, should recall that Ginger did everything Fred did, only backward and in high heels.

Literature for Amateur Astronomers

By L.S. Kudashkina & I.L.Andronov

Department of Astronomy, Odessa State University
T.G.Shevchenko Park, Odessa 270014 Ukraine

The experience of working with amateur astronomers in the countries of the Commonwealth of Independent States and in Ukraine shows a noticeable lack of literature, especially educational and methodological. The amateurs, possessing an observational base, do not know what best to observe at a given moment, and those, who are not yet ready for practical work in astronomy, do not know how to be prepared.

A series of brochures under the title "The Atlas of Amateur Astronomy" has been prepared, which pursues the purpose of delivering to amateurs a minimum of the necessary information on the following items:

1. Popular scientific reviews (lectures) on various directions in astronomy and astrophysics.

2. Methodological articles on the bases of observations and their processing.

3. Programs of observations, finding charts of variable stars, short information on comets, meteor showers etc.

4. Help material (tables, ephemerides, items of information from the General Catalogue of Variable Stars etc.).

5. Observations made by the amateurs themselves.

Five issues of "The Atlas of Amateur Astronomy" have been published. Together they contain information on about 60 objects, for which finding charts and comparison stars are given.

Part I contains the introductory articles, description of a structure of the atlas, which is repeated in the other issues, finding charts for 20 variable stars, recommendations for observations and the table of Julian dates from 1980 till 1995 (the atlas was issued in 1990).

Part II contains the first lecture from a cycle "Variety in the world of variable stars" on a theme "Long-period variables". In this part the finding charts with comparison stars for 30 variables are given. Instruction is also given here on how to make observations and how to reduce moments of time to the center of the Sun for short-period stars.

Part III contains a lecture on a theme "Binary stars" (general and theoretical aspects). In this part only two finding charts (for the stars R Tri and RU Cam) are shown, but the standard NPS (North Polar Sequence) is given and a lot of help material, requested by amateur astronomers, e.g. the names of variable stars, Greek and Latin alphabets, list of constellations. Also, in the "short communications" there are placed observations of the Mira-type star R Aql and a drawing of sunspots.

Part IV contains a continuation of the lecture about binary stars, namely "Determination of the physical characteristics of eclipsing binary stars". This issue also contains information about the 2nd Congress of the Astronomical Society of the Former Soviet Union in Moscow. The most interesting reports are briefly described. Also in part IV are placed the electrophotometric standards IC 4665 and Coma Berenices. A finding chart of the Mira-type star R Cas is given, and in the short communications - observations of the bright bolide above Saratov city.

Part V was issued in 1996 after a long gap (parts II-IV are dated 1991). It contains a lecture from the same cycle on a theme "Supernova stars". It is opened by an article about the Crab Nebula. The detailed information about the bright supernovae 1987 and 1993 is also included. Finding charts for 4 variable stars are placed. In the short communications, observations of the supernova 1993 in M81 are indicated.

Another series of publications has been prepared by I.L.Andronov and L.L.Chinarova for the amateurs who are members of the section of Astronomy of the Small Academy of Sciences "Prometheus" working at the Odessa Regional Station for Young technicians. This "Academy" gathers scholars interested in different branches of science - astronomy, mathematics, physics, chemistry, biology, ecology, PC programming, air plane- and auto-modeling, humanitarian sciences et al. It works under the supervision of the associate and full professors from the Odessa State University and sometimes other institutes. Every year competition conferences are organized starting in individual schools, progressing through regional heats with the final covering the whole of Odessa State. The winners are recommended to all-Ukrainian and international conferences. The best scientific works made by scholars are being published in the annual volume of the proceedings booklet "Pervye shagi" ("First steps"). These publications are of course not so important for science as the professional journals, but they are really first steps and some of the authors improve these contributions and resubmit articles to more "serious" journals. In the section of astronomy this takes place usually if a variable star is observed visually or photographically. One may note that the Sky Patrol plate collection of the Astronomical Observatory of the Odessa State University contains about 100,000 negatives and thus is the third in the world after Harvard (USA) and Sonneberg (Germany) and attracts attention of professionals and amateurs from many countries. Such articles were published in "Variable Stars", "Astronomical Circular" (Russia), Information Bulletin on Variable Stars (Hungary), Bulletin de l'AFOEV (France), "The Astronomer" (United Kingdom), "Variable Star Net (VSNET)" (Japan), "Odessa Astronomical Publications" (Ukraine). Among the most active observers making not only first but many other subsequent steps are A.V.Halevin, V.I.Marsakova, Yu.V.Beletskij, K.A.Antoniuk.

For amateurs, summer schools are organized, and also a "correspondence course" which was the only one in the former Soviet Union and was initiated by Professor V. P. Tsessevich (1907-1983) in 1977. Now it is the only one in the Ukraine. In Russia a similar course started in Ekaterinburg. For the amateurs and students several years ago a series of booklets was published in Odessa by I.L.Andronov. Among them: "Visual and photographic observations of variable stars" (1991, 84pp.), "Structure and evolution of variable stars" (part 1, 1991, 86pp, Part 2, 1992, co-author L.L.Chinarova) and 16 smaller ones.

Some popular articles by I.L.Andronov were published in the journals "Priroda" (Moscow, 1987), "Die Sterne" (Leipzig, 1988, 1989), "L'Astronomia" (Milan, 1995), "Koszmos" (Prague, 1996).

Popular articles on various fields of Astronomy in the Ukrainian language are published in "Short astronomical calendar" issued in Kiev.

We hope that our attempts contribute a little to the popularization of astronomy in our country and will be happy if this could help somebody "to find his own star".

Desktop Space Exploration

By M. Martin-Smith[1] & R. Buckland[2]

[1] General Practitioner, Hull, UK

[2] Department of Design and Innovation, The Open University, UK

The Humble Space Telescope project aims to launch a small space telescope for educational and recreational purposes, in time for the New Millennium.

The arrival of the 3rd Millennium, accompanied in the United Kingdom by a Millennium Commission distributing 250 million per year of National Lottery funds for good causes and imaginative projects which would otherwise require direct funding by the taxpayer, provides a unique opportunity to design, build and operate a small but capable version of the pioneering Hubble Space Telescope.

In July 1994, a leading British newspaper with a long history of covering developments in science, launched a competition for members of the public to propose science projects to be funded by the Millennium Commission. The idea of a small satellite telescope, fitted with a CCD detector package was submitted by Dr. Martin-Smith, and won a share of the top prize. Meanwhile, Rodney Buckland, a Trustee of the National Science Centre project, took up the idea as an ideal new field site for the Centre, and has become its Project Manager.

It is well established that specialised and initially-expensive technologies - for example Schmidt-Cassegrain optics, CCD cameras, computers and the Internet - began as the advanced tools of professionals, and in time become accessible to amateurs, educators, and the public, for learning and recreation. The advent of small capable low cost satellites, as developed by Professor Martin Sweeting of Surrey University, simpler more flexible management structures promoted by Rodney Buckland at the Open University , "off the shelf" telescopes, such as Questar (a model of which is already space qualified) the phenomenal growth of the Internet, and the expansion of the small satellite launcher market combine to provide an excellent window of opportunity for placing access to space within the reach of desktop explorers at school and armchair ones at home.

Remote control of observatories, whether ground based, or in Earth orbit, is an established fact. Humble offers all this to amateurs and educators at a cost of 1/300th that of Hubble, which is used predominantly by professionals .

The Millennium Commission, looking for visionary projects to celebrate the New Millennium, thus provides the chance to launch a popular educational space and astronomy project without calling on the usual sources of public funds.

1. The Astronomy Option

In the UK as elsewhere, astronomy forms a part of the National Curriculum and yet students' observations are hampered by many problems. As in all industrialized countries, enjoyment of the night sky, our legitimate heritage, as well as the ability of astronomers both amateur and professional to carry out serial observations and measurements, are hampered by increasing light pollution and built-up skylines. Terrestrial weather, as any Briton will tell you, hardly helps. Furthermore, for school students already heavily loaded with study for their national curricula, astronomy, while an option, requires out-of-hours work, being mostly a nocturnal pursuit, while school classes are in the daytime. High quality amateur telescopes, equipped with CCD detectors and remote control software and facilities have been available off the shelf for the past few years.

The round-the-clock, all-weather, high-quality, multispectral images from Humble will allow serious contributions to be made by students and amateur astronomers to a science which has large gaps left behind by the advancing wave-front of discovery. Again, it is a truism that when a problem meets an opportunity, great advances result. The threat and opportunity posed by Near Earth asteroids is perhaps a most pertinent example, forcing us to choose between annihilation and building a space faring civilization. It is generally accepted that young people find education in astronomy and space science intellectually exciting and challenging - more so, perhaps, than many other fields. Many countries, for example India and America, consciously use space programmes to stimulate pupils to take up science and technology for future careers. In the UK we must do the same.

2. Astronomy On Line

A taste of things to come is being promoted by The European Association for Astronomy Education (1). In November 1996, schools across Europe will be offered access, via the Internet, to ground-based observatories and other sources of astronomical data in a programme called Astronomy On-Line.

3. The National Science Centre (NSC)

The National Science Centre is a project to develop a national centre for public understanding of science. It will attract one million visitors a year who will be able to see real satellite operations and data acquisition and processing in action. It is also planned that school pupils will be able to participate in the Pupil Researcher Initiative programme operated by NSC partner Sheffield Hallam University (2), whereby pupils will carry out research programmes using Humble, write papers for a Humble journal, and present results to a Humble Conference. The British National Space Centre has agreed to circulate British schools nationwide with information on Humble and an invitation to participate in the user requirements capture process.

4. Small Mission Architects and Builders

Surrey Satellite Technology Ltd. (3), a world-leading group of small space mission architects and builders, has pronounced itself keen and capable of designing and fabricating the Humble spacecraft as funds become available, and following a feasibility study projected for September 1996. In addition, the Mullard Space Science Laboratory, Satellites International Ltd, D.R.A Farnborough, Birmingham University and the University of Kent have signified interest in providing instrumentation and other facilities for Humble.

The potential exists to equip Humble with multi- spectral capabilities, allowing observations in ultra-violet and near infra-red wavelengths from above the atmosphere. This would give non-professional users access to hitherto unfamiliar views of celestial objects, and open up the advantages of space-based astronomy to new groups.

The growth of the small launcher market over the next few years enhances the affordability and attractiveness of the project to potential supporters.

5. Faster, Cheaper, Better

The Atmospheric and Mesospheric Particle Transfer Experiment (AMPTE) went from concept to operational status in 3 years, as did the Clementine mission in 1994. The NEAR spacecraft, likewise achieved a three-year gestation period, within budget. It is against this background that a launch for Humble, before the year 2000, at a cost of 6-8 million is a reasonable proposition.

Very preliminary soundings of possible launchers have been made. Depending on the results of feasibility studies, possibilities include Ariane and Orbital Sciences Corporation's Pegasus launcher.

The National Science Centre has the management framework for success. The Centre's Trustees have considerable experience in the management of scientific and educational programmes, and plan to handle data collection and distribution from the Centre. Work is proceeding on the architectural and financial aspects of the NSC, with a view to submitting a detailed technical and business review in the autumn.

The growth of the Information Super-highway allows effective distribution of images and data from a space-based observatory - Hubble has shown the way. Humble will provide for desktop exploration in schools and universities, and armchair exploration by amateur astronomers. Many schools and institutions, as well as an increasing number of homes, have access to the Information Superhighway, with the prospect of considerable growth by the proposed launch date of Humble in late 1999. Indeed, the likely merging of large screen flat television sets and Internet linked computers in the same console makes Armchair Space Exploration an inevitable development, as shown by plans for emplacing lunar rover vehicles equipped with telepresence for entertainment as well as exploration purposes.

6. Space for Education

Ready access by the public to Humble imagery at the National Science Centre, on the Internet, or on participating television science programmes , will enhance peoples' appreciation of the beauties and wonders of the Heavens, as well as advance the cause of astronomical education.

In addition to giving schools and the public the opportunity to view the heavens in comfort and do real research, the Humble Space Telescope programme will allow participants to gain experience in satellite management, data acquisition, processing, and presentation of results to peer groups. Thus, a relatively small and inexpensive space programme, realisable in three years, has the potential to give millions of people access to Space.

7. Education for Space

The variety, quantity, and quality of observations made possible by Humble to people from many different backgrounds and walks of life will allow Humble to become more than just a small satellite in Earth orbit: all being well, it will become a Millennial institution providing many young people with the inspiration to become space professionals and creating considerable public awareness of benefits arising from space activities which has not been forthcoming in Britain.

The late Professor Hermann Oberth always hoped that space exploration and development would be carried out with popular enthusiasm. The authors hope that, in addition to its educational potential, the Humble project will foster public support for enhanced astronomical and space activities. In allowing students to conduct on-going studies of our neighbouring planets with high quality imagery, it is hoped that that Humble will inspire members of the rising generation to consider becoming space professionals and explorers.

The National Science Centre has authorised a user requirements capture process to be carried out starting in July 1996. It is expected that a full feasibility study will be carried out in September, as part of the National Science Centre's progression through the DAR (Detailed Appraisal Review) process for Millennium Commission funding.

British capabilities in small satellite mission execution are well established. Surrey University and its commercial affiliate Surrey Satellite Technology Ltd, have built 12 satellites to date, each within 2 years. Humble is expected to weigh about 250 kilogrammes, and to cost about 6-8 million. There will be several opportunities for demonstrating and using British space technology. Possible candidates include ion propulsion from the DRA and passive radiative cooling from the Open University.

8. Publicity

It is planned that Humble's potential users become aware of the project now, and begin considering its possible utilization . To this end, articles and notifications of the project have already been widely circulated. A site has been set up on the World Wide Web (3), while readers of astronomy magazines and members of societies have been notified and asked to contribute to the definition of user requirements. The international space community will also be introduced to Humble at the annual IAF gathering in October 1996 (4).

Local radio stations and several newspapers have carried features on Humble, resulting in considerable publicity. Several schools and universities have been contacted by the authors, as have conferences of physics teachers and astronomy educators in Britain and Europe. International magazines have carried articles on the project this year by one of the authors. The UK Secretary of State for Education, and her Opposition Shadows, have been acquainted with the project and its educational potential , and contact has been made with the London and Armagh Planetaria.

Many potential users are now aware of the project, and developments on the funding and feasibility studies should hasten participation. The authors plan a Humble Space Telescope Working Committee to co-ordinate the development of a user community. Through the astronomical and educational communities, the authors hope to realise a new valuable European institution, with the potential to raise seeds on Earth and beyond well into the New Millennium.

REFERENCES

1. The European Association for Astronomy Education web site is at:
   http://www.eso.org/astronomyonline/
2. The UK Pupil Researcher Initiatives web site is at:
   http://www.shu.ac.uk/schools/sci/pri/index.htm
3. A summary of Surrey Satellite Technology Ltd missions can be found at:
   http://www.ee.surrey.ac.uk/EE/CSER/UOSAT/missions/
4. Desktop Space Exploration, Dr. Michael Martin-Smith and Rodney Buckland, IAF-96-Q.1.08, presented at the 47th International Astronautical Congress, Beijing, China in October 1996.

Teaching Astronomy in the Schools

Current Trends in European Astronomy Education

By Richard M. West

European Southern Observatory, Karl-Schwarzschild- Strasse 2, D-85748 Garching, Germany

1. Introduction

1.1 A European Initiative

Astronomy knows no geographical borders – the sky is the same over all countries. However, while professional astronomers have long established bi- and multilateral collaborations, many of which take place under the auspices of the IAU, few similar schemes exist within astronomy education.

Now, following the establishment in 1995 of the European Association for Astronomy Education (EAAE), this situation is about to change on that continent. This new association offers an efficient platform for astronomy educators at various levels - in particular at the approx. 7000 secondary schools in this geographical area – to interact in all related matters, e.g. curricula, all kinds of teaching materials, student exchanges and other events. Together with the European Commission and some of the professional institutes, EAAE is now planning a major, international event in November 1996.

1.1. *Geographical disparity*

Astronomy is taught at secondary school level in most European countries, but there are enormous differences from area to area. In some places astronomy plays an important role within the physics curriculum, in other places CCD-equipped telescopes of medium size are available for observational studies, and in some places astronomy is barely visible or the connected matters are spread over many different subjects. With some notable exceptions, it cannot be said that the teaching of astronomy in Europe is satisfactory, and it would appear that the great potential inherent in this science with so many connections and of such an outspoken interdisciplinary nature is very poorly exploited.

The reasons for this are numerous and vary from country to country. If one common factor is involved at all, it may be the personal devotion by some teachers with an astronomical background. In some European regions, teachers of physics may have had comparatively little contact with modern astronomy at the time of their graduation; in others, advanced astronomy courses are obligatory for physics students at the universities.

Another point which is apparent when surveying the teaching of astronomy in Europe is the comparative lack of interaction between teachers in the various countries. This is due, partly to the different school systems, but also to language problems. However, the great variety and enormous experience of teaching astronomical subjects which is available on this continent is a very valuable asset that ought to be exploited much more efficiently than has been the case so far.

Now, however, there is a definite trend towards internationalisation of astronomy teaching in Europe. This is not an entirely new development, but a solid foundation is presently being created which will ensure that future interaction among educators will become much more vivid and useful, for the benefit of all involved and, not the least, to the advantage of astronomy in general.

2. Establishment of the European Association for Astronomy Education

Some years ago, Commissioner Ruberti of the European Commission (in a position equivalent to the Minister of Science and Technology of Europe) took the excellent initiative to organize the "European Week for Scientific Culture" aimed at bringing new developments in this field closer to the European public. The first such Programme was organized in 1993 and the participants were mainly science-related institutions from many parts of Europe.

The European Southern Observatory used this welcome opportunity to organise a competition for secondary school students in 17 European countries; the national winners were invited to visit ESO and were taken to the La Silla observatory to perform observations with some of the professional telescopes there (West, 1994).

This Programme being an outstanding success, it was decided to organise an event for their teachers in 1994. ESO thus invited 100 physics teachers with interest in astronomy from 18 European countries to a one-week workshop at the ESO Headquarters in Garching, Germany. This was the first time a meeting of this type had ever been held and it proved extremely productive, especially for the exchange of information about the teaching of astronomy in different countries. There is no doubt, however, that the most important outcome was the decision by the participants to establish the European Association for Astronomy Education (EAAE) which would henceforth provide the platform for the attempted unification of astronomy teaching all over Europe (West 1995).

The EAAE held its first General Assembly in Athens in November 1995, under the auspices of the Third European Week for Scientific and Technological Culture. On this occasion, the statutes and by-laws were drafted, various procedures for the efficient running of the Association were defined and the first Executive Council was elected, together with seven Working Groups. The Association has since become officially registered in Germany as a non-profit organisation and national representatives in the individual countries are now busy setting up local structures. The membership of EAAE is rapidly increasing and many activities are planned for the coming years. These include in-service training for teachers, exchange of students, summer schools at observatories with the opportunity to perform observations etc., adaptation of courses with input from several countries etc.

There is little doubt that the EAAE is developing into a very useful tool for efficient teaching of astronomy. By bringing teachers of different nationalities and from different school systems closer to each other, new horizons have opened for many. This process is providing new stimuli, not only within the teaching curricula, but also because of the personal contacts established with colleagues in other countries.

Since the beginnings of the EAAE, ESO has supported this effort in various ways, by advice and with materials. For instance, the first meeting of the EAAE Executive Council with the chairpersons of Working Groups and all national representatives took place at the ESO Headquarters in June 1996.

3. Astronomy On-Line

3.1. *The goals*

Astronomy On-Line represents the first large-scale attempt in the world to bring together pupils from all over one continent (and, to a limited extent, from others as well) to explore challenging scientific questions, using modern communication tools both for obtaining and for communicating information.

The programme is a collaboration between the EAAE and ESO. It is sponsored by the European Union (EU) via the European Commission (EC) through its Directorates XII (Science and Technology) and XXII (Education). The programme will take place in conjunction with the 4th European Week for Scientific and Technological Culture.

The active phase of Astronomy On-Line will start on October 1, 1996, and reach a climax in the period November 18 - 22, 1996.

In Astronomy On-Line, a large number of school students and their teachers, mostly from Europe but also from other continents, together with professional and amateur astronomers and others interested in astronomy, will participate in a unique experience that makes intensive use of the vast possibilities of the World-Wide-Web (WWW). Although the exact number of participants will not be known until the middle of November, it is expected to run into the thousands, possibly many more. The size and scope of Astronomy On-Line will contribute to make it an important media event.

Through the WWW, the participants will "meet" in a "marketplace" where a number of different "shops" will be available, each of which will tempt them with a number of exciting and educational "events", carefully prepared to cater for different age groups, from 12 years upwards. The events will cover a wide spectrum of activities, some of which will be timed to ensure the progression of this programme through its three main phases (see below). It is all there: from simple, introductory astronomy class projects to the most advanced on-line access to some of the world's best telescopes, from discussions with peer groups to on-line encounters with leading professionals.

In fact, Astronomy On-Line will be the first, internationally organised and fully-structured programme which offers a very large number of students the possibility to familiarize themselves with the use of modern communication tools, unequalled possibilities for fruitful international communication, and at the same time to learn much about the science and technology of astronomy, including the scientific methods. Moreover, they will be able to actively contribute to co-ordinated sub-programmes that will draw on the combined forces and ingenuity of participants from all areas.

Astronomy On-Line is not just about "trivial" data retrieval or about enhancing the seductive drive into virtual reality. For example, through the possibility of designing and conducting a real observing run on professional telescopes, it offers the opportunity of hands-on experience to students even in the most remote areas. In addition, they will be able to "meet" some of the professional astronomers at the participating observatories on the WWW and discuss subjects of mutual interest.

Apart from its astronomical component and the opportunity for students to familiarize themselves with one area of the natural sciences, a particularly fascinating perspective of the project is that it significantly contributes to an understanding of the usefulness and limitations of the communication technologies that will increasingly govern all our daily lives.

There are many other side benefits, of course, such as stimulating schools to go on-line and prompting international cooperation among young people. Another important aspect is that the programme will lead to natural involvement of business and industrial partners in the local areas of the participating groups. Also its unique character and international implications will be very inviting for extensive media coverage, both in human and scientific/technological terms.

3.2. *Steering Committees*

A preparatory meeting of the Executive Council of the EAAE and National Representatives in 17 countries was held in Garching (Germany) on June 15 - 16, 1996. An International Steering Committee (ISC) of seven persons was established for the programme. The ISC is responsible for the planning of the main activities in Astronomy

On-Line. At the same time, it was decided that the EAAE National Representatives will set up National Steering Committees (NSC) which will coordinate the Programme in their respective countries.

The NSCs consist of educators, scientists (many with an interest in dissemination of science), representatives of leading planetaria, internet specialists and as far as possible also representatives from sponsors (internet providers, PC hardware suppliers etc.). Most NSCs have established good connections with their National Ministries of Education.

The ISC has prepared a provisional programme description together with basic guidelines that will serve to coordinate the work of the NSCs. They in turn will provide organisational and technical information (e.g. computer and communication link specifications) to the participating groups, sponsors and supporters of the programme.

This information is available on the two central computer nodes of the programme and will be continuously updated as the elements are specified in increasing detail. The Astronomy On-Line WWW Homepages can be reached under:
http://www.eso.org/astronomyonline/ and
http://www.algonet.se/~sirius/eaae.htm

3.3. *Participation*

The first task of the NSCs was to issue a call for participation to interested schools, astronomy clubs and other astronomy-interested groups in their respective countries. The deadline for registration was first set as October 1, 1996, i.e. the day when the active, first phase of the Programme will start and some sub-programmes will become accessible on the WWW. However, in view of difficulties for many schools to achieve internet connectivity before that date, the deadline was later shifted to mid-November.

The participating groups consist of at least one teacher and his/her students or of one or more astronomy enthusiasts. Each group must have access to the WWW. A summary of requirements for access to the WWW was published. If access is not yet available at the school, this has been arranged by "sponsors" in the local area. These are planetaria, science institutes, business undertakings (e.g. in the field of electronics, computers, communication, etc.), industrial firms or private benefactors.

All communication via the WWW will take place in English. Some groups may decide to include participants with particular WWW and language qualifications. The local language may of course be used at the national level.

Only registered groups can participate actively. Groups must register with their NSC under a name or a designation and provide basic information about themselves (who they are, information about their school etc.).

3.4. *Computer Nodes*

The NSCs are responsible for the establishment of national computer nodes for the Programme. In many places, this has been done in collaboration with a national university/observatory or with a (sponsoring) internet provider. The National Astronomy On-Line Home Pages have two main features: 1) A national component, dedicated to the activities in the specific country, and 2) A mirror of the "ESO Astronomy On-Line Home Page" which acts as the central "European Homepage".

The preparations for the establishment of these Homepages began as early as possible and most will be available by the end of September 1996.

In parallel, ESO has provided a limited number of VHS video tapes (PAL/English) which can be used to promote the Astronomy On-Line programme. Beta-SP copies will also be available on request for interested TV channels. ESO has also produced a colour poster to be used by the NSCs in schools, planetaria etc.

4. The Astronomy On-Line Concept

The Astronomy On-Line Programme is based on the concept of a WWW "market-place" with "shops" that will be consulted by the participants. These shops will open at specified times, some from the beginning on October 1, 1996, and others later. They will display a variety of "goods" (activities) at different levels of complexity in order to attract participants of different age groups. At this moment, the following shops are foreseen:

4.1. *General information*

Information about the Programme as such and the overall schedule. Guidelines and Help facilities. List of participating groups. Links to all related Web sites.

4.2. *Collaborative projects*

Projects which require observations by many groups, all over the continent, thereby leading to "joint" results. For instance, observations of the Moon and Sun, auroral activity and meteors; parallax measurements of nearby objects (probably of the asteroid Toutatis when it passes near the Earth); degree of light pollution in cities, etc.

4.3. *Astronomical observations*

Preparation of a real observing programme, to be submitted and executed by telescopes at participating observatories. The data obtained will be transmitted from the telescopes to the groups via the Web. Reduction and evaluation of data by the groups. Publication of the results on the Web before the end of Phase 3 (see below).

4.4. *Astronomical software*

Use of a variety of general astronomical software which can also be taken over for future use at the schools. For instance, ephemerides and orbits, eclipse predictions, etc.

4.5. *Use of astronomical data on the WWW*

Retrieval of data, available on the WWW at different sites – images, texts, astronomical data in observatory archives, etc. Thoughtful combination of data will make specific projects possible, ranging from the preparation of an exhibition of astronomical imagery at the school to the solution of more complex tasks. Links to other useful Websites, all over the world. This shop will also include educational "Treasure Hunts" on the Web.

4.6. *Prepared exercises (Try your skills)*

A variety of prepared, astronomical exercises of different levels to be solved by the groups.

4.7. *Talk to the professionals*

Talk over the WWW to professional astronomers and educators at participating institutes. Question and Answer sessions, open to all.

4.8. *Group communication*

Talk over the WWW to other Participating Groups about astronomy and other subjects.

4.9. *Newspaper*

Publication on the WWW of the results of the various activities, etc. Announcements about the Programme and its progress.

5. The schedule

Astronomy On-Line will be divided into three phases, lasting from early October to November 22, 1996, and reflecting the gradual development of the associated activities. During this period, a number of special projects will take place, for instance in connection with the Partial Solar Eclipse on October 12, and the amount of information on the Astronomy On-Line Webpages will continue to grow.

During all three phases the ESO Astronomy On-Line Home Page, as well as the National Homepages, will be the anchors and they will serve as the media for international exchange among the groups.

5.1. *Phase 1 (October 1 - November 17, 1996)*

Phase 1 of the actual project will last about six weeks, starting on October 1st, and continuing until about a week before the start of the 4th European Week for Scientific and Technological Culture. It will mark the beginning of the project and lead up to the subsequent "hot phases".

During this period, the Participating Groups will have the possibility of preparing themselves for active participation. The time may be used by the participants to familiarize themselves with the hardware and software, to become aware of the potential of astronomical observations and to consider specific programme opportunities, as they become available on the Astronomy On-Line WWW pages. It will also be the time for the groups to begin to communicate actively with each other. This may lead to the establishment of regional clusters or even larger constellations of Participating Groups.

5.2. *Phase 2 (November 18 - 19, 1996)*

Phase 2 will start on Monday, November 18, and last until Tuesday, November 19, 1996. On these and the three following days during Phase 3 (see below), the 'active period'will be in the six-hour interval 15h - 21h UT. This period has been chosen to allow students to participate outside the normal school hours, and by taking into account the time zones across Europe (from UT in the West to UT + 2 hours in the East). On these two days, the Participating Groups will interact intensively with each other and consult the available Home Pages for the detailed information about the various projects that have now become available.

Various events will be planned to happen at certain times and in certain places, keeping the programme lively and ensuring continued attention and expectation by the participants.

It is not expected that all groups will be "on the WWW"all the time; this will be taken into account for the planning.

5.3. *Phase 3 (November 20 - 22, 1996)*

This Phase will follow from Wednesday, November 20 to Friday, November 22, 1996 and will mark the climax of the Programme. During this period the participants will continue their work on the various projects and bring them to a successful end. Where applicable, they will prepare their concluding reports. These will be published in the Newspaper.

6. Follow-Up Activities

The results from Astronomy On-Line will remain available on the established Home Pages for a while, and the main conclusions and statistics will be summarized by the ISC in a written report to the European Commission. The EAAE Working Group on

Research on Teaching will evaluate the Programme and submit a detailed report at a later stage.

It is quite likely that some groups will not be able to make full use of the proposed activities during the limited duration of the present Programme. However, it is of course possible to use the offered information, programmes, exercises, etc. in the classroom, also after the formal end of this Programme.

Astronomy On-Line is a complex programme and it serves as a pilot project that is expected to pave the way towards a more permanent network of astronomy educators. While such a network will undoubtedly have a particular importance for the future work of the European Association for Astronomy Education, it may also constitute a useful basis for the subsequent development of a more global tool.

REFERENCES

WEST, R.M., 1994, *Sky & Telescope*, September 1994, p. 28.

WEST, R.M., ed. 1995, Proceedings of *ESO/EC Workshop on the Teaching of Astronomy*, ESO, Garching.

Project ASTRO: A Successful Model for Astronomer/Teacher Partnerships

By M. Bennett, A. Fraknoi & J. Richter

Astronomical Society of the Pacific, 390 Ashton Avenue, San Francisco, CA 94112

1. Overview

Project ASTRO is designed to improve astronomy education and science literacy in grades 4-9 by creating effective working partnerships between teachers/youth leaders and astronomers (both professional and amateur). Key elements of the program include:

- training the teachers/youth leaders and astronomers together in inquiry-based "hands-on, minds-on" learning activities
- encouraging an active working partnership between the astronomer and the teacher/youth leader
- encouraging multiple visits by the astronomer to the classroom or youth group meetings.

The ASP conducted a Project ASTRO pilot in California from 1992-1995, funded primarily by the National Science Foundation. The success of the pilot led to a second grant from NSF (1996-1998) to expand Project ASTRO to several other cities in the United States.

In the 3-year pilot project a total of 104 astronomers and 150 teachers formed 96 teams. More than 85% of the astronomers visited their adopted classrooms four or more times, with 46% making 5-10 visits during the school year. Approximately 10,000 students were involved. The Project ASTRO staff developed an extensive set of astronomy activities and tested resources, now available from the ASP as The Universe at *Your Fingertips: An Astronomy Activity and Resource Notebook*. The staff also published the *Project ASTRO How-To Manual for Teachers and Astronomers*.

In the first years, astronomers and teachers received stipends for attending training workshops. All participants either volunteered their time or were given release time by their employers for all time involved in planning and implementing their partnership visits and activities.

The independent evaluator rated the project "extremely successful", noting that 91% of the teachers felt they were teaching more astronomy as a result of the project. Most of the astronomers felt that they had learned a great deal about how children learn and how to develop problem solving and reasoning ability in children. As another indicator of the project's success, 62% of the teams continued the partnership on their own after their initial year.

In the expansion phase, Project ASTRO programs will be formed in 6-10 communities in the US. Science centers, planetaria, universities, school districts, or other astronomy and education institutions will take the lead in replicating the project locally. The goal of the project is to learn the most effective methods for forming and maintaining Project ASTRO programs that are locally self-sustaining.

2. Background — The Astronomical Society of the Pacific

The ASP is a worldwide scientific and educational organization founded in 1889 that brings together professional astronomers, educators, amateur astronomers, and interested laypeople. Technical publications include the well-known *ASP Conference Series* and *Publications of the ASP*. The ASP also publishes a popular bi-monthly magazine, *Mercury*.

Educational activities include an annual two-day summer workshop in astronomy, generally attended by 100-200 teachers, held since 1980, and the quarterly *Universe in the Classroom*, a free education newsletter currently reaching over 12,000 teachers worldwide. The newsletter is currently translated into 14 languages.

As part of the ASP's annual meeting, held in various North American cities, the society for the past several years has also been cosponsoring with *Astronomy* magazine a public weekend event of lectures, demonstrations, and commercial exhibits, which often attracts over 2,000 people.

3. The ASTRO Pilot Project 1993-1995

The pilot project consisted of the following major components:
- Recruitment and application process
- Selection and matching process
- Initial training
- Development of resource materials
- Implementation
- Evaluation of results

4. Recruitment and Application Process

Two application forms were developed-one for teachers and one for astronomers. These were distributed through appropriate local channels and networks. Teachers were recruited through local school districts, local branches of professional associations like the California Science Teachers Association, science education reform projects, and the state systemic initiative. All teacher applicants were required to demonstrate their school's commitment to the project by obtaining a signed release form from the principal of their school.

Astronomers were recruited through the ASP's mailing list, university astronomy departments, and amateur astronomical societies. Astronomers were required to indicate their time availability and commitment to the project.

Response to the program by both teachers and astronomers was high; nearly 10 teachers applied for each available slot.

5. Selection and Matching Process

Application forms were reviewed by project staff using criteria such as geographic area, balanced representation of grade levels, willingness to use activity- oriented teaching methods, ethnic and gender representation, and representation of urban, suburban, and rural districts.

Approximately 70% of the teachers were from elementary schools, 24% were from middle schools, and 5% were from high schools. Sixty-five percent of the teachers indicated they had minimal previous astronomy training or had not taught astronomy prior to their involvement with Project ASTRO.

Twenty-four percent of the astronomers were professional astronomers, 13% were professional astronomy educators (in planetaria or community colleges), and 63% were amateur astronomers. Few of the astronomers had any experience teaching any subject at the elementary grade levels.

Project ASTRO staff selected teams based on geographical proximity, common interests, prior experience teaching astronomy to particular age groups, and ethnic and gender considerations. In 62% of the teams one astronomer was paired with one teacher ; the remaining third consisted of one astronomer paired with two teachers.

6. Initial Training

All participants were required to attend a two-and one-half-day training workshop, usually held during the summer, where, in most cases, teammates met their partners for the first time. The primary objective of the workshops was to train both astronomers and teachers in activity-based, age-appropriate classroom methods that engage students in problem solving and critical thinking. The workshop also provided the teams with help and guidance in beginning their individual planning processes.

Astronomers' and teachers' responsibilities and expectations were discussed and defined. Some astronomers, particularly some of the amateurs with less knowledge of astronomy, were relieved to discover that they were not expected to always know the answer to every question or to know all the constellations. Teachers were encouraged to be active members of the teaching team while continuing to be responsible for class organization and student behavior.

7. Development of Resource Materials

During the pilot project, the staff, assisted by the project participants and consulting master teachers, compiled, tested, revised, and finally published *The Universe at Your Fingertips: An Astronomy Activity and Resource Notebook*. More than 800 pages long, it contains more than 90 classroom and group activities, drawn from many existing curriculum projects and other publications. In addition, content notes, articles about teaching and learning, bibliographies, and resource lists were included. The publication has been favorably reviewed in several astronomy and educational periodicals.

The Project ASTRO How-To Manual for Teachers and Astronomers, a 32-page booklet with many tips and suggestions for how to create and maintain partnerships, was published in 1996.

Both publications are available from the Astronomical Society of the Pacific.

8. Implementation

Although Project ASTRO had recommended that astronomers visit their classrooms a minimum of four times, including an initial observation visit, most astronomers visited more frequently. More than 46% visited between 5-10 times and another 13% visited more than 10 times. Only 15% visited fewer than 4 times.

Astronomers and teachers engaged in a number of different activities during the visits. In many cases, the teacher took the lead in suggesting and developing partnership activities, and in helping astronomers gain confidence in working with young people. Conducting activities from the resource materials, particularly those which had been modeled at the workshop, was the most common activity. Organizing and conducting star parties, often held on the school grounds and involving families and other classes, was the second most common activity. Other activities included assembly (large group) presentations, slide show and video presentations, using school computers and network connections, working with student clubs, organizing field trips, daytime telescope viewing, refurbishing the school planetarium, and various planning activities.

During the first two years of the project, groups of teams were formed in several California cities. Separate workshops were held in the San Francisco Bay Area (northern California) and in the Los Angeles area (southern California). In the third year, new teams were formed and workshops held only in northern California. Project staff elected not to formally continue the project in southern California because it was decided that the project works best if managed locally. A few partnerships, however, did continue on their own in southern California. A total of 104 astronomers and 150 teachers formed 96 teams. Approximately 10,000 students participated in Project ASTRO over the three-year period.

To improve communications and long-term networking between teams, the Project ASTRO staff publishes a quarterly newsletter, the ASTROgram, which is mailed to all teachers and astronomers who have participated in the program. It contains case histories and "success stories" written by participants, announcements of upcoming events, etc. All participants with email addresses (more astronomers than teachers) are placed on the project's email "exploder", but so far use of email has been light.

A few times a year, participants from both current and past years are invited to attend follow-up workshops, often featuring a combination of astronomy content and additional training in hands-on activities.

9. Results

A major goal of Project ASTRO was to increase the amount of astronomy taught in classrooms, as well as to increase the use of inquiry-based activities. More than 90% of the Project ASTRO teachers report that they are teaching more astronomy as a result of the project. In many cases, Project ASTRO seemed to empower teachers to use more creative approaches to teaching, as well as to focus on more hands-on, constructivist activities.

There also was evidence that the program influenced teachers' attitudes towards teaching girls about science. The workshops included a session sensitizing participants to issues related to gender, and one female teacher, partnered with a male amateur astronomer, reported that the Project ASTRO experience had changed her whole perspective on teaching, particularly teaching girls. She felt that she was far more sensitive to the importance of encouraging girls in her classes.

Project ASTRO also had a significant impact on the participating astronomers, exciting them about astronomy teaching, training them to teach effectively, and providing a valid and respected entré into the school systems by improving their abilities to be effective science education advocates and teachers in their communities. There were no differences in the success rates of professional astronomers, amateur astronomers, or astronomy educators.

Although design of the project did not include direct measurement of student learning, there is considerable evidence that impact on the students was substantial. More than 80% of teachers and astronomers indicated they believe that students had become more interested in astronomy as a result of the project, and 74% said that the students had actively learned more astronomy. Impacts included exciting students about the topic, providing opportunities for them to be active learners, and using their discretionary time by involving them (and often their families) in field trips, after-school clubs, and star parties. One fifth-grade class won a city-wide Mind Olympiad that focused on astronomy, and the teacher attributed their victory to Project ASTRO.

In three California communities (Stockton, Santa Barbara, and Sacramento), small groups of astronomers and teachers have formed their own "mini" Project ASTRO groups, recruited and trained new participants, and continued with limited financial support from the main project.

It is clear that the success of Project ASTRO is the result of the combination of three key factors – the formation of true working partnerships between teachers and astronomers, the emphasis on multiple visits by the astronomers, and the increased use of activity-based teaching methods over lecture-based techniques.

10. The Project ASTRO Expansion Program

In 1996 the NSF funded a three-year expansion grant for Project ASTRO, consisting of three main elements:

- continuation of the San Francisco Bay Area program
- expansion to 6-10 other cities in the United States
- development of a training program for individual astronomers

The primary goal of the new project is to explore methods for replicating self- sustaining Project ASTRO efforts in other communities. One replication strategy being encouraged is for expansion sites to form local coalitions of cooperating institutions to support the program. Part of the three-year project is to explore effective strategies to make the expansion sites self-sustaining.

In the expansion phase, increased emphasis will be placed on including youth group leaders, rather than just classroom teachers, in partnerships. Continuing emphasis will also be placed on reaching young people typically underrepresented in the physical sciences, particularly girls, children of color, and children in lower socioeconomic status groups.

11. The San Francisco Bay Area Program

The Bay Area Project ASTRO Coalition (BAPAC) consists of nearly 20 representatives of local school districts, amateur astronomy groups, university astronomy and/or science education departments, science centers, planetaria, and museums. The purpose of the Coalition is to support the continuation of the local ASTRO group and to eventually find resources and funding to support the modest administrative costs of continuing the program.

In the spring of 1996 20 new astronomers and 23 teachers were recruited to form a new Project ASTRO group for the 1996-97 academic year. A two-day workshop, to be held at one of the local county education offices, is scheduled for August, with the teams expected to begin working together in the fall 1996 semester.

12. National Expansion

How can the Project ASTRO model be successfully transferred to other localities? What are the key elements needed to create and sustain a local Project ASTRO? These are the questions that will be explored in the next three years.

The Project ASTRO staff intends to identify and recruit "lead institutions"in a total of 6-10 US cities. These institutions will be responsible for recruiting, matching, and training ASTRO partners, for generating interest and support among other local institutions, and for finding sources of ongoing support. The lead institution will receive "seed"funding to support project coordination, training, and purchase of materials, but within three years it is hoped that the local Project ASTRO program will become self-supporting.

The Adler Planetarium in Chicago, Illinois is the lead institution of the first such expansion site. They are in the process of recruiting participants and will hold their first training workshop in August, 1996. Project ASTRO staff expects to identify at least two other institutions in other cities by the end of 1996, so they can begin operations in the spring of 1997.

13. The Astronomer Partners Program (APP)

A "second strand" of the ASTRO Expansion grant is to find a way whereby individual astronomers (usually in non-ASTRO cities), with an interest in becoming more personally involved in science education, can use the Project ASTRO model to channel their efforts more effectively.

Approximately three one-day workshops per year are planned, to be held in conjunction with major national or regional professional and amateur astronomy meetings. Participants will be given an introduction to current issues in science education reform, and to the importance and effectiveness of activity-based age- appropriate learning techniques. They will also be given some tips on how to identify and recruit interested local teachers or youth group leaders with which to partner. Participants also receive a full set of resource materials. These astronomers, and the teachers and youth group leaders they work with, will become part of the national Project ASTRO network, receiving the ASTROgram, having access to email communications, etc.

The first APP workshop was held at the ASP annual meeting in June 1996, with 19 participants. The next is scheduled to be held in conjunction with the AAS meeting in Toronto in January 1997.

14. Conclusion

Many astronomers - researchers, educators and amateurs - wish to share their enthusiasm and energy with young people in schools and youth groups. Project ASTRO has demonstrated that this wonderful spirit of volunteerism, this desire to help and inspire children, can be channeled, leveraged, and made even more effective by combining three key elements forming ongoing partnerships with teachers and youth group leaders, learning how to conduct age-appropriate inquiry-based activities, and making multiple visits. The benefits of Project ASTRO extend far beyond the students themselves, giving the teachers and astronomers themselves new skills and understanding they will use throughout their careers. As astronomers become more personally involved with the realities of science education in U.S. schools, and more informed about the alternatives presented by the science reform movement, they will themselves become more effective agents for change, both in the schools they visit and beyond.

Acknowledgments
The Project ASTRO pilot and expansion programs have been funded by the NSF office of Informal Science Education, NASA Office of Space Science and Office of Education, the Banbury Fund, and the Bart Bok Fund of the ASP. The evaluation of the pilot project was conducted by Lynn Dierking of Science Learning, Inc.

REFERENCES

DIERKING, L. & RICHTER, J., 1995, *Project ASTRO: Astronomers and Teachers as Partners* in *Science Scope*, Mar. 1995, p. 5.

FRAKNOI, A., ed., 1995, *The Universe at Your Fingertips: An Activity and Resource Notebook for Teaching Astronomy.* Project ASTRO, Astronomical Society of the Pacific.

FRAKNOI, A., RICHTER, J. & HILDRETH, S., *Project ASTRO: Partnerships Between Astronomers and Teachers* in J. Percy, ed., *Astronomy Education: Current Developments, Future Coordination.* 1996, Astronomical Society of the Pacific Conference Series.

MARINO, K., *Education Programs Let JPL Give Back to the Community: Project ASTRO Links Astronomers and Elementary Students Statewide* in *JPL Universe*, Dec. 29, 1993, p. 1.

RICHTER, J. & FRAKNOI, A., *Matches Made in the Heavens: The A.S.P.'s Project ASTRO* in *Mercury*, Sep./Oct. 1994, p. 24.

RICHTER, J. & FRAKNOI, A., 1996, *Project ASTRO How-to-Manual for Teachers and Astronomers.* Project ASTRO, Astronomical Society of the Pacific.

The Training of Teachers

By L. Gouguenheim[1] & M. Gerbaldi[2]

[1]Université de Paris XI-Sud and Observatoire de Paris, 92195 Meudon, France

[2]Université de Paris XI-Sud and Institut dAstrophysique de Paris 98bis Brd Arago 75014 Paris, France

1. Introduction

Informal and formal astronomy education is present through many channels: newspapers and TV; amateur associations; clubs and science associations; at school at any level. The teachers are not only the main agents of the educational process at school, but they are also very active in extra-curricular activities: they run clubs, educational projects etc.

These activities are present everywhere in the world, as can be seen from the reading of the National Reports published every 3 years by Commission 46 "Astronomy Teaching"of the International Astronomical Union and published in its Newsletter.

A quick look at these reports shows that there is a huge variety of educational systems from one country to another: some countries have a specific curriculum in astronomy, others are just beginning to develop it; in other places, astronomy has been considerably reduced in the newly created curricula. One more difference: in some countries, education has a national curriculum; in others the responsibility for teaching is left entirely to each Province, a term used here to refer to the local situation. Such a situation and its consequences was were depicted by Wentzel (Williamstown IAU Colloquium 105, 1986).

1.1. *Why Astronomy in the curricula?*

In spite of these differences, a general trend can be drawn: it is very rare that astronomy is considered as a separate subject; it is nearly everywhere part of the programme either of Mathematics, Physics and Chemistry or Natural Sciences. The objective to be reached with the introduction of Astronomy in these fields was to widen the students knowledge of science, giving him (her) an idea of how to work scientifically from observations and known physical laws in order to get new knowledge within another field of science.

Astronomy can inspire students to study sciences, can help to fight against "math phobia"(see Appendix 1), but it is efficient only if the science teachers have a sufficient specific knowledge in the field of astronomy, in order that they feel competent enough to teach this subject.

1.2. *The Training of Teachers*

There is everywhere a huge demand for training school teachers in astronomy, whatever their level: elementary school teachers seldom receive much training in science in general and astronomy in particular; junior and senior high school teachers receive an initial training in mathematics, or physics-chemistry or biology, but with astronomy courses included in very few cases.

There is a general lack of teacher training in astronomy; the way it is organised varies from one country to another; and according to the many different local situations different strategies were defined, but very few of the teachers had initial training in astronomy at the university.

All teachers, primary as well as secondary, whether of physics, mathematics, earth science or geography, should thus be educated in astronomy both in initial and subsequent

256

in-service training. They should be instructed in scientific matters as well as in teaching methods; they need access to scientific research, to educational materials and methods and the possibility for exchanges of experiences.

It is obvious that a strong impetus was given during the Williamstown IAU Colloquium: a number of collaborations and activities in this field started there.

We have also to mention the International Conferences on Teaching Astronomy held at the Polytechnical University of Catalonia under the responsibility of Dr Rosa Maria Ros. The history of this conference started in 1984 and 1985 with the first and second Meetings on Teaching Astronomy at Secondary School for teachers from Catalonia; the number of participants increased every year and the following year, in 1986, the third National Conference on Teaching Astronomy in primary and secondary schools took place, open to teachers from all over Spain. After that, the conference became international, in 1990, with the 4th International Conference on Teaching Astronomy to teachers and professors representing all educational levels, ranging from primary schools to universities. The 5th one took place in March 1995: it brought together around a hundred participants, mainly teachers, from 15 different countries; they had the opportunity to attend five general lectures and to present their own ideas, experiences and contributions. All of them are published in the Proceedings. All those who attended one of these Conferences can attest how personal links were developed among participants and further collaborations started.

2. Present European situation

In what follows, we develop the present situation in Europe as we know it through the Workshop on the Teaching of Astronomy in Europe's Secondary Schools held at the ESO headquarters in November 1994 under the auspices of the Second European Week for Scientific Culture.

During this meeting, 100 school teachers coming from 17 western European countries gathered, as well as officials from the Ministries of Education of these nations.

The present status of astronomy teaching everywhere in western Europe was presented and an important exchange of information took place also during this meeting, for the first time in Europe, on that subject. The "European Association for Astronomy Education" (EAAE) was initiated.

It was recognised that there is a crisis for the teaching of scientific subjects everywhere; in that context, astronomy was felt as being a unique opportunity for promoting scientific teaching. It was generally emphasized that less formalism should be used in the teaching of physics, in order to concentrate more on experimentation and on the various representations of natural phenomena. In such a framework, astronomy fits well. Such a conclusion was reached independently and under various conditions in each country.

The EAAE goes further, pushing the idea that an overall holistic view of astronomy should be presented to all European children. One strong point of the "Declaration on Teaching Astronomy in Europe's Schools", which was unanimously adopted by the participants of the EU/ESO Workshop is the following:

"By the end of compulsory education, students should have been involved in observation, experimentation and discussion of the following ideas from astronomy:
- Our place in the solar system, progressing to our place in the Universe;
- The nature of objects we see in our sky, for instance, planets, comets, stars, galaxies.
- Examine thinking from the past ages and more recent times to explain the character, origin and evolution of the Earth, other planets, stars and the Universe.

In initial training of teachers and their subsequent in-service training, these ideas should be introduced and reinforced.

Recent studies of students' misconceptions and ideas in astronomy provide a useful basis for the further development of teaching methods."

3. Examples of Activities Developed in Europe

The training of teachers is organised in many different ways, but the needs are similar: (i) need education in order to be able to teach astronomy, (ii) need educational material, (iii) need educational methods and the possibility for exchanging experiences. It is interesting to note that nowhere was anything in that domain organised on a nationwide basis by the national authorities. Instead, several actions were undertaken spontaneously by various groups of professional astronomers and/or didacticians in sciences, in close collaboration with school teachers.

We mention in the following a few examples.

The **Italian Astronomical Society**, a free society of professional astronomers, amateurs and teachers, which established an Educational Committee many years ago: this committee is a sort of nucleus for activities concerning the teaching of astronomy: it regularly interacts with the public authorities, the other scientific societies, the teachers associations; it ensures the presence of astronomy in the science curricula; it works out curricula proposals; it offers modular courses for teachers; it arranges summer courses for teachers; it promote competitions in schools and text-books analysis and it bring out educational materials. It works in close collaboration with several didactic laboratories of sciences such as in Turin, Rome or, more recently Reggio Calabria.

The educational system in Germany is ruled by the authorities of the 16 regions (Länders) which constitute the German-Federal Republic; the situation has changed after the German reunification in 1989, because astronomy has been a regular teaching subject in the schools of the 5 new regions of East Germany. In this situation, a Working Group was established in the German Physical Society, which cares for teaching of "astronomy and astrophysics at school". In the last 5 years, this group has been continuously organising workshops. Interesting didactic material is being developed by Prof. Roland Szostak in the Didactic laboratory of Münster University. We mention also an interesting package of 31 programmes in Turbo-Pascal available from Dr. Rainer Gaitzsch in Bavaria: their intention is the physical point of view in many selected examples of astronomical problems.

The long-standing activity of the **Finnish Association of Amateur Astronomers, URSA**, which publishes its Newsletter Tahdet ja Avaruus (Stars and Space), a number of books, generally written by astronomers and organises locally public conferences given by professional astronomers.

The long-standing activity of the **Association for Astronomy Education** in the United Kingdom and its Newsletter "Gnomon" and that of the **French Liaison Committee between Teachers and Astronomers**, CLEA and its Newsletter "Les Cahiers Clairaut" are well known. '

It is worth noting that many exchanges have taken place among members of these different associations: Les Cahiers Clairaut have already published many contributions from Italian colleagues (Lidia Nuvoli, Nicoletta Lanciano, Mascellani...), Spanish (Rosa M. Ros, Estrallela, Anglada,...), German (Roland Szostak), Finnish (Pekka Teerikorpi), Polish (Cecylia Iwaniszewska)...

Joseph Nussbaum (Israel), Cecylia Iwaniszewska (Poland), Roland Szostak (Germany), Christoffel Waelkens (Belgium) have already given lectures to CLEA summer universities for school teachers and several participants came from Spain and Belgium. Two workshops have also been organised in Paris in which Darrel Hoff (USA) and CLEA members exchanged their experiences. Foreign astronomers have been invited to participate in the annual meeting of the Educational Committee of the Italian Astronomical Society, in teacher training sessions in Barcelona or Münster.

4. Conclusion

It is interesting to note, that though the situations in different European countries, and even within a given country, are somewhat different, similar conclusions are reached concerning both the contents and methods. Emphasis is being put everywhere on the importance of making observations, or at least studying observational documents, on constructing models and instruments.

Many school teachers prefer to construct for themselves, preferably with their pupils, relatively simple models, rather than buying more sophisticated ones ready to use, provided that the simple device illustrates sufficiently well the concept it is aimed at. However, they manifest also a strong interest in using modern techniques, such as CCDs.

We have also to mention the problem of the language, and not only the language, but also of the culture. Because astronomy is not only a science, cultural aspects are particularly important. The exchange of pedagogical documents may thus have different trends. Sometimes, they are simply translated from one language to another; sometimes, the same practical activity is presented in a different way.

It is clear that such teacher training can be made neither rapidly or massively especially when astronomy is part of the curricula. It needs progressiveness and feedback. This is the way the CLEA is working in France since 1977.

APPENDIX 1: THE POINT OF VIEW OF A FRENCH MATHEMATICS SCHOOL TEACHER (Josée Sert)

As a Mathematics teacher, I have been teaching Astronomy to students in the last year of their curriculum and who have chosen the Literary option of the Baccalaureat. Until June 94, the official instructions stipulated that about one third of the curriculum could be chosen among several optional topics, Astronomy being one of them. The students were to go through an oral examination for the *Baccalaureat*, during which they could present a personal project related to this optional subject. The syllabus consisted mainly of the History of Astronomy up to Newton's gravitation law and the measurement of time (sundials and calendars).

As years passed and I gave lectures on these subjects, I could assess how important it was to teach Astronomy, and this in several ways:

First, Astronomy is a means of reconciling those students who had often been "negatively" oriented towards Literary studies because they were not good at mathematics, with scientific subjects. Their curiosity towards such topics emerges at once and bursts into numerous questions, especially if you are lucky enough to take them to an evening observing session. It increases when they are confronted with science, a field still fraught with uncertainties, but where knowledge progresses fast and it leads them to other fields that help them to get answers. You can then notice that students are more active and develop skills (such as searching for information, delivering a lecture ...) and that they even accept using the mathematical tools which they have so far considered as artificial. As an examiner, I could see rich and thorough studies presented by students on topics they were particularly interested in (for instance the origin of the Universe, the possibilities of extraterrestrial life in the solar system ...) as allowed by the particularly flexible syllabus.

Then, this brought them elements of a scientific culture which they had never approached before: the evolution of our ideas on the Universe through History, how and why this evolution occurred, what an experimental process consists of (the relationship between observation and the model), the difference between a model and a theory, how ideology or religion can influence the advancement of knowledge and stimulate their reflection on Man and his place in his environment.

For that reason, throughout these years, because of the students' queries or hints, I have more particularly dwelt on some particular points, sometimes with the help of the Philosophy teacher. I developed Aristotle's ideas; I made the students concretely study the equivalence between the Ptolemaic and the Copernican models. I emphasized the importance of observation in the process that led Kepler to his laws, the fundamental reflection of Galileo on inertia. I opened the issue of the influence of the Copernican revolution in other fields than Astronomy. I tried to bring to light the qualitative leap that the theory of gravitation represented and at the same time I encouraged the students to apply their own research method, not only to books but also to their immediate environment: getting pictures from observatories, searching for calendars different from the Gregorian one, thus discovering that their grandmother, their neighbour or the grocer at the corner of the street may practice another religion (which is of great interest because it involves contact with other people and respecting them, and instructive in so far as beliefs may be an obstacle to exchanges, for example if you cannot obtain a calendar because any document that bears the name of God may not be thrown away, or because you don't pledge to burn the sheet of a block calendar after it has been used).

So I discovered that the great interest of Astronomy was that it could lead these nonscientific students towards processes that made them discover the surrounding world, in order to understand its physical reality and become aware of some of its human dimensions. Is that not a promising start with respect to the education of the European citizens of the year 2000?

New Trends in Astronomy Teaching

By L.E. Abati

Società Astronomica Italiana, Unità di Ricerca CNR "Asiago", c/o Department of Astronomy
Vicolo Osservatorio, 5 - 35122 Padua, Italy

We are all aware of the fact that Astronomy teaching is not an easy task for many different reasons which we are going to examine during this Colloquium. The present contribution focuses on one of these reasons we consider of major importance for Astronomy in the school: Teacher Training.

Teacher training has been debated extensively for a long time and discussion is being presently livened up.

Institutions and associations are promoting research, studies and comparisons on this issue. For instance, the Osnabrück conference "Teacher Education in Europe: Evaluation and Perspectives" (June 1995) – the International Forum of Rome (September 1995) and, specially devoted to Astronomy, the EU/ESO Workshop "Astronomy teaching in the European secondary school "(Garching, 1994), SAIt Workshop in Reggio Calabria "European Science Teacher Training "(September 1995), Conferences of Teaching Astronomy in Spain, the Constitutional Conference of the European Association for Astronomy Education (EAAE, Athens, 1995).

It is difficult to treat Astronomy teacher training without including it in a more general context. Teacher training does not only mean providing teachers with suitable teaching skills for each subject. First of all, teachers should bear in mind the interaction with a social and cultural reality that may affect learning processes. And the educational (and teaching) system is not neutral to the external framework. European and non-European countries have their own national differences with different school systems and choices made in the field of teacher training. Time does not allow us to go in detail into a comparison of the various solutions adopted in different countries. However there are some common elements in teacher training both in terms of organization and content (Osnabrück Conference, ESO Workshop, EURIDICE '90, EURINS), see Fig. 1.

Common organizational elements†

- Training often in post-secondary institutes, not necessarily in universities
- Possible training both in specialized institutes and universities
- Curricula of specialized institutes mainly with an educational background, lead to teaching in compulsory schools
- University curricula mainly with a specialized background in one or two disciplines, often leads to teaching in secondary schools (Dutch model)
- No specific curricula leading to teaching in Universities
- In-service training and requalification linked to initial training often in different centres
- Refresher courses compulsory almost everywhere
- Common changes of the institutional mechanism referring to teacher training

FIGURE 1.

† with the exception of Italy

COMMON ELEMENTS

UNIVERSITY	///???
SECONDARY SCHOOL (High School)	///\\\\\\\\\\\
SECONDARY SCHOOL (Junior H. S.)	///////////////////////////////\\\\\\\\\\\\\\\\\\\\\\
PRIMARY	///////////////\\\\\\\\\\\\\\\\\\\\\\\\\\\\\\\\\\\\
NURSERY	///////////\\

///////////   subject knowledge

\\\\\\\\\\\   educational skills

FIGURE 2. Taken from the presentation of G. Luzzato, President of the Italian Conference of the Interdepartimental Centres for Educational Research (CONCIRD) given at the GIREP '95 held in Udine, Italy, August 1995.

Concerning content, Fig. 2 is extracted by considering teachers' degree of specialization and teachers training curricula.

It shows the relative proportion of subject knowledge and educational skills at various school levels. Almost everywhere, outside this conference, it is still believed that at university level the qualification for teaching is automatically guaranteed by a competence in research.

We all agree that good astronomy teachers should be competent both in subject knowledge and educational skills. Therefore, the professional profile of teachers should combine the two aspects. This implies that, in teacher training, disciplinary courses and educational courses should converge in subject didactics which then interface with school practices as shown in Fig. 3, also taken from the above mentioned work.

Astronomical didactics becomes the valuable bridge between the disciplinary themes and the educational ones. In the subject didactics, competences deriving from different areas should converge.

Features emerging from an analysis of the teacher training system show a general dissatisfaction concerning the actual integration between subject and educational science which are located in separate departments often scarcely communicating with each other.

UNIVERSITY

SUBJECT DEPARTMENT   EDUCATION DEPARTMENT

Disciplinary   Educational
courses   courses

Subject
didactics

SCHOOL SYSTEM

School practice

FIGURE 3.   Suggested Devision of Teacher Education

For teaching purposes, one must consider that scientific knowledge is increasing and changing very rapidly, too. Most of what we are going to hear at the next IAU Assembly in the field of Astronomy could hardly be imagined at the time when a person, today in his forties/fifties, was attending University. The same can be applied to students. Therefore, the need is to focus on skills and on the development of abilities, on flexibility and basic training, as also stressed by George Greenstein, and astronomy seems to be particularly suitable for this purpose. Moreover, efficient refresher courses play a key-role in teaching. It should be noted that the institutional training and requalification system is subject to change almost everywhere. But the discussion would shift towards matters of educational policy, scientific, cultural and administrative independence of schools. This is not the most suitable place to tackle such issues. Our aim is to consider the role that Astronomy can take on in an educational system.

SOCIAL AND POLITICAL ELEMENTS

- Cultural
  Scientific   Independence of schools
  Administrative
- Legal status of teachers
- Careers depending on the quality of service
- Flexible curricula
  curricula updating
  curricula adjusting to different situations

FIGURE 4.   Social and Political Elements

Astronomy is a discipline rooted in the early culture of "man-anthropos". Therefore, tracing its history means tracing the main stages of the development of human thought and knowledge. Since the most ancient evidence of man's interest in the universe, such as primitive graffiti as well as theoretical approaches of the oldest schools of thought (Chaldean, Egyptian, the Pythagorean School and pre-Columbian civilizations), it is possible to find a constant reference to two paradigms: 1) a rational knowledge based on the "scientific"observation of the phenomenon; 2) a "magical", "mythological"aspect involving the emotional, irrational sphere, the individual and collective imaginary world. In front of the starry sky, man feels "sub divo", immersed in mystery, in the arcane. The first paradigm has lead to the technical applications of astronomical knowledge, first of all in time measurement based on the cyclical trend of physical and biological phenomena (from tides to recent biorhythms). Historically, these applications precede the theoretical elaboration and systematization which is continued in modern astronomical research up to its most complex fomulations. Just think of cosmology, of the theory of evolution of

Universe, until the recent discoveries in the solar system. After all, in the sciences, practical aims always come before theory (a typical example is geometry which was invented as a response to the need for measuring distances on the ground). The second paradigm led to all the intuitions, interpretations, experiences which converged in the pre-Galilean culture of the so-called pseudosciences, among which astrology (compared to astronomy) represents the attempt to establish a direct connection between the cosmic phenomenon and the individual and social existence. The reference to these two paradigms allows us to place astronomy in a privileged position in a modern scientific context. Until Positivism, in fact, what seemed to be an insurmountable distinction between "science" and "myth" existed and was based on Cartesian and Galilean assumptions: scientific knowledge excluded, by definition, the whole mythological heritage.

However, more recently, after questioning some "scientific" certainties (in mathematics and physics) and revaluing what is "indeterminate" and "probable" with respect to the strict "cause and effect" relationship (determinism, probabilism, the laws of chaos), scientific and mythological approaches no longer clash. Nowadays, they are considered two separate but integrated ways of experiencing and exploring reality. This vision is also supported by a biological approach, as more recent studies in neuroscience have proved: the two cerebral hemispheres are entrusted with different but integrated functions and are not subordinated one to another, according to the obsolete principle of "cerebral dominance". For example, in listening to music, the perception of notes and their sequence depends upon the function of the left hemisphere, but our perception of melody and impression is due to the right hemisphere.

In the creation of scientific knowledge and in that "breeding ground" called school, of course, both ways of knowledge are always present. The teaching criteria, in fact, should represent and implement the two approaches.

At this stage, we can say that astronomy is included with a special educational value in school curricula, since it is one of the few, if not the only, subject that maintains a double link:

a) with the most advanced research modalities complying with modern epistemological criteria, according to the well known Popperian statements, which have enabled achievement of present technological results and applications;

b) with an attractive and "magical" background that has an immediate hold over irrational but all the more important aspects in the acquisition of knowledge. Let us think of science fiction, which does not mean a "fantastic science", but a science combined with fantasy, where astronomy has a leading position; think of the appeal of equipment, telescopes, radiotelescopes, space probes, satellites, etc.

From a strictly educational point of view, we must bear in mind that teaching is above all communication and, as such, must keep in touch with the different aspects of personality, with the cognitive-rational sphere and the emotional one. Modern neuroscience is developing identification of these aspects in their biological substratum (cortex and sub-cortical structures - limbic lobe). School curricula should follow and take into account this basic ability to communicate and interact. In this interaction, there is space for all theories of communication (rational sphere) and relational dynamics studied by psychological sciences (emotional sphere).

Astronomy is suitable for such modular knowledge and its introduction in all school curricula should promote the most complete education of the individual. Therefore, we are allowed, on a scientific basis, to say that the value of astronomy is twofold:

• Informative: indispensable knowledge in a modern culture which is concerned (for biological reasons, too) about protecting the planet;

• Formative: it is a subject that involves different levels of individual and social knowledge, thus representing a model of integration between cognitive-rational modalities and structures and emotional-irrational ones. Recognizing these two values and clarifying the function of astronomy in school, that is interacting with students at an emotional level too, and using the appeal of a starry sky to guide them towards a scientific understanding of the world, can be considered a very successful new trend in astronomy teaching, that no other school subject can achieve.

Experience from cooperation in school programme projects has pointed out that a teaching programme is justified if, besides the content, the aims, the learning objectives and educational methodology are made clear. Of course it is always important to take into account the natural ideas of the students.

We believe that the aim of teaching astronomy in all schools is to understand that we are on the Earth but we live in the Universe. As generally astronomy is not an independent subject at school, there should be astronomy "modules"during teacher training in didactics of natural science and physics, the aims of which could be:

• the consciousness of Earth as a complex system in dynamic balance within the solar system (science) as well as

• understanding the universality of the physical laws from human level to microcosm and macrocosm (physics) with the objective of

• locating the Earth in space and time and setting the fundamental steps of its evolution.

As for methodological indications we think that, at all school levels, we do not have to teach the history of astronomy, but with the cognitive instruments of our students, try to explain the world around us. However, the formal introduction of astronomy in school curricula is not sufficient to guarantee the teaching of astronomy and particularly a correct teaching of it. The greatest effort in teaching astronomy has to shift attention from the curricula to the training of teachers. It is also a common experience that young people are interested in the phenomena surrounding them, rather than in their academic explanation. Students read newspapers, watch TV, sit at workstations watching the WWW, visit science museums, planetaria. In other words, we have to consider that school is not the only educative agency nowadays.

In Italy a lot of activities concerning astronomy teaching are organized by many observatories and astronomy departments and by the Italian Astronomical Society, of which I would like to give you some examples. For a long time the Italian Astronomical Society (SAIt), a free association of professional astronomers, amateurs and teachers, has set up an Educational Committee consisting of university professors and teachers. This commission is a sort of concentration nucleus for the activities concerning astronomy teaching; it regularly interacts with the public authority, the other scientific societies, the teachers associations; it keeps astronomy to the fore in the science curricula; it works out curriculum proposals, it offers modular courses for teachers. It arranges summer courses and promotes competitions in schools, and text-books analysis; and it brings out educational material.

Problems in teaching astronomy: there are many problems concerning astronomy teacher training which we are not going to treat here exhaustively, but just mention a few of them common to other disciplines: initial teacher training, in-service teacher training, curricula, etc. For Astronomy we have to add some other specific problems:

• Astronomy is present in other university curricula, physics and natural science

• Astronomy teachers need broad competence

• Research in astronomy education is very poor

• Time and Space are peculiar.

The latter could be considered as a drawback, but in my opinion it represents a great advantage for astronomy since it could guide students to respect time, space and, more generally, the modality of the natural world.

I hope I have been able to convince you of the fact that, as mentioned at the beginning of this presentation, teacher training does not only mean providing teachers with suitable teaching skills for each subject. Above all, teachers should be aware of the interaction with students in a social and cultural reality which can affect learning processes. The basic objective should aim at recovering an interpersonal relationship that reproposes the fundamental values of true humanism in the scientific universe. Historically, these values have never been set aside. Therefore we believe that if we want astronomy in the school, we have to take care not only of curricula and structures, but, most of all, of teachers, in their initial training, in their in-service training and in practical training. Institutions and parties in general which care for astronomy in the school and with the school have to move in this direction for the future of our society and of man.

REFERENCES

LUZZATO, G. 1995, Proceedings of GIREP '95, Udine.

Teacher Education in Europe, Evaluation and Perspectives, 1995, Proceedings of the Osnabrück Conference.

Workshop on European Science Teacher Training, 1995, in press, SAIt, Reggio Calabria.

Astronomy teaching in European secondary schools, 1994, Proceedings EU/ESO Workshop, Garching.

Scuola di Specializzazione all'Insegnamento secondario, 1994, CIRE, Bologna.

Itinerari astronomici, 1993, Itinerari Educativi, Venice.

ABATI, L. 1995, Il Giornale di Astronomia no.2, SAIt.

Initial Teacher Training, 1991, Eurdid Station Unit.

Coping with a New Curriculum: The Evolving Schools Program at the Carter Observatory, New Zealand

By K. Leather, F. Andrews, R. Hall & W. Orchiston

Carter Observatory, PO Box 2909, Wellington, New Zealand (Wayne.Orchiston@vuw.ac.nz)

1. Introduction

Carter Observatory is the National Observatory of New Zealand and was opened in 1941. For more than ten years the Observatory has maintained an active education program for visiting school groups (see Andrews, 1991), and education now forms one of its four functions. The others relate to astronomical research; public astronomy; and the preservation of New Zealands astronomical heritage (see Orchiston and Dodd, 1995).

Since the acquisition of a small Zeiss planetarium and associated visitor centre in 1992, the public astronomy and education programs at the Carter Observatory have witnessed a major expansion (see Orchiston, 1995; Orchiston and Dodd, 1996). A significant contributing factor was the introduction by the government of a new science curriculum into New Zealand schools in 1995 (*Science in the New Zealand Curriculum*, 1995). "Making Sense of Planet Earth and Beyond"comprises one quarter of this curriculum, and the "Beyond"component is astronomy. As a result of this exciting innovation, within just a few years, astronomy will be taught at almost every school in New Zealand – from entry primary school through to final year secondary – at eight distinct levels. This, in turn, will eventually lead to the emergence of one of the most astronomically-aware nations on Earth.

In 1995 the Ministry of Education also introduced competitive funding for museums, science centres, observatories and other institutions wishing to offer "Learning Experiences Outside the Classroom", and the Carter Observatory was successful in negotiating a three-year contract. As a result, a second full-time Education Officer was appointed, and the Observatory's schools program was totally revised in order to cater to the evolving needs of students, teachers and trainee teachers under the new astronomy curriculum. In addition, other programs and resources, such as "Overnight Extravaganzas", Astrocamp Booklets, and special holiday programs, were introduced to enrich the after-hours astronomical education experiences of school students, while a number of training programs were introduced for teachers and trainee teachers aspiring towards astronomical literacy.

This paper focuses on new resources developed for school visits to the Carter Observatory and other initiatives that have been taken to support the new curriculum.

2. Schools Program

There are two quite distinct components to the Carter Observatory's schools program: visits by groups to the Observatory, and outreach visits by Observatory staff to schools. Those visiting the Observatory can take full advantage of the facilities in the Visitor Centre, including the Zeiss planetarium and, on evening visits, the historic 23cm Cooke refractor (see Andrews and Budding, 1992; Orchiston *et al*, 1995).

One of the aims of the new astronomy curriculum is to assist students and teachers "... to investigate and understand relationships between planet earth and its solar system, galaxy and the universe"; this is typically achieved by using a combination of class discussion, practical demonstrations, selected videotape excerpts, slides, audio-visual shows, and planetarium programs. Material in the Observatory's Resource Centre is freely drawn on in preparing classes, and Unit Booklets are often used during the classes themselves. Information on both of these appears below.

2.1. *Unit Booklets*

In response to demand, a decision was made to develop Unit Booklets that could be used by teachers as a basis for planning their classroom work in astronomy. Furthermore, because of the novelty of astronomy as a discipline, and its perceived "difficulty" (e.g see Richter and Fraknoi, 1994), it was decided to adopt a cross-curricular approach and demonstrate how astronomy could be imaginatively linked with other curriculum areas (c.f. Whitehouse, 1994). The first batch of Unit Booklets focussed on linkages between astronomy and other physical sciences, and between the astronomy, mathematics and technology curricula. Work is now in progress on developing Unit Booklets from a Maori and Pacific Island perspective, in order to recognise the *tangata whenua* (original pre-European inhabitants) and take account of New Zealand's unique multicultural ethnic mix.

As at 1996 June, 45 different Unit Booklets had been produced, and each of these deals with a particular topic at a specified curriculum level. Between them, the 45 Booklets span a wide range of topics. In each Unit Booklet the curriculum objectives are clearly defined, and a range of options and guides for implementing the unit are given. Any necessary background information is included. The intent of these Booklets is to take the drudgery out of planning astronomy lessons and where possible to instil creativity in the teaching. We also prepared three associated Assessment Booklets and these have photocopyable pages, which make the task of setting up assessment schedules far less arduous.

All of the Unit and Assessment Booklets were produced in-house on the Observatory's computers, and then photocopied. Unit Booklets range in length from 6 to 17 pages and are sold at $NZ3:50 each. They have been very well received by New Zealand teachers and by colleagues in a number of overseas astronomical institutions.

2.2. *The Resource Centre*

In order to provide a diverse range of resource material for Observatory staff, teachers, and students undertaking our training programs we have been building up a Resource Centre. This contains the following range of material, drawn from national and international sources:

- books, booklets, leaflets and pamphlets
- reprints and photocopies of articles on astronomy and astronomy education
- teaching and resource material developed by overseas observatories, science centres, planetariums, etc.
- kits, toys, and equipment
- slides
- videotapes
- computer programs
- CD-ROMs and video disks

As at 1996 June, material in the Carter Observatory Astronomy Education Resource Centre occupied the equivalent of one four-drawer filing cabinet, and about 40 linear metres of bookshelf space. Carter Observatory in-house publications in the Resource Centre include the latest *Astronomical Handbooks*; full sets of Unit Booklets, Astrocamp

Booklets (see below), Activity Booklets and selected Information Sheets (see Orchiston *et al.*, 1997). We are eager to expand further the number of non-New Zealand items in the Resource Centre and would welcome the offer of material on an exchange basis.

During the building up of the Resource Centre, staff have also made a point of identifying items suitable for sale in the Observatory's Space Shop, and details of these reach schools throughout the country via our Summer and Winter Mail Order Catalogues.

3. Other Related Initiatives

3.1. *Holiday Programs*

In 1995 the Observatory ran two different two-week school holiday programs for 5-11 year old children (see Leather *et al.*, 1997). One focussed on Mars and the other on the Sun. Enrolments were limited to 30 children per day, and most days were fully subscribed. In addition to providing the children with astronomical games, quizzes, video screenings, art and craft activities (including mural painting), story- telling, drama sessions, and practical and design projects, staff also prepared two different 12-page Activity Booklets titled *Spaceship Mars and Journey to the Sun* (after the names of the respective holiday programs). These contain basic information and an assortment of activities, games, puzzles, and astronomical recipes, and were sold at cost ($NZ3 each).

Although both holiday programs were highly successful, staffing limitations and space constraints led to a much regretted decision not to schedule any further programs in the foreseeable future.

3.2. *"Overnight Extravaganzas"*

One of the Carter Observatory's most successful innovations are the "Overnight Extravaganzas" (see Orchiston and Andrews, 1995). On average these are held fortnightly, on a Friday evening, and groups of pre-teenage children and accompanying adults spend a night at the Observatory and sleep under the stars in the planetarium chamber (after the chairs have been removed).

The evening to some extent follows the program set out for the Observatory's twice-weekly "Public Nights", with two different planetarium programs ("The Southern Night Sky" plus whatever happens at the time to be the current feature program), the current audio-visual show, videos and discussions on astronomy, and sky-viewing through the 23-cm refractor (weather-permitting). The "Orbits" computer program is also popular. An additional optional activity towards the end of the evening program is a 10-minute walk through the Wellington Botanic Garden to view an area crowded with glow-worms.

Since their introduction in 1993 November "Overnight Extravaganzas" have proved extremely popular, and attracted groups from all over New Zealand. There is a strong demand to expand the client base by including teenage and adults groups, but the limited number of staff available for "Overnight Extravaganzas" thus far has precluded this.

3.3. *Astro-camp Booklets*

School camps and camps organised for brownies, cubs, scouts, guides and other groups are particularly common in New Zealand, and over the past few years Carter Observatory staff frequently have been asked to supply night sky information and suggest observing projects and other activities.

In responding to this obvious demand, during 1996 May and June one of the authors (K.L.) prepared six different Astro-camp Booklets. These deal, respectively, with general projects, constellations and mythology, galaxies, the Sun, and the Moon (two different

Booklets), and were produced in-house on the Observatory's computers and then photocopied. They range in length from 18 to 28 pages, and are sold "at cost"(i.e. from $NZ5:50 to $NZ7.00).

The Astro-camp Booklets are an important addition to the Observatory's range of publications (see Orchiston *et al*., 1997 for details of others), and are now listed in our twice-yearly Mail Order Catalogues. As a new, unique type of astronomical education resource specifically designed for New Zealanders, they are proving popular.

Given the continuing demand, further Astro-camp Booklets will be produced as time permits.

3.4. *Training*

With the introduction of the new astronomy curriculum, large numbers of secondary science teachers and primary school teachers suddenly faced the challenge of becoming "astronomy experts"virtually overnight. The Carter Observatory has responded to this demand by introducing a number of different training options.

After experimenting with an introductory astronomy course in 1995, we decided to replace this with four specialised courses, all set at about first year university level. They are:

- Hitchhikers Guide to Cosmology
- Hitchhikers Guide to our Galaxy
- Hitchhikers Guide to the Solar System
- Observational Astronomy

All involve eight successive 2-hour weekly evening classes, and provide up-to-date largely descriptive overviews. And in order to provide a national perspective, research by past and present New Zealand astronomers is introduced, where relevant. These courses are taught by three of the authors (F.A., R.H., and W.O.), and "lecture guides"with text, half-tones and line-drawings are prepared for most sessions. Students have access to a "loans section"of the Resource Centre, which contains an extensive collection of journal articles and other relevant material. Although actual observing sessions are encouraged and slotted into the program as optional extras, at least one lecture in each course is held in the Observatory's planetarium in order to circumvent the vagaries of Wellington's weather.

Students can take each course on a stand-alone basis, or by paying a surcharge and successfully passing set assessment work they can complete all four courses and obtain a Carter Observatory Diploma of Astronomy. All Diploma students undertake two different library-, observation-, or laboratory-based projects, and present their results in the form of essays and/or reports.

At the time of writing (1996 June), both the Cosmology and Solar System courses had already been taught (in 1995 and 1996, respectively), and the Observational and Galaxy courses were due to be piloted during the second half of 1996. Around 20 students enrolled in each of the completed courses, although most of them were amateur astronomers or those with a general interest in science rather than teachers or trainee teachers. Feedback received from the teaching profession indicated that teachers were "shell-shocked"by the new curricula and other educational restructuring, and lacked the energy or commitment to undertake evening courses, no matter how useful they may prove to be. We anticipate that this situation will improve with the passage of time, and that we will witness increasing teacher support for our courses.

Two other training initiatives specifically for teachers were introduced. In 1995 and 1996 the first two authors offered a number of 2-hour seminars that examined possible linkages between the new astronomy curriculum and other curricula (technology, mathematics, social studies and English) at mainly primary and intermediate teaching levels. Relevant resource material was examined and discussed, and appropriate teaching methods were explored and evaluated. These seminars generated considerable teacher interest, but because they were conducted after school hours and at personal expense, they too received less than satisfactory patronage.

Finally, in 1996 one of the authors (R.H.) began offering planetarium-based workshops designed to introduce teachers and other astronomical beginners to the night sky, as seen from New Zealand. The emphasis was on learning the main constellations, and learning about notable stars (including double and multiple stars), spectacular clusters and appealing nebulae. Four different suites of workshops were planned, dealing with summer, autumn, winter and spring skies respectively, with four different 2-hour evening sessions for each season. The summer and autumn workshops are now over, and once again teacher patronage has been minimal, even though surveys reveal that this is precisely the sort of practical program most needed by New Zealand teachers!

4. CONCLUSION

Since its restructuring in 1992, Carter Observatory has developed a range of dynamic education programs for school-age students, teachers and trainee teachers, as well as amateur astronomers and members of the general public. Further initiatives are planned as additional funding and space become available.

Acknowledgements

We are grateful to the British Council and the International Astronomical Union for grants that made it possible for one of us (W.O.) to visit London and present this paper at IAU Colloquium 162.

REFERENCES

ANDREWS, F., 1991, The Carter Observatory education program. *Publications of the Astronomical Society of Australia* 9, 174.

ANDREWS, F. & BUDDING, E.,1992, Carter Observatory's 9-inch refractor: the Crossley connection. *Southern Stars* 34: 358-366.

LEATHER, K., ANDREWS, F., BUCKLEY, D., CARTER, B., HALL, L., HALL, R., HALL, S., LEATHER, K., LEATHER, N., MATLA, P., ORCHISTON, W., & SULE,C., 1997, Special school holiday programs at the Carter Observatory, *Publications of the Astronomical of Australia*, in press.

ORCHISTON, W., 1995, Public astronomy and education at the Carter Observatory: some recent developments. In Orchiston,W., Carter,B., and Dodd,R.(eds.). *Astronomical Handbook for 1996*, 59-69, Carter Observatory, Wellington.

ORCHISTON, W.& ANDREWS,F., 1995, A cloudy night under the stars: "Overnight Extravaganzas"at the Carter Observatory, *The Planetarian*, 24, 15-19.

ORCHISTON, W.& DODD, R., 1995, *A Brief History of the Carter Observatory*, Carter Observatory, Information Sheet No. 1, Wellington.

ORCHISTON, W.& DODD, R., 1996, Education and public astronomy programs at the Carter Observatory: an overview, *Publications of the Astronomical Society of Australia* 13, 165-172.

ORCHISTON, W., ANDREWS,F. & BUDDING,E., 1995, *A History of the 23cm Refracting Telescope at the Carter Observatory*, Carter Observatory, Information Sheet No. 13, Wellington.

ORCHISTON, W., CARTER,B., DODD,R. & HALL,R., 1997, Selling our Southern Skies: recent public astronomy developments at the Carter Observatory, New Zealand. This volume.

RICHTER, J. & FRAKNOI, A., 1994, Matches made in heaven, *Mercury*, 23(5), 24-27.

Science in the New Zealand Curriculum, 1995, Learning Media, Wellington.

WHITEHOUSE, M.L., 1994, Geography with an astronomical slant in Nova Scotia schools, *Journal of the Royal Astronomical Society of Canada* 88, 277-279.

U.S. Science education reforms: is astronomy being overlooked?

By D.B. Hoff

Department of Physics, Luther College, Decorah, IA, USA

1. Recent history of science education reform in the USA

In 1981, in response to growing concerns that the United States was falling behind the rest of the world educationally, the federal Secretary of Education created a national commission on excellence in education. This commission was charged with gathering data about the status of U.S. education compared to the rest of the developed world and to define the problems which would have to be faced to successfully pursue the course of excellence in education.

In 1983 this commission issued its report, *A Nation at Risk*, (Secretary of Education, 1983). The release of this book produced a flurry of activity by schools, political entities and professional groups representing various educational disciplines. These groups included, the National Council of Teachers of Mathematics, the National Governors Association and the National Science Teachers Association and others. By 1989, the American Association for the Advancement of Science (AAAS), a major American organization representing a broad spectrum of the sciences, produced its own call for an improved educational climate for science and engineering. Their book, *Science for All Americans*, attempted to produce a comprehensive expression of the scientific community as to what constitutes literacy in science, mathematics and technology (Rutherford and Ahlgren, 1990). The release of this report, coming from a credible, broad-based and nationally recognized organization of scientists and engineers produced a great deal of interest in the American press and calls came for developing strategies for action.

Following the publication of *Science for All Americans*, with financial support from a number of public and private agencies and foundations, the AAAS followed up this broad call to action by establishing a large task group of teachers, administrators, scientists and science educators charged with establishing benchmarks by which literate students in the 21st century would be judged. The task group was collectively called Project 2061. They realized that achieving improved science literacy would to be a slow process. The year 2061, marking the year of the return of Halley's Comet, was chosen as the project name to denote the long time likely required to acquire uniform science literacy in the U.S. The many participants in the benchmark planning process were divided into discipline teams and four years later, in 1993, produced *Benchmarks for Science Literacy* (AAAS, 1993). This volume listed the benchmarks for the various sciences for children from kindergarten through grade 12. More about the general benchmarks and those specific to astronomy will be given below.

Other action was brewing on the reform movement as well. The prestigious National Research Council, chartered by the United States Congress in 1863, got into the act. Starting in 1991 this council began planning for the development of a document that would, not only examine science education as defined narrowly by science content, but would develop standards by which one could judge not only children's understanding of science content, but also set the standards for the professional development of teachers, for science teaching itself, for assessment of learning and for the entire science education system from kindergarten through graduate school–and beyond. Their document

National Science Education Standards was released just this year (National Research Council, 1996).

2. Astronomy content in the reform documents.

We now come to the question in the title of this talk. In the United States reform movement in science education has astronomy been overlooked? Certainly as one examines the astronomy content of the NRC standards document , you would have to agree with Jay Pasachoff who wrote a clear indictment on the lack of astronomy subjects and content in the last issue of the *American Astronomical Society's Newsletter* (Pasachoff, 1996). Let us quickly look at the NRC list in its entirety. As Pasachoff notes, there is no astronomy outside the solar system listed for grades 5-8. That nuclear energy is the driving force within stars is not present, and it is a major leap to go from the solar system to the origin of the universe. (I have copies of these standards if anyone is interested.)

But what about the AAAS document, *Benchmarks for Science Literacy*? Here the bench- marks for astronomy content is found in Chapter 4, "The Physical Setting". This section is further divided into sub-sections on the universe and one on the earth. These sections cover ten pages and are much more extensive than the standards found in the NRC book. For example by the eighth grade, students following a curriculum based on the AAAS benchmarks would have an introduction to the world of stars and that nuclear processes are the driving force in stars. Also the link between the stellar world and the origin of the universe is much smoother than in the NRC work. (The AAAS list of topics is too extensive to put on transparencies, but I have extracted the benchmarks themselves from the book and have a number of copies to distribute to anyone interested.)

3. Why does it appear that astronomy has been overlooked in the NRC document?

Typically, astronomy has not been a major topic in American science curricula. While astronomy as a subject did appear in American high school curricula in the late 1800's, it had virtually disappeared after the beginning of this century, not to reappear until after the beginning of the Space Age. This disappearance can be traced to an educational reform movement designed to standardize the high school curriculum. I shall not trace the reasons for that drive to reform, but will only note that in 1893, the so-called "Committee of Ten" produced recommendations for what course work a high school student should have completed prior to admission to college. These recommendations grew out of recommendations from a number of task forces from a variety of disciplines, including the sciences. While the committee was referred to as "The Committee of Ten", there were really a number of conferences held that reported back to the main committee. The main committee's final report recommended coursework in both physics and chemistry, but not astronomy. If offered at all at the high school level, the report further stated, astronomy need only be a 12-week elective course.

The denigration of astronomy as a subject in schools a century ago may have a modern counterpart. There were no astronomers on the committee charged with developing recommendations for "Physics, Astronomy, and Chemistry" at the beginning of this century. (In fact, at their first meeting, this group renamed itself the sub-committee for "Physics, Chemistry , and Astronomy", moving astronomy to a third, and minor role.) In today's reform, an examination of the names listed in the "Working Group on Science Content Standards" for the *National Science Education Standards* shows a similar lack of involvment from the astronomical community. In the appendix of the book listing contributors we find the names of biologists, chemists, geologists and physicists but no astronomers. I have no way of telling if this was the result of lack of interest on the part of astronomers or an oversight by the planning committees, but nonetheless none were on the science content standard's group.

It may be more than mere speculation that the reason astronomy appears to be treated more fairly in the AAAS work is that in the equivalent contributor listing in the AAAS document there are two astrophysicts listed, along with biologists, chemists, geologists, meteorologists and physicists.

4. What will be the effect on astronomy education as a result of the current reform movement?

In answer to that question, it is probably too early to tell. First, change comes slowly in education. Second, direction for science education in the U.S. is highly decentralized. Education in the U.S. is under the direction of the individual states. Under the U.S. Constitution, any function of government not specifically spelled out as being the function of the federal government is reserved to the states. Education is therefore described as a "state's right". This means that there really are 50 different loci for educational leadership in the U.S. In fact, it is more complex than that, as there is a great deal of local control within the states themselves. There are approximately 16,800 separate school districts in the U.S., each with a fair degree of autonomy in constructing and choosing curriculum materials. It is true that some states are producing educational frameworks, based on the reform documents, that are to be used by local districts as they construct their own science curricula. Add to that, the U.S. has no centralized, uniform testing in the sciences. Some textbook publishing companies are tailoring their products to conform to the AAAS benchmarks and likely will pay lip service to the NRC documents as well but it remains to be seen whether these products will have a major impact on student achievement.

What about curriculum development or dissemination projects funded by the federal government? Are they incorporating topics that reflect the guidance of the reform documents? The American Astronomical Society is conducting a very agressive teacher resource agent program whose purpose is to disseminate materials that have been developed with National Science Foundation funds. This program currently has about 220 teacher-agents located in nearly all of the United States whose function it is to conduct workshops for other teachers sharing materials developed by federally funded projects. This program is selecting those topics for the appropriate grade level that best reflect the guidance of the reform movements. But this is *post facto* as most of the curriculum development projects themselves do not appear to have paid much attention to the work of the AAAS, in spite of the fact that it has been out in final form for three years.

REFERENCES

AAAS, 1993, *Benchmarks for Science Literacy*, New York, Oxford University Press, New York.

NATIONAL RESEARCH COUNCIL, 1996, *National Science Education Standards*, National Academy Press, Washington D.C.

PASACHOFF, J. M., 1996, "Astronomy and the New National Science Education Standards: Some Disturbing News and an Opportunity", *AAS Newsletter* (80), June.

RUTHERFORD, F.J. & AHLGREN, A., 1990, *Science for All Americans*, Oxford University Press, New York.

SECRETARY OF EDUCATION, 1983, *A Nation at Risk*, U.S. Government Printing Office, Washington, DC.

"Plaza del Cielo" Complex: its state of evolution

By N. Camino

Department of Physics, Faculty of Engineering, National University of Patagonia, 9200 Esquel, Chubut, ARGENTINA.
e-mail: nestor@unpate.edu.ar

1. Our motivations for the creation of "Plaza del Ciel "

We consider that one of the most important aspects in the harmonic development of a person is his relationship with the natural environment in which he lives: in that environment he initially trains and enlarges his curiosity and the capacity for astonishment, both innate properties of human beings. It is so much so that we could locate the germ of all creative future activities, scientific or otherwise, in what happens with those characteristics during the years of childhood, that if fully developed will help children to become sensitive and critical adults. So we consider it necessary to generate the mechanisms to reinforce the sensibility, the critical observation and the interaction with natural phenomena, in a systematic way, as we propose in "Plaza del Cielo".

Education by means of Astronomy is the principal nucleus of our proposal. This is so especially because we consider that those aspects of Nature studied by Astronomy and the way this science works is a powerful tool to motivate children and adults, that it is one of the best ways to study Nature, and that it brings us as educators new means to show the evolution and full integration among the many ways, not only scientific, Humanity has constructed throughout History in order to comprehend not only the Universe, but ourselves as well.

1.1. *Geographical location and zone of direct influence*

The Plaza will be situated at Esquel, a small city of 23,000 inhabitants, close to Los Andes mountain range in central Patagonia, Argentina (Fig. 1). The zone we have defined as of direct influence has an area of approximately 80,000 square kilometres, with Esquel as the most important city, which includes 65,000 inhabitants; 12,000 of them are primary students, 2,200 secondary students, 950 primary teachers and 450 secondary teachers.

1.2. *Institutions supporting the Plaza*

There are three institutions that bring the main support to Plaza del Cielo Complex: Municipality of Esquel (the project has been declared "of community interest"), National University of Patagonia (for education and investigation reference) and "Educándonos" Foundation (for the economic aspects). Furthermore, the "Social Educational Plan", a Program of the National Ministry of Culture and Education, which assists nearly ten thousand indigent schools in the whole country (like those in Patagonia), brings its economic, material and human support to this project. We want to thank all of these four institutions for their support, without which this project could not have been developed.

2. Synthetic Description of the Project

The Plaza could be thought of as a huge interactive didactical tool, without any restriction, neither social, economic, nor by age of those who will go to play and to learn in it. The complex (see Fig. 2) includes many modules designed for the teaching of Astronomy,

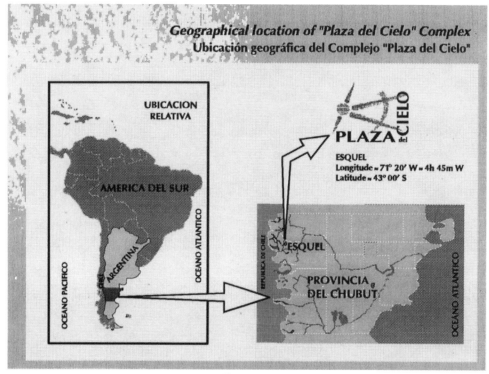

FIGURE 1.

FIGURE 2.

capable for their utilization outdoors or even indoors, which could be built with simple, low cost and low maintenance materials. In the design of all the modules we kept in mind two aspects: the educational and the ludic, the two foundations of the proposal.

The Plaza modules have been organized in four main "conceptual streets"(Fig. 3): the "Sun's diagonal", to understand the relative motion of the Sun with respect to our topographical location; the "South's street", in order to understand our southern location on a planet with a definite rotational movement; the "History's street", to understand that science, and particularly Astronomy, is a cultural product, dynamically

"Plaza del Cielo" Complex – Esquel – Argentina
Complejo "Plaza del Cielo" – Esquel – Argentina

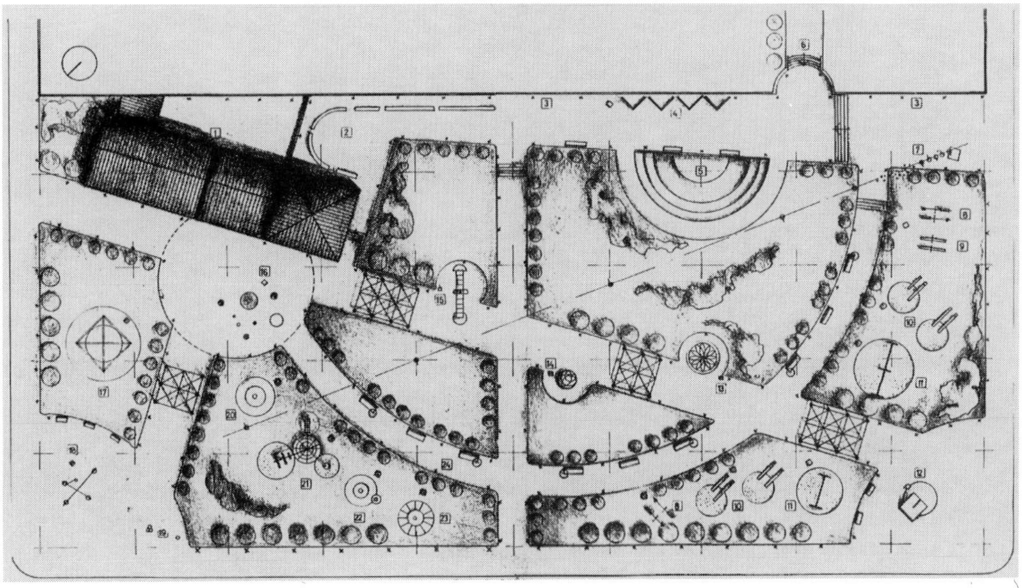

SAENZ PEÑA

FIGURE 3.

changing during time, a human tool to describe the Universe in which we live; and the "Culture's street", a connection with cultural aspects in a young and active community like Esquel and Patagonia. Furthermore, there are two kinds of square games: the "Gravitational" ones, to "live" consciously our relation with the Earth's gravitational field; and the "Planetary" ones, in order to dramatize, by playing with them, the gravitational relations among Solar System bodies, natural or artificial.

The modules included in each street and set of games are the following ones.

"Sun's diagonal": equatorial sundial, climbing armillary sphere (not indicated in Fig. 3), parallel Earth globe, Solar System traverse board game, Solar System distances mock-up, Solar System sizes mock-up.

South's street: Southern Cross mock-up, Foucault's pendulum (now inside the building).

History's street: statues of those astronomers that gave humanity a transcendental cosmic vision, considered each of them as the "convergence of an epoch"; they are, in our personal epistemological conception (we can talk about it...), Aristarco, Tolomeo, Copernico, Galileo, Newton, Einstein, and there will be an empty bench, waiting for who will sign the next cosmic vision, the next convergence.

Culture's street: cultural center (with the Foucault's pendulum and a planetarium, and specific audiovisual media, library, etc., conference room, and a classroom), an artisans fair, an open theatre, and the Poplar's mall.

Gravitational games: see-saws for one child, see-saws for two children, slides of different heights and equal inclinations, slides of Earth-Mars-Moon, swings of different longitudes.

Planetary games: climbing Voyager mock-up, asteroids' round, Earth annual movement merry-go-round, Earth-Sun-Moon system merry-go-round.

FIGURE 4.

3. State of Evolution of Plaza Del Cielo Complex

At present, the beginning of the construction of this complex is in a "hard stand-by", only because of economic reasons; so, we don't know when it will be possible to inaugurate it.... Notwithstanding this, and as we are very tenacious, the "Plaza team": Inés Irigoyen (a primary teacher), Mariano Kasanetz (a physicist) and myself, is developing an ambitious and intense activities plan for the 1996-1998 period (Fig. 4), as follows.

Lectures for primary and secondary students: many lectures and activities are developed each year, indoors and outdoors, observational, etc., in schools of the region, as required by teachers or even by the students.

Teacher training program: activities or courses to train primary and secondary teachers in those astronomical concepts included in the curricula, with a strong focus on the observational aspects.

"El rastro del Choike": a monthly publication in the local newspaper, directed to the general public, but especially to children and teachers. Its name is a homage to those ancient Tehuelches, native people of central Patagonia, who used to name in that way the group of stars we, nowadays, call the Southern Cross. The "Choike" is a big bird, an American ostrich (ñandú), and this native contellation means "its footsteps (el rastro) in the night sky" (Fig. 5). The "main" section has the objective to explain many phenomena or events, focussing on their local observational aspects; the "children's questions" answering those questions produced in the planetarium performances or in the lectures; "I ask you..." in which children answer our questions; "Astronomy and Expression" to show connections between our science and the Arts; "shooting news" is astronomical news in order to show the present and future evolution of Astronomy. "El rastro del Choike" is being reedited by the "Social Educational Plan" in order to extend this educational effort on the teaching of Astronomy to other geographical regions.

Scientific communication articles: many articles in radio, TV, and newspapers, related to phenomena, news, etc.

Planetarium program: GOTO Mfg. Co. has donated an EX3 planetarium for the development of Plaza del Cielo Complex (we want to thank them very much, because it was a real lift-off for us). We have developed two performances: "Below the Southern Cross", oriented to know the sky of Esquel and Patagonia, to learn how to determine cardinal points and to know some constellations (Orion, Taurus, Southern Cross, Scorpius); and "The Zodiac, the path of the Sun among the Cultures", oriented to discuss

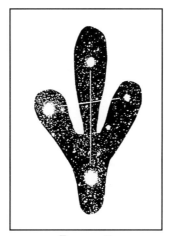

FIGURE 5.

FIGURE 6.

both the astronomical explanation of the Zodiac and how we can know past Cultures by learning about the mythology of those constellations. Weekly, three performances for primary schools and one for the general public are held. The planetarium was inaugurated in May, 1995, and at present approximatelly 2,500 persons had come to participate in them, 90% primary students, which means an intense response to this activities program.

Audiovisual lectures: lectures held once a week for the general public, in which, after a video or slides, an informal talk is developed, related to the interests of the public.

The Astronomy week: the 24th of October is Astronomy day in Argentina, so we have planned an intense week, full of activities : audiovisual lectures; planetarium performances; special issue of "El rastro del Choike"; observational activities; expositions of stamps, photographs, instruments; astronomical puppets (by Horacio Tignanelli, an astronomer and an Astronomy educator, specializing in puppets and an author of many books); primary childrens' drawings exposition; and a talk about "how to become an astronomer" designed for secondary students.

CREDITS OF PLAZA DEL CIELO COMPLEX

The staff of the Plaza is the one we describe in Fig. 6; without them all, the Plaza could not be working like it is actually now. But there are many other credits, especially these:

Arq.Alberto Teszkiewicz and Ing.Eduardo Averbuj, specialists in the architectural and conceptual design of interactive places for the teaching of Science, for they made possible the materialization of a dream; Prof. Nicoletta Lanciano, for she is a strong source of inspiration and of learning; Lic. Horacio Tignanelli, for he is a natural creative person who helped me to respect my fantasy; Lic. Hilda Weissmann, for she motivated me to create the Plaza; and my family, for they are a living support.

Astronomy as a School Subject

By J-L. Fouquet

18, Rue de Puits de Fev, 17630-La Flette, France

1. A brief survey about the teaching of astronomy in France at Primary and Secondary School level

a) From 9 to 11, a "Science and Technology" curriculum is extensively provided by primary schools :

"Pupils, through some aspects of the scientific approach, are encouraged to develop their abilities to formulate questions and logical answers through a wide scope of investigative and practical observations and other scientific activities such as collecting measures, linking data, interpreting documents" – extract from the official syllabus which includes the following chapters :

1. Unity and Diversity of the Living World :
   - the development of living creatures,
   - animal reproduction,
   - human sexuality,
   - ecology,
   - nutrition.
2. The Human Body and Health Education :
   - sports,
   - hygiene,
   - nutrition,
   - first aid.
3. The Sky and the Earth :
   - the earth movements / revolutions,
   - the solar system,
   - the universe,
   - light and shadow,
   - calendars,
   - sundials.
4. Energy and Matter :
   - the cycle of water,
   - the quality of air,
   - energy : providing and saving energy.
5. Objects and Technological Projects :
   - wire connections,
   - mechanical and electrical devices,
   - models.
6. Computer Science :
   - using software within the scope of school subjects.

Officially, this programme should be divided into 72 periods of one hour and a half each. In actual practice, about 200 hours are devoted to partly cover the teaching of science, among which 20 hours, spread over two years, deal with astronomy.

In most primary schools, teachers are obviously the main suppliers of information, and they do so according to their own preferences and abilities. Nevertheless, teachers, as individuals or as members of the teaching team, can always ask a qualified speaker from outside the school to take part in the teaching / learning process. It is within this specific framework that I have been able to teach astronomy in various schools in the area where I teach physics, and this for more than 10 years, now, mostly dealing with children from 9 to 11.

b) At Secondary School level :

• from 11 to 13 : physics and astronomy are no longer taught.

• from 13 to 15 : physics is taught by a qualified teacher 2 hours a week. During those two years, no more than 10 hours can be devoted to astronomy.

2. What is the common ground between Primary and Secondary Schools?

The first thing we should emphasize is the major importance of inquisitive thinking and scientific questioning at the earliest age possible ...

... as early as nursery school, when children are full of enthusiasm, eager to communicate and imaginative, when they are always ready to listen to fairy tales, sing a "ring-a-ring-a-roses", dance and play in circles, create stories and draw.

... and until they are about 10 and have still got a natural curiosity that they will unfortunately lose a few years later when they will start feeling conscious and ashamed of their body, under other people's eye.

Then, the astronomer should go to school and meet the children there, on a very regular basis, for a period of at least a few weeks.

There should be, well in advance, some kind of preparation shared with the teacher in charge, so that the long term objectives of the project, its different steps and the final evaluation of the achievements are outlined long beforehand.

The various sessions can be varied :

• indoors : movies, slides, models, balls, lights can be used.

• outdoors : observation of the moon, shadows, sunspots; dancing in rings to imitate the moon, earth revolutions and the solar system.

• in magic meeting points : such as a planetarium, an exhibition or "discovery classes" where magic meetings can happen with a real genuine astronomer for example.

We should take into account the initial interpretation of children. We can, at a certain point, accept a child's inaccurate representation with some kind of reservation, as long as we can slowly improve its accuracy through discussions, observations... Otherwise, pupils will become passive.

Let's take an example with the annual earth revolution and the cycle of the seasons : A set of slides illustrating the above activities at school were shown.

• number 1 : Experiment with the help of a globe and a spotlight to exclude possibilities. The northern hemisphere would always experience summer! See Fig. 1.

• number 2 : Statement : the Pole Star is fixed as are the other stars.

• number 3 : 4 tennis balls and 4 parallel strings long enough to show that the earth axis always points towards the Pole Star. See Fig. 2.

Finally, we should offer a lot of books to read to young children. Avoid encyclopedias, which are too compact, and books with too many beautiful illustrations or already solved problems. We should always prefer books or, for younger ones, tales that lead to new tracks, awaken curiosity and initiate discussion. Reality and fiction will be divided as such, later on, when and if necessary.

A set of slides illustrating the above were shown.

polaris

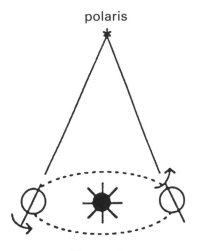

FIGURE 1.

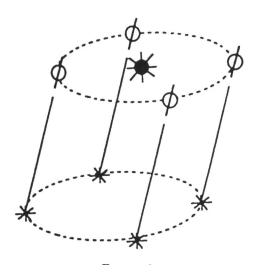

FIGURE 2.

Teaching Astronomy at Secondary School Level in Europe

By R. M. Ros

Universitat Politècnica de Catalunya, Barcelona, Spain.

1. Introduction

The status of teaching Astronomy in European countries is variable. Sometimes Astronomy appears as a compulsory subject or as an optional subject, but on many occasions Astronomy appears within another subject, depending on the country. It is even possible for Astronomy not to appear anywhere in the curriculum. But of course the position here is better than in other less developed places. In Europe there are various topics which can be organized into two main groups: aspects related to relative motions and aspects related to properties of light. Some examples of teaching activities and materials in various countries will be described.

It is also necessary to emphasize several initiatives such as the review of Astronomy curricula, the publication of general books on Astronomy for secondary schools and the organisation of new journals to promote Astronomy in schools.

It is essential to mention the new European Association for Astronomy Education (EAAE) founded last November in Athens. This meeting was attended by 100 teachers and astronomy professionals from 17 European countries. It is hoped that this, in conjunction with the other initiatives, will do much to encourage the study of Astronomy.

2. Relative Motions.

In this field, as in others, there is some very interesting material promoted by the Comité de Liaison Enseignants et Astronomes (CLEA) in France. Denise Wacheux has produced a special umbrella which is used to study the movement of the Sun and celestial sphere in relation to the horizon, and which has very interesting didactic applications in secondary schools. It is possible to change the latitude and to move the umbrella around its axis. Designed for a similar purpose, Roland Szostak (University of Münster, Germany) has interesting material to study the Sun's movement through a half-sphere of plastic on a horizon (Fig. 1). Students use this to plot the sun's trajectory by putting stickers on the half-sphere at regular intervals of time.

There is also a model made using photography from the Permanent Astronomy Seminar of Catalonia (PASC) (Polytechnical University of Catalonia, Spain) made in collaboration with Nicoletta Lanciano (La Sapienza University, Rome) who has wide experience in this field in primary schools. To do this activity the students have to take a collection of photographs of their horizon at school (Fig. 2). With this material they make a cylinder which is supported on a piece of wood. In this oriented model according to the meridian line, they have to put several astronomical concepts such as the local meridian, the equator, the Earth's rotation axis, etc. This model is very important in order to have a vision of the celestial sphere from outside and inside. When the teacher draws the celestial sphere on the blackboard the students have a view from outside. But when the students do a practical activity of course, they are inside the sphere. This kind of change is clarified with this model.

Some simulations of several astronomical aspects have also been developed. Some planetaria (Vittorio Mascellani, Italy) organize activities in which students participate

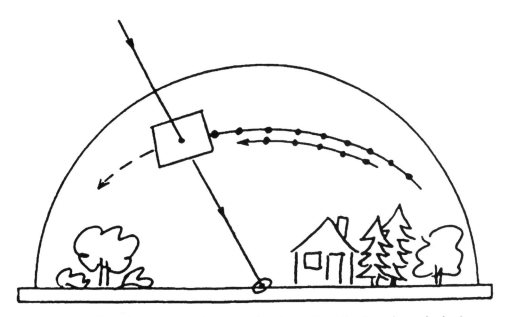

FIGURE 1. *The plastic sphere for measuring the path of the Sun above the horizon. (Germany).*

to study solar phenomena or to construct a small model of the solar system in the garden. In this type of simulation, members of the Permanent Astronomy Seminars of Euskadi and Catalonia, (PASE and PASC), prepared a model to observe the different kinds of conics on a horizontal plane obtained from the Sun's diurnal movement. This material enable us to explain why the end of the gnomon's shadow describes a conical line. The Sun's ray is simulated by a thread which starts from the Sun's position (in the celestial sphere), passes the end of gnomon and crosses the horizontal plane at the same point as the end of the gnomon's shadow on the horizontal plane. We repeat this simulation at various solar positions. If we observe all the intersection points, we obtain the conical line (ellipses, hyperbolae, parabolae or straight lines). The type of conical line depends on the latitude. The students can only observe one of these conical lines in their city. A similar model was made with small lights which it is possible to switch on and off, one by one, and to observe the shadows on the horizontal plane. This last model has the advantage that it is possible to change the latitude and obtain all the different conical lines in the same model. With this model, the teacher and the students can make an imaginary journey to another country with a different latitude. So they are then able to observe the conical line according to this new latitude.

Still on simulation but this time using computers, there is the work done by Francis Berthomieu (CLEA, France) concerning interactive programs of the Moon's phases and the planets' movements.

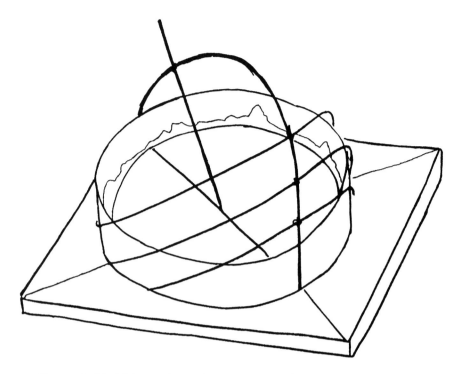

FIGURE 2. *Model for studying the specific school's horizon. (Italy and Spain).*

3. Properties of Light

In the field of Astrophysics and the properties of light, there are interesting experiments developed by Laura Abati (Italy), for example, one to determine the Sun's temperature and another related to Hubble's law. Also a group of activities were developed by Lucienne Gouguenheim (CLEA, France) about spectra and the construction of a spectroscope to observe the spectrum lines. In the same area, Lucette Bottinelli presented a collection of practical experiments which use photocopies of a spectrum to determine its characteristics. She covers very easy cases, for example assigning the spectral type in the Harvard classification (Fig. 3), up to others which are more difficult and elaborate.

The PASE also made a spectroscope constructed with a piece of compact disc placed in a match-box. Along these lines, members of PASC produced a general set of practical activities about the H-R diagram which include the construction of a spectrocope which can be used to read the wavelength of several lines. Another group of professors from the same PASC used photographic activities related to the H-R diagram, magnitudes and variable stars.

In particular this last activity was done in collaboration with members of the Permanent Astronomy Seminar of Aragón (PASA, Spain). The activity on the variable star is very easy to run and has interesting results. The original idea is to test the luminosity curve of a regular variable star, knowing the star's period and epherimedes. If it is necessary, the students have to prepare a simple programme to obtain the timetable to make the observations. For example to observe $\beta Persei$ it is necessary to know when the star will be at its minimum brightness, but if we observe $\delta Cephei$ or $RRLyrae$ it is not necessary because these stars do not have such a spectacular change. In every case the

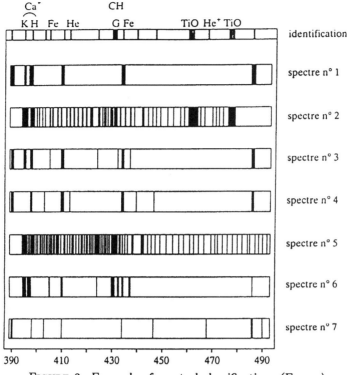

FIGURE 3. *Example of spectral classification. (France).*

students have to take some photographs (colour slides) of the star and its constellation, and note the time. When the teacher has some of these photographs, in the classroom the students can calculate the phase and can assign the magnitude for each case, using the method of Argelander. Then, the students obtain some points of the luminosity curve (Fig. 4). Of course it is possible to use the same process to obtain directly the curve without previously knowing it, but this takes longer and does not provide any new information and is boring for students.

The Open University of the United Kingdom offers several projects which can be useful in secondary schools. For example, they have interesting material concerning visual stellar magnitudes and luminosity of the Sun. In particular their study on limiting magnitudes in and around Orion is very appropiate because this constellation is very easy for school pupils to observe when the day is very short.

4. Aspects of Curricula and Publications.

In Greece, the Ministry of Education decided to establish a working group, presided over by Panayioti Niarchos and comprising specialists in Astronomy, in order to review the current curriculum for the subject of Astronomy suitable for secondary schools. To encourage and increase the knowledge of Astrophysics among school students, a first summer school in Astronomy this September, will be organized sponsored by the Academy of Athens.

In Portugal, a group of teachers from secondary schools coordinated by Felisbela Martins, is organizing its national association on teaching Astronomy. At present, its first

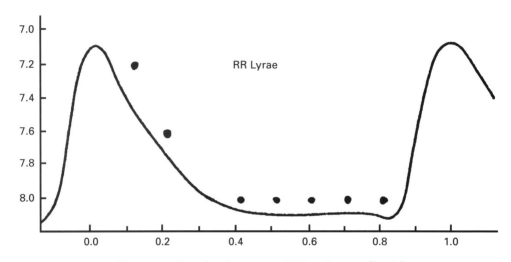

FIGURE 4. *Luminosity curve of RR − Lyrae. (Spain).*

and principal objective is to organize the 1st National Meeting on Astronomy Education next year.

In some eastern European countries a text book on Astronomy for secondary schools has been recently published. It is necessary for each country to have a book of this kind adapted to its own conditions as it is a vital support in the teaching of the subject. For example in Latvia, Ilgonis Vilks (Riga Observatory, Latvia) finished an Astronomy book for secondary schools and Nikola Nikolov (Sofia University, Bulgaria), Vytautas Straizys (Institute of Theoretical Physics and Astronomy, Lithuania) and E. Kharadze (Tbilisi State University, Georgia) have recently finished other textbooks for the same level.

I do not want to finish my talk without mentioning Edward Kononovich (Moscow University, Russia). He organized an almanac called "The Universe and Ourselves" to cover school problems such as teaching aids, school telescopes and various materials. The main goal of this publication is to help students to understand the place of Mankind in the Universe.

It is necessary to talk about the Blossoms of Science in Israel and their publications and studies on the conceptual frameworks in Astronomy. In particular they observe the secondary school students' progress towards heliocentric concepts, the solar system and astronomical phenomena from a sidereal perspective. Space Science and Astronomy were introduced lately into the curriculum of primary and secondary schools in Israel. Tel Aviv University and the Centre for Educational Technology have produced books and computer software.

5. International Contacts.

To close this paper about teaching Astronomy in Europe it is necessary to speak about the new European Association for Astronomy Education (EAAE), which will improve and promote scientific astronomical education all over Europe in schools at all levels and other institutions involved in teaching Astronomy. The existance of this new EAAE proves the vitality and dynamism of teaching Astronomy in Europe. We can hope this initiative will bear fruit soon.

The interchange with other countries is very enriching. Before finishing I would like to talk a little about the 4th and 5th International Conferences on Teaching Astronomy held in Spain, in 1990 and 1995. Both conferences were very important to promote and stimulate Astronomy in Spain. Teachers and professors from different countries participated in these activities and this promoted several collaborative projects, students exchanges and, of course, a lot of new personal contacts. For example the last meeting, in 1995, gave encouragement to the teachers of secondary schools who were organising the 1st National Meeting of ApEA (Spanish Association for Astronomy Education) held in Caceres last year. It also encouraged us to set up the 1st Summer School of the teaching working group of SEA (Spanish Association of Professional Astronomers) which will be held in Alicante this September.

The work carried out by Kurt Locher from Switzerland can be used as a basis to promote activities between several countries. His studies on Astronomy throughout the history of ancient cultures encourage observational Astronomy today. This common past for a lot of countries is a nexus between them, which it is possible to explore.

Obviously many aspects concerning teaching Astronomy in Europe have not appeared in this brief talk. The intention was not to provide an exhaustive description but simply to give you a taste of what is happening in Europe.

REFERENCES

Astronomy and the New Technologies, 1995, European Association for Astronomy Education, Eugenides Foundation, Athens.

Proceedings of 5th International Conference on Teaching Astronomy, 1995, Institut de Ciències de l'Educació, Universitat Politècnica de Catalunya, Barcelona.

A High School Astronomy Course for a Wide Range of Student Abilities

By G. E. Sampson

Wauwatosa West High School, Wauwatosa, Wisconsin 53222 U.S.A.

1. Introduction

Astronomy is, inherently, a high-interest subject. However, at the high school level there is a tendency to teach astronomy using higher- level abstractions and complex mathematics. This teaching approach thus eliminates a large number of students who have difficulties with abstractions and complex mathematics, thereby restricting the study of astronomy to a rather select group of students.

The astronomy course offered since 1976 at Wauwatosa West High School was developed to reach a wide range of students with differing abilities. The prerequisite for this one-semester elective course is the successful completion of one year of high school science. Most of the students enrolled in this course are high school juniors and seniors, ages 16-18. Since algebra is not a prerequisite, the "math phobic"students have been attracted to the course. The higher ability students enjoy the challenges posed by astronomy and often take this course as a supplement to their physics classes. Students who normally have difficulty with science suddenly discover that they can succeed in astronomy, and we have introduced a whole new group of students to this high-interest subject.

2. Activities

Our astronomy course focuses on hands-on activities, which can illustrate higher-level abstractions in a more concrete manner. In particular, we use Project STAR activities, which were developed at the Harvard-Smithsonian Center for Astrophysics in Cambridge, Massachusetts. Our text is also the Project STAR textbook (Coyle, et al, 1993). The following subsection will illustrate some of these hands-on activities.

2.1. *Orientation: Celestial Sphere*

[Slide: exterior of Wauwatosa West High School] Wauwatosa West High School is located in a suburb of Milwaukee, Wisconsin, on the western shore of Lake Michigan. As a public school, we serve a wide range of students. Thus, as we begin the astronomy course, orientation (both on Earth and on the celestial sphere) become important. [Slide: student assembling globe] Students plot star maps on the inside of two plastic hemispheres. [Slide: student assembling globe] An ecliptic strip is added, and the two hemispheres are joined together on a shaft made from a surplus coat hanger. [slide: student measuring horizon collar] A horizon collar is added. [Slide: student using celestial globe] A map pin is placed onto the ecliptic to represent the Sun, thus making it possible to simulate the Sun's daily motion for any day of the year. The globes are tilted at an angle equal to our latitude of 43° North. These globes can be duplicated for other areas of the Earth, although there was a bit of a dilemma when we constructed celestial globes with American and Australian students in Wollongong, Australia at latitude 35° South. We resolved this dilemma by "inverting"the globes, so that south was at the top.

2.1.1. *Diurnal Path of the Sun*

[Slide: student using sun-tracking hemisphere] We also plot the diurnal path of the Sun directly with these Sun-tracking hemispheres. First the students draw their predicted daily path of the Sun on the hemisphere. A common misconception is to show the Sun's path passing through the zenith, which never happens at our latitude. Then the students plot the actual path of the Sun, with the tip of a pen acting as a gnomon. When the pen's shadow falls on the center of the hemisphere's base, the Sun's current location can be plotted on the outside of the hemisphere. At least five points are plotted, and when joined to the sunrise and sunset points for that day, it is possible to plot the diurnal path of the Sun.

2.2. *Earth-Sun-Moon*

2.2.1. *Ratios and Proportions*

[Slide: students with spheres] Ratios and proportions are very important, especially in our unit on the Earth, Sun, and Moon. One of our activities involves the students in choosing from a series of assorted sizes of spheres everything from lead shot to basketballs. Students work in pairs, with the first student selecting a sphere to represent the Earth. The second student selects a sphere to represent the Moon in its correct proportion to the Earth on the same scale, with the Earth-Moon ratio being roughly 4:1. Then the students move to the correct proportional distance between the Earth and the Moon, using 30 Earth diameters as the proportional distance.

2.2.2. *Lunar Phases*

[Slide: student with styrofoam sphere] Moon phases are simulated by using a light bulb to represent the Sun, a student to represent the Earth, and a styrofoam sphere, held at arm's length, to represent the Moon. A second student observes the changing "phases" of the Moon as the first student slowly rotates 360° while holding the "moon" at arm's length. [Slide: student using lunar phase dial] These lunar phases are recorded on a chart, and the names of the phases and a time-of-day dial are added. With this lunar phase dial, students can determine the time of day when a given phase of the Moon rises, sets, or when it is highest to the south.

2.2.3. *Lunar Surface Features*

[Slide: students working with lunar maps] Inexpensive lunar maps can be used to locate and describe lunar surface features. [Slide: student observing lunar rocks] The National Aeronautics and Space Administration (NASA) loans out samples of lunar rocks for study in the classroom. Here a student is observing lunar rock samples under a stereoscopic microscope.

2.3. *Pinhole Tubes*

[Slide: student with pinhole tube] Students construct pinhole tubes from empty paper towel tubes. Aluminum foil is placed over one end of the first tube, a pinhole is made in the foil, and tracing paper is placed over the end of the second tube. The pinhole tube is aimed at a light bulb, and an image of the light bulb is formed on the tracing paper. By making a series of calculations, this activity can be used to illustrate similar triangles, among other uses.

2.4. *Refracting Telescopes*

[Slide: student constructing telescope] The pinhole tube is then converted into a simple refracting telescope. The tracing paper is replaced by an eyepiece and the aluminum foil is replaced by a 16-cm focal length lens. [Slide: student testing telescope] Students check out their telescopes in the classroom and often fear that they have done something wrong when they see an inverted image. [Slide: student calibrating telescope] The telescope can be calibrated to determine width of field of view.

2.5. *Three-Dimensional Constellation Models*

[Slide: students measuring strings for three-dimensional constellation model] Students construct three-dimensional constellation models. A photograph of a constellation is attached to a piece of cardboard. Beads are threaded onto strings suspended from the cardboard. [Slide: students attaching strings to washer] Then the strings are fastened to a washer at a distance of 56 cm from the cardboard, since the camera that photographed the constellation had a 56-cm focal length. The washer represents the Earth, and the beads, representing stars, are adjusted for their correct scale distance from the Earth, using a scale of 1 cm = 100 LY. Constructing these three-dimensional constellation models can be quite tedious and involves a great deal of teamwork and cooperation.

2.6. *Observational Astronomy*

Observational astronomy is enhanced by the plotting of constellations, use of star finders, regular observation in the planetarium, and direct observation whenever possible. Five major constellation groups are studied: circumpolar, the summer triangle, autumnal constellations, the winter hexagon, and the spring diamond.

3. Course Content

The astronomy course is divided into seven units as follows;
- **Astronomy and the Beginning of Science**: the origins of science through astronomy, interpreting sky phenomena, methods of astronomical inquiry, including the uses of technology.
- **Solar System Dynamics**: measuring size, distance and scale in space, modeling the solar system, and observing the planets.
- **Observing Earth Motions**: apparent daily motion of the Sun, timekeeping systems, time zones, revolution of the Earth, the seasons, and precession.
- **Earth-Sun-Moon**: scale of Earth-Sun-Moon system, lunar phases, solar and lunar eclipses, the Apollo moon voyages, and lunar observation.
- **The Sun and Stars in General**: measuring and observing magnitudes of stars, magnitude scales, physical properties of the Sun, and solar activity.
- **Stellar Evolution**: spectral classification, the Hertzsprung-Russell Diagram; energy production in stars; and evolution of low mass, medium mass, and supermassive stars.
- **Galaxies and Cosmology**: observing the Milky Way and deep space objects, types of galaxies, large-scale structure of the universe (the Bubble Theory), and theories of cosmogeny.

4. Basic Skills

Application of basic skills is another major thrust of the course. As new vocabulary terms are introduced, their use is integrated into laboratory, audiovisual, planetarium and other observational activities. Scientific notation and unit conversion activities are used extensively throughout the course. The use and understanding of time zone changes is studied thoroughly in the unit on Earth motions. Use of outside reference material is integral to each unit. Interpretation of charts, graphs and coordinate systems facilitate basic understanding of potentially complex material. Models are used extensively. Computer technology is used on a classroom basis, particularly to study constellation maps. There are provisions for students to do independent study, and even for a separate independent astronomy course.

5. Assessment

Formative assessment includes regular quizzes on constellations and basic skills. Summative assessment is based on unit tests, a constellation unit test, a final examination, as well as written homework and laboratory procedures. The grading scale is deliberately widened at the C and D range to accommodate those students who truly work hard, but have difficulties in academic situations.

6. Conclusion

Astronomy has become a very popular course in the Wauwatosa high schools, but it is not widely taught in high schools elsewhere. Perhaps the present model could be used to provide an incentive for more widespread teaching of astronomy at the high school level.

REFERENCES

COYLE, H. P. et al, 1993, "Project STAR: The Universe in Your Hands," Kendall/Hunt Publishing Company, Dubuque, Iowa.

Measuring the Eccentricity of the Terrestrial Orbit: An Experiment in the Classroom

By R. Szostak

Universität Münster, Institut für Didaktik der Physik, Germany

1. The eccentricity is very small

Most textbooks of physics present the terrestrial orbit by a drawing which shows an ellipse of substantial eccentricity. This suggests a remarkable variation of the distance between Sun and Earth during the year up to a value of about 3:1 and more. Imagine the dramatic variation in size of the radiating area of the Sun seen by the terrestrial inhabitants with all the terrible consequences for temperature. All this is not true. There is no obvious change in size of the solar disc.

In nature the numerical eccentricity of the terrestrial orbit is only $\epsilon = e/a = 0.01675$ (a = major or long axis, b = minor or short axis, e = focal length). This value is so small, that this ellipse cannot be distinguished from a circular orbit in a drawing when using a normal pen ($a/b = 1.00014$). The deviation would be 1/20 of the width across the line of the pencil. By what procedure would it be possible to measure this small eccentricity using only simple means in the classroom?

2. Observe the varying size of the solar disc

A first approach could be the idea to take photographs of the Sun throughout the whole year. The angular width of the solar disc varies by about 3% within this period. The focal length $f = 50$ mm of a normal camera produces an image of the Sun, which is 0.4 mm in diameter on the film. Trying to determine the eccentricity from these pictures better than 10% would mean ability to measure difference of 1μm in size on the film. So this procedure will not work well. It will be necessary to use a camera with a considerably larger focal length or a telescope with a magnifying factor which has to be kept reliably constant throughout the year.

But we can be independent of this condition by measuring the angular width of the solar disc via the daily rotation of the earth. In this case it is sufficient to measure the motion of the solar image on a screen, produced by a telescope at rest. The Sun exhibits its angular size by the time elapsed in passing a line on the screen. This happens within about 2 minutes. For determining the variations of the angular width to better than 10%, it is necessary to measure the time of this passage to within 0.3 s. This must be done carefully. In addition there is the problem that the brightness of the solar disc drops near its contouring edge. So we have some little problems to determine the eccentricity of the terrestrial orbit.

3. Variation in solar time

But there is another way to determine it; this method gives results with a few percent error which can be obtained with almost no technical equipment: there is a phenomenon which was well known already the Greeks in ancient times: the seasons, whose beginning and ending are given by the dates of the equinoxes and solstices, are not equally long. They differ by up to 5 days, as one may find in any calendar. In ancient times people

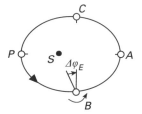

FIGURE 1.

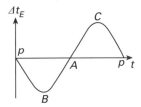

FIGURE 2.

thought that the Sun was moving around the Earth, which was at rest. So these deviations were understood as an eccentric motion of the Sun. Changing to the Copernican heliocentric view this is caused by the eccentricity of the Keplerian elliptical orbit.

The eccentric motion of the Earth has an interesting consequence. Look at Figure 1: it produced a periodic deviation of the Sun's position for a terrestrial observer compared to the case of a circular motion. The figure shows the Earth on its orbit in four positions belonging to intervals of a quarter of a year each. At the positions of perihelion P and aphelion A there are no deviations. But at B and C a terrestrial observer perceives the Sun in a position which deviates by the angle $\Delta\phi_E$ compared to the case if the terrestrial orbit is circular. By the daily rotation this results in a deviation in time of day . At the positions P and A the Sun is passing the meridian at 12 h noon. But at the position B the Sun has not yet reached the meridian at 12 h; this means that the Sun is somewhat "late". Correspondingly at position C the Sun will pass the meridian earlier than 12 h. Figure 2 shows this time deviation throughout the year. Because we have good clocks now, which measure exactly the continuous flow of the time, we can measure these differences of "solar time" against the independently running time easily.

4. Procedure for the classroom

This variation of solar time can be checked by simple means in the classroom by observing a sun beam which passes through a little hole in a cardboard sheet which has been fixed at a window facing South. We register the time, when the centre of the Sun's image passes a line on the floor (Figure 3). This time can be read on a radio controlled clock, which is available at low cost in the shops now and which gives the time to 1 s exactly. By observing the Sun's image carefully when it passes the measuring line, we' may obtain this moment to about 3 s exactly. We need no lenses, only this piece of cardboard.

It is sufficient to repeat this measurement in the classroom once a week in a sequence of a few times, in order to obtain remarkable deviations of several minutes. This demonstrates the deviation convincingly. It shows that the Sun is not crossing the reference line at the same time each day. A longer extension of these measurements will not be

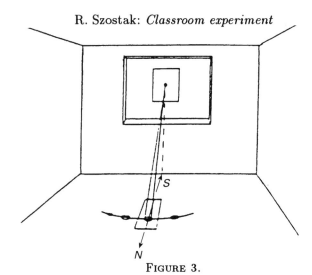

FIGURE 3.

suitable for school practice. But when the students have been convinced of the Sun's delay by their own experience, they may collect the residual data from other sources.

There is a data block cheaply available to everybody: in a normal calendar you may find the data for sunrise and sunset, at least once a week. By taking the mean value between sunrise and sunset one obtains the time, when the Sun passes the meridian. This way one finds the deviation throughout the year as shown in Figure 4, which is called the "equation of time ". But unfortunately this curve is not the same as expected using Figure 2. With some imagination the expected curve can vaguely be recognized in the two big extrema at left and right. Apparently there is another effect of double periodicity superimposed as shown in Fig. 5.

In order to find out the origin of this additional effect let us consider a normal sundial. If we install a gnomon vertically on a horizontal ground, the full hours will not occur equidistantly on the clock face. An equidistant division will be obtained only, if the clock face is inclined into coincidence with the equatorial plane and if the gnomon is in coincidence with the direction of the terrestrial axis. We see, that the nonequidistant characteristic was an artefact of oblique projection, when the sundial was inclined to the Earth's axis. This deviation has evidently four zeros: at noon, at 06.00, at 18.00 and (virtually) at midnight. So it has the expected double periodicity.

A corresponding artefact, but with regard to the period of a year, results from the inclination of the equatorial plane to the orbital plane of the Earth. For a terrestrial observer the Sun does not move during the year in the equatorial plane but in the ecliptic plane which is the plane of the Earth's orbit and is inclined by 23.5° (Fig. 6). This produces the following deviation: if the Sun were continuously moving on a circle in the equatorial plane, an observer on the daily rotating Earth would see the Sun passing the meridian all the year in equidistant intervals of average solar days. But as the Sun moves along the inclined plane, it passes the meridian of the observer somewhat later on its way between F and S in Figure 6. Correspondingly the Sun passes the meridian somewhat earlier than the "average time", if it is on its way between S and H.

This projection error $\Delta\Phi$ has four zeros during the year: the two equinoxes F and H for springtime and for autumn, and the solstices S and W for summer and winter. These are the beginnings of each of the seasons. So this function $\Delta\Phi_P$ possesses the double periodicity, which we looked for. From Figure 6 one can find by geometrical rules

$$\Delta\Phi_P = \Phi_{\ddot{A}} - \arctan(\cos\epsilon \cdot \tan\Phi_{\ddot{A}}).$$

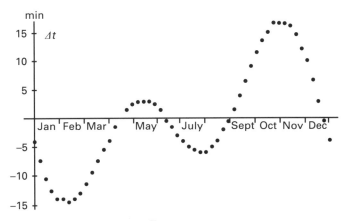

FIGURE 4.

This function can be completely calculated ($\epsilon = 23.5°$) and converted to the time difference Δt_P by the angular velocity of the terrestrial rotation, where $1°$ corresponds to 4 min. As the zeros are well known by the dates of the equinoxes and solstices, the values of this function can be subtracted from the empirical data given in Fig. 4. By this we obtain the curve in Fig. 7, which corresponds to the expected behaviour of Fig. 2. This now contains only the influence of the eccentricity. It is the amplitude in Fig. 7 which depends directly on the eccentricity. The zeros in Figure 7 reveal immediately the dates of perihelion and of aphelion which can be identified with an error of less than 2 days. This is surprising in the case of small eccentricity.

Let us now calculate the value of the numerical eccentricity. Consider a point C in Figure 8 which is reached exactly at half of the time between A and P. Due to the second law of Kepler the areas of the elliptical parts SAC and SCP must be equal. But the calculation procedure of these areas is tedious. We exchange the shadowed parts in Figure 8 and obtain two areas of equal size of a quarter of an ellipse each. These two shadowed areas are congruent triangles except for the little curved part which in case of the small eccentricity causes an error of only 10^{-4}. The triangle containing the angle $\Delta\Phi_E$ offers the relation

$$\tan\Delta\Phi_E = \frac{2e}{p} = \frac{2e}{b^2/a} = \frac{2e}{a}\left(\frac{a}{b}\right)^2 = 2\epsilon\frac{a^2}{a^2 - e^2} = \frac{2\epsilon}{1 - \epsilon^2} \approx 2\epsilon$$

where the term $\epsilon^2 \approx 10^{-4}$ can be neglected. $\Delta\Phi_E$ can be related to Δt_E by $1° = 4$ min as before. Comparison with Figure 8 shows that the date of C will be close to the maximum of the curve in Figure 7. So it is justified to take the amplitude $(\Delta\Phi_E)_0$ in Figure 7 for calculating

$$\epsilon = \frac{1}{2}\tan(\Delta\Phi_E)_0.$$

The amplitude in Figure 7 gives $\epsilon = 0.01768$. Comparing this with the exact value $\epsilon = 0.01675$ shows that we have a result which is exact to 5%. Remembering that we use only dates of sunrise and sunset in a cheap calendar, this result is quite satisfying. If we take any ordinary annual data book for amateurs, we obtain results with an error of about 1%.

There is still another correction to be made which is rather trivial but to be considered at the school level. The dates of sunrise and sunset given in the calendar are valid only for a certain geographical position, in our case for the town of Kassel, which has a longitude

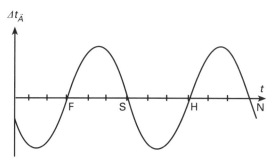

FIGURE 5.

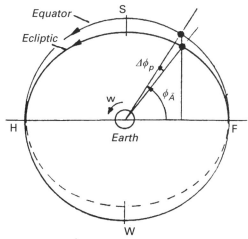

FIGURE 6.

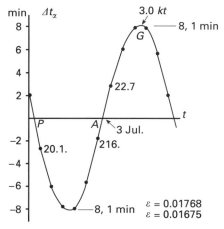

FIGURE 7.

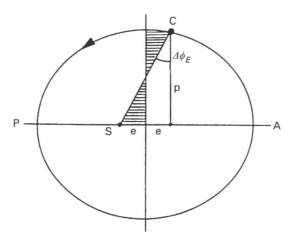

FIGURE 8.

of 9.5° East. At this longitude the Sun passes the meridian generally somewhat later than the reference meridian for the Central European Time which is located at 15° East. The difference 15° − 9.5° results in a delay of 22 min. this had been taken into account already in our data given in Figure 4 without mentioning. If this correction is not made accurately, the zero line in Figure 7 will be shifted up or down. But for calculating ϵ from this figure this shift is not important. We can omit this shift in the end because we need only the amplitude which can be read between the minimum and the maximum of Figure 7.

A Program Incorporating Physics, Astronomy and Environment

By Kononovich, E.V.[1], Fadeeva, A.A.[2], Kiselev, D.F.[1] AND Zasov, A.V.[1]

[1]Department of Physics, Moscow University, 119899, Moscow Russia

[2]Institute of General Middle Education, Russian Educational Academy, 119906, Moscow, Russia

A cultural role of astronomical education at all levels is well known and it is needless to repeat the corresponding arguments. Nobody denies it, but nobody can propose any universal way of introducing Astronomy on a level this branch of science deserves.

There is a good tradition to appreciate Astronomy in Russian schools. For more than a century part of the natural history science in school dedicated to the Universe has been considered as a separate part of the school curriculum in Russia. Before 1917 it was named Cosmography, and Astronomy thereafter. And up to now there is no decision or prescription to rule it out of the school program.

Nevertheless the teaching of Astronomy becomes less and less. Astronomy is taught only then and there where the enthusiasts of this subject are in existence. But the recent process of liberation of the educational system demands different approaches. Up to the present time several attempts to integrate astronomy with physics have not been very successful. The reason is the difference between their educational purposes.

The main purpose of this report is to emphasize the advantage of a somewhat more balanced program incorporating physics, astronomy and environment. Such a choice provides a more natural reason for integration, based on the ideal common to all three parts, to consider the world we live in as our home and property. The most general and fundamental ideas should be emphasized in such a course. As for astronomy, whose social importance is enormous in spite of the negligible teaching time, the necessary requirement is to elaborate a certain school minimum of astronomical knowledge.

As a first approximation we use our earlier elaborated and recently revised short textbook in Astronomy (Zasov and Kononovich, 1993) as a sum of basic astronomical information, principal ideas, and main teaching methods to introduce into this new type of project. The recently organised almanac "The Universe and Ourselves" (Kononovich, 1993, 1995) might be very useful for supporting this program.

During the last decades an opinion has developed that Astronomy should be integrated with some other sciences. This works rather well for the primary school education, e.g. in Natural Science (Suravegina et al., 1994) where the Earth as a celestial object is successfully presented. But when we try to incorporate Astronomy into Physics usually we fail. Unfortunately it takes place even in the best courses (e.g. see Pinsky et al., 1993, 1995). Theoretically this merger is quite possible but only when a special course of physics is concerned. Strictly speaking it is relatively easy to integrate only astrophysics with theoretical physics and no more. The traditional course of school physics differs from that of astronomy like, e.g. a modern theatre building from a Gothic cathedral.

Time table of Physics and Astronomy in Natural Science complex

Form	Year	Physical Level	Astronomical Level

Elementary (Initial) School
"Surrounding World"

Form	Year	Physical Level	Astronomical Level
1	7		
2	8	Air Temperature	Time. Seasons.
3	9	Objects. Matter. Solid. Air. Liquid. Water.	The Earth
4	10	Observations. Experiments. Tools. Measurements.	The Earth from Space. Globe

Middle (Main) School
"Natural Science"

Form	Year	Physical Level	Astronomical Level
5	11	Mass. Density. Particles. Diffusion. Interactions.	Celestial objects. Solar System. Ptolemeus and Kopernicus. The Planet Earth.
6	12	Motion. Forces. Heat. Light. Sound. Magnetism.	Origin of the Solar System.

The level of existing experiment: textbooks are available

"Physics. Astronomy. Environment

Form	Year	Physical Level	Astronomical Level
7	13	Molecular physics.	Astronomy. Universe overview. Matter in space. Gas in the Universe. The stars and the Sun.
8	14	Mechanics	Apparent sun, stars and planet motions. Gravitation in the Universe
9	15	Electricity and magnetism. Atomic structure	Solar, planetary and stellar magnetic fields. Stellar interior

The level of existing experiment: experimental textbooks

High (Advanced) School

Form	Year	Physical Level	Astronomical Level
10	16	Mechanics. Thermodynamics. Electrodynamics	Celestial mechanics. Physical properties of stellar and interstellar gas. Cosmic electrodynamics. Physics of Solar System
11	17	Wave, atomic and quantum physics	The Sun and Stars. Interstellar medium. Stellar Systems and the Evolution the Universe.

The textbooks are in preparation

FIGURE 1.

Certain aspects should be taken into account for a successful integration of astronomy with other subjects. We suggest as the most important among them the following:

1. Astronomy treats the largest space-time scales in the Universe.

2. Astronomy presents the Universe as the most general conception of the Environment and gives evidence of tight connection between Man and the Universe.

3. Astronomy proves the unity and the universal character of all fundamental laws of Physics.

These aspects may be treated as astronomical milestones on the educational road.

There are many attempts to present astronomy and/or physics using some other background. This is especially successful in fiction, science fiction, historical essays or textbook chapters and so forth. The course in Natural Science is also an example.

Positive results were obtained using poetry during lessons on Astronomy. Also for young children didactic material was prepared to support lessons on astronomical topics by means of paintings and drawings including those made by the pupils themselves. The main purpose of all these attempts is to stimulate childrens' interest and to make the education more humanistic.

Now we propose a new idea to use the Environment as a generalising background. It suggests treating the Universe as a whole as our living home, the widest subject of Environment we belong to. On the other hand Physics is a main tool to investigate the relations between the Universe and Ourselves.

The general outline of this project is summarised in the table (Fig. 1). It covers all three educational levels: Elementary (Initial), Advanced (Main) and High (Middle) school. The corresponding Russian titles of these educational levels are enclosed in brackets.

This program has been experienced in the frames of the two first levels and this is marked grey in the table (Fig. 1). It means that the permanent or experimental textbooks are available. Some of them are presented on this Colloquium as a poster.

Each educational period incorporates General Physics at the corresponding level: The course "Surrounding World" presents propaedeutical level for Elementary School (the first 4 years of education).

The course "Natural Science" presents an introductory level for the beginning of Middle School and for the time being is prepared in two variants:

1) as a two year program (Suravegina et al., 1993, 1994) and

2) as a three year program (Kalinova et al., 1991; Khripkova et al., 1992, 1994).

The course "Physics. Astronomy. Environment" presents an advanced level for the remaining years of education. It also consists of two independent cycles: one at the end of Middle School, the other during the whole High School.

The main problem of compiling such a course is to choose the right information corresponding to a given level. We try to follow the well known pedagogical principles, which make together a scientifically correct approach and a necessity to take into account pupils' abilities. Also we try to accumulate the educational material around fundamental physical conceptions, laws and theories to draw a general picture of the World in its integrity and unity. Astronomy presents an outstanding possibility to illustrate the evolution in the World over the widest space and time scales and extremely different stages from elementary particles to Homo Sapiens and further to cosmology.

REFERENCES

KALINOVA, G.S. et al, 1991, "Natural Science 5", Prosveschenie, Moscow.

KHRIPKOVA, A.G. et al, 1992, "Natural Science 6", Prosveschenie, Moscow.

KHRIPKOVA, A. G. et al, 1994, "Natural Science 7", Prosveschenie, Moscow.

KONONOVICH, E.V., 1993, "The Universe and Ourselves", Astronomical Society, Editor. Almanac, No 1, Moscow.
1995, "The Universe and Ourselves", Astronomical Society, Editor. Almanac, No 2, Moscow.

PINSKY, A.A. et al, 1993, "Physics and Astronomy 7", Prosveschenie, Moscow.

PINSKY, A.A. et al, 1995, "Physics and Astronomy 8", Prosveschenie, Moscow.

SURAVEGINA, I.T. et al, 1993, "Natural Science 5. A Human Being and Natural Environment", Ecology and Education, Moscow.

SURAVEGINA, I.T. et al, 1994, "Natural Science 6. A Human Being and Environment", Ecology and Education, Moscow.

ZASOV, A.V & KONONOVICH, E.V., 1993, "Astronomy 11", Prosveschenie, Moscow.

Classroom Activity: Kepler's Laws of Planetary Motion

By Y. Tsubota

Keio Senior High School, 4-1-2 Hiyoshi, Kouhoku-ku, Yokohama 223, JAPAN

1. Introduction

This activity was originally developed by a group of teachers in Japan during 1960s under the influence of the American Curriculum-Reform Movement. This was used in Earth Sciences in order to develop the students' cognitive skill. Kepler had been trying to analyze Tycho's observations of Mars, fitting them into the Copernican orbital system. It simply would not work. The problem is with the circular orbit that the Copernican system still used. Mars obviously did not have a circular orbit about the Sun. So Kepler tried a variety of other geometrical shapes, until he finally found the ellipse. This worked for all of the planets as well. Kepler's law of planetary motion is cursorily mentioned in most secondary Earth Science and Physics textbooks as shown in Table 1.

Textbook	Chapter	Chapter Title
SPACESHIP EARTH Earth Science Houghton Mifflin, 1984	4	Solar System
Modern Earth Science Holt, Rinehart and Winston, 1989	29	The Solar System
Investigating the Earth Houghton Mifflin, 1991	1 20	The Earth in Space The Solar System
Earth Science Scoot, Foresman	23	The Solar System
PHYSICS Addison Wesley, 1991	6	Circular Motion and Gravitation
PHYSICAL SCIENCE Scoot, Foresman	4	Graviton

Table 1. Correlation of this experiment with American science textbooks.

But the objectives of this activity covers Kepler's Law in greater detail. After completing this activity, students will be able to understand Kepler's method to determine an orbit of Mars; understand Kepler's laws of planetary motion; and describe the planetary motion. Students will also learn about the history of science and nature of the scientific belief. This 90-minutes activity is divided into three parts: drawing the orbit of Mars - sixty minutes; analysis - fifteen minutes, and post-lab discussion - fifteen minutes.

2. Student Activity

2.1. Materials For Each Student

Each student needs following materials: a ruler, a circular protractor, a compass, graph paper, observational data as shown in Table 2.

Mars	Earth	Date	Sun	Mars	Mars-Sun
M1	E1	7/28/69	124.8	244.4	119.6
	E1'	6/15/71	83.3	317.8	234.5
M2	E2	9/16/71	172.5	312.3	139.8
	E2'	8/3/73	130.5	25.4	-105.1
M3	E3	11/4/73	221.4	28.5	-192.9
	E3'	9/22/75	178.4	79.4	-99.0
M4	E4	12/24/75	271.5	78.1	-193.4
	E4'	11/10/77	227.5	125.4	-102.1
M5	E5	2/11/78	321.9	114.7	-207.2
	E5'	12/30/79	277.7	163.5	-114.2
M6	E6	4/1/80	11.4	146.0	134.6
	E6'	2/17/82	328.8	199.0	-129.8
M7	E7	5/21/82	59.6	180.9	121.3
	E7'	4/7/84	17.4	238.2	220.8

Table 2. Observational data of Mars.

Positions of the Sun and Mars in Table 2 are given by the angles from the vernal equinox as shown in Figure 1. The positions of the planets can be described only by the ecliptic longitude if we take the ecliptic coordinate since planetary motions are confined along the ecliptic. It is clear from the small value of the angle inclination of orbit to ecliptic as shown in Table 3. The positive angle of the positions is measured counter clockwise and negative angle is measured clockwise. Since Mars' orbital period of 1.8809 Earth years corresponds to the 678 Earth days, Mars to traverse the orbit by 687 days. The date on the bottom row(E1', E2', E3' and so on) is 687 days later from the date on the top row(E1 E2, E3 and so on) in Table 2. Therefore, Mars is located at same position on the orbit.

	Mercury	Venus	Earth	Mars	Jupiter	Saturn
Distance (AU)	0.3871	0.7233	1	1.5237	5.2026	9.5549
Eccentricity of orbit	0.2056	0.0068	0.0167	0.0934	0.0485	0.0555
Inclination of orbit to ecliptic (.)	7.006	3.395	0.001	1.85	1.303	2.489
Sidereal period (Y) Year=365.2422Days	0.2409	0.6152	1	1.8809	11.862	29.458
Synodic period (day)	115.9	584	------	779.9	398.9	378.1
Ecliptic longitude of the perihelion (.)	77.441	131.564	102.908	336.018	14.311	93.003

Table 3. Basic data for the planets

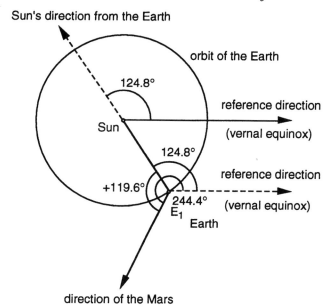

FIGURE 1. The relationship between the Sun's direction on the celestial sphere and the Earth's position on the orbit.

2.2. *Background*

Students should know certain keywords in order to study planetary motion. The term 'aphelion' and 'perihelion' are the farthest and closest point in an elliptical orbit of a planet, asteroid, or comet from the Sun, respectively. The axis between the aphelion and perihelion is defined as a major axis of the orbit. The half length of the major axis is called as mean distance between a planet and the Sun. The Astronomical Unit (AU), roughly 150 million kilometers, is the mean distance between the Earth and the Sun.

2.3. *Procedure*

(1) Draw a circle with radius of 5 cm on the center of a sheet of graph paper.

(2) Draw a line from the center of a circle to any direction (e.g., right). Let us assume a circle be the orbit of the Earth, the center the location of the Sun, a line the direction of vernal equinox.

(3) Locate the Earth's position on the orbit using Sun's position in Table 2, which is an angle from the vernal-equinox, as described in Figure 1.

(4) Draw the direction of Mars from the Earth's position using the angle of Mars from vernal equinox in Table 2.

(5) Repeat (3) and (4) for another date within a pair. The intersection of the two lines is the position of the Mars.

(6) Repeat (3) through (5) using other six pairs of observations.

(7) Determine the Martian orbit by connecting seven locations of Mars. Try to draw the Martian orbit as a circle using compass. The center of a circle can be determined by the intersection of the perpendicular bisector of the chords.

(8) Measure and fill the Table 4 using the orbital data drawn in this exercise.

2.4. *Analysis*

(1) Confirm that the center of the Martian orbit is not the Sun.

(2) Locate the perihelion and aphelion of Mars and measure the counterclockwise angles from the vernal-equinox direction. Determine the mean distance between the Sun and Mars in AU.

    Perihelion: degrees from the vernal equinox.
    Aphelion: degrees from the vernal equinox.
    Mean distance: (AU)

(3) Calculate areal velocity from Table 4 and confirm the Kepler's second law. $\bar{r}$ is the distance between the Sun and the Mars computed as: $\bar{r} = 0.5(r(M_i) + r(M_{i+1}))$ A and V are the area of the fan-shape sector (SM_iM) and areal velocity of the Mars respectfully computed as:

$$A = \pi\bar{r}^2 \frac{\theta}{360}, \; V = \frac{A}{t}$$

where θ is the center angle of fan-shape area.

(4) Fill Table 5 using your results and calculate the third column in order to confirm the Kepler's third law. R and P are mean distance between the planet and the Sun and orbital period of the planet, respectively.

Mars	Earth	days(t)	$r(M_i)$	$r(M_{i+1})$	$\bar{r}$	θ	A	V_{area}
M$_1$-M$_2$	E$_1$'-E$_2$	94						
M$_2$-M$_3$	E$_2$'-E$_3$	94						
M$_3$-M$_4$	E$_3$'-E$_4$	94						
M$_4$-M$_5$	E$_4$'-E$_5$	94						
M$_5$-M$_6$	E$_5$'-E$_6$	94						
M$_6$-M$_7$	E$_6$'-E$_7$	94						

Table 4. Worksheet for the results

Planet	R (AU)	P (Year)	R3/P2
Earth	1 AU	1.00 Year	1
Mars		1.88 year	

Table 5. Verification of Kepler's Third Law

2.5. *Discussion and Conclusion*

(1) List the assumptions applied in this exercise.
(2) Consider the source of the error of this exercise.
(3) Summarize the Kepler's Law of Planetary Motion.

3. Teaching tips

There are some things to follow to teach Kepler's Law more effectively. First, the radius of the earth's orbit should be two thirds of the radius of the circular protractor. This means that the orbit of Mars corresponds to the size of the circular protractor and hence, it should be easy to identify the center of orbit. The Law of Ellipses states that the Sun's position differs from the center of the orbit. Second, the time intervals between the Mars' positions which are for example, two dates E1 and E2, should be the same in order to compare the Law of Equal Areas. Mars' areal velocity will be measured by the use of computing a fan-shaped area within Mars' orbital positions about the Sun. An alternative method of measuring the area of a sector is the use of graph paper and count the number of squares in the sector. We may use the simple assumption that all partial squares are count as 0.5 square. Another way uses a accurate balance scale to weigh the paper that fills the sector. We may use the simple assumption that paper's thickness and density are constant. The best method to demonstrate it is by the weights of the fan-shape Mars' trace. Fan-shaped areas will be cut out from the huge cardboard diagram. A visual effect can be created to give an impact to students using a simple balance. Third, the pre-lab activity should include the derivation of the orbital period. The orbital period cannot be measured directly from observation. We can determine only the synodic period which determines the orbital period and hence, is most important period. This exercise is a good problem for algebra. Another pre-lab activity should cover the apparent motion of the planets which include direct and retrograde motions and stationary points since the appropriate theory must explain these planetary motions. Fourth, the post-lab activity should include the analysis of the results. The Law of Period can be confirmed by comparing the results with the Earth. Also, the variation of the planets' orbital velocities should be applied to comets. Halley's comet return every 76 years but the best observation period lasts only a year. That is because of the Kepler's second law of planetary motion. The Earth's orbit was assumed to be circlar in this activity and hence, contradict the results that identifies it as elliptical. However, the eccentricity of the earth is only 0.0167 corresponding the 0.84 mm in the diagram as compared with Mars' 0.0934 corresponding the 7.12 mm. Therefore, we can neglect the eccentricity of the earth in this activity. There are revelations in this activity. Students will learn the history of science and the process of a geocentric system versus a heliocentric system. That is not an easy process because the idea is counter to our common sense. Scientific belief is often a hypothesis. The hypothesis should be the most-reliable idea at that time. However, it might be replaced by a newly-developed hypothesis. That is the nature of scientific belief.

4. Summary

Instructors should bear in mind that this is a kind of 'integrated curriculum'. Students need to utilize and apply their knowledge of mathematics, physics, earth science, and history of science. Data was computed by Stellar Navigator (a commercial software in Japan). Moreover, diagrams presented here were drawn by Mac Draw Pro on a Macintosh computer. Therefore, this is not only teaching astronomy but also scientific method. It requires simple materials (ruler, circular protractor, compass, and graph paper) and hence can be done at any school all over the world. The most recent observational data will be supplied on the Internet from Keio Senior High School(www.ifet.or.jp/ keio).

REFERENCES

CHURTTRAND, M. R. & TIRION, W., 1991, *The Audubon Society Field Guide to the Night Sky*, Alfred Knopf, Inc., New York.

MORRISON, D & OWEN, T., 1988, *The Planetary System*, Addison-Wesley, New York. Many Japanese Standard Textbooks

5. Appendix

Grade ___ Student ID# _____ Name _____

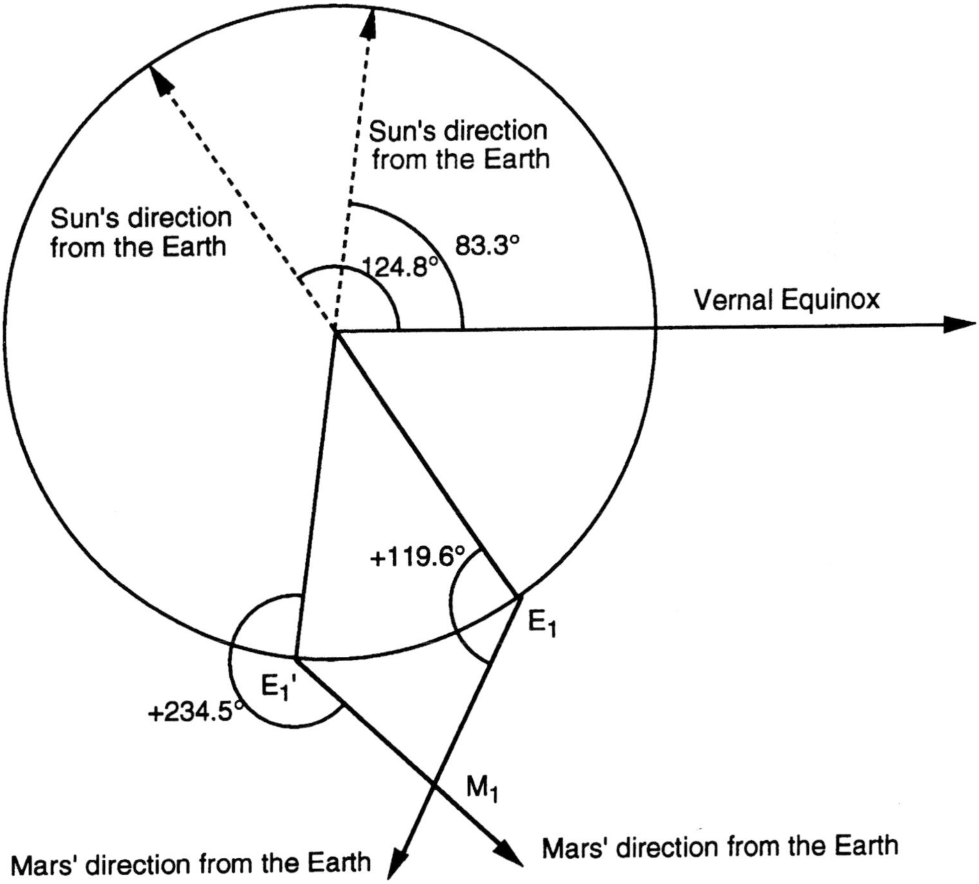

FIGURE 2. Worksheet for classroom use.

Collaboration As a Viable Approach for Making Astrophysics Research Accessible to the K-12 Community Through the Internet and the World Wide Web

By ISABEL HAWKINS[1], ROBYN BATTLE[1,2]
AND MARLENE WILSON[3]

[1] UC Berkeley, Center for Extreme Ultraviolet Astrophysics, 2150 Kittredge Street, Berkeley, CA, 94720, USA
[2] UC Berkeley Graduate School of Education, Berkeley, CA, USA
[3] Fruitvale Elementary School, Oakland, CA, USA

1. Introduction

We describe a partnership approach in place at UC Berkeley's Center for Extreme Ultraviolet Astrophysics (CEA) that: (a) facilitates the adaptation of astrophysics data and information from NASA and other sources for use in the K-12 classroom, (b) facilitates scientists' participation in astronomy education, and (c) engages a sustained collaboration typically including personnel from research institutions, centers of informal science teaching such as museums and planetaria, university-based schools of education, and K-12 schools. We are investigating several ways of engaging scientists in partnerships for the purpose of making their research results accessible in appropriate ways to the K-12 community via Internet and World Wide Web technologies. Our investigation addresses the hypothesis that the transition of scientific data and research results from the workplace to the classroom can be facilitated by the joint creation of curriculum materials by teams of cognitive experts, subject-matter experts, and teachers. In particular, we are investigating how space science, astronomy, and Earth science research results can be adapted through a partnership approach into more effective representations for use in the classroom. Our strategy for evaluating our partnership approach engaged the participation of personnel from scientific research institutions, centers of informal science learning, and schools. We describe several projects led by UC Berkeley's Center for EUV Astrophysics: "Science On-Line," "Science Information Infrastructure," and "Satellite Operations Class for Teachers." Our projects have two primary and complementary components, namely, implementation in school districts serving students with wide-ranging socio-economic backgrounds and a science education research component based on in-depth project evaluation. We describe several specific advantages that arise from a partnership approach in the context of using the Internet and the World Wide Web as relatively new authoring and representation media for K-12 curriculum materials.

2. The Importance of Developing Collaborative Efforts in Astronomy Education

The astrophysics community is increasingly interested in becoming involved in the education of today's students and is eager to share exciting space science results with the general public. Although it is sometimes feasible for institutes to create self-contained educational programs, this method is probably not the most efficient nor the most productive way of participating in educational processes. Often we find that institutes do not have a complete range of expertise in the fields of science, research, and education that responds to the needs of the varied K-12 and general public communities. Thus it is imperative for quality space science education that organizations join together to build collaborations containing expertise in all of these essential fields. This important collaborative strategy (a) avoids duplication of effort, (b) allows each partner to operate primarily in their area of expertise, (c) provides a natural mechanism to translate cutting-edge research and technology into resources that respond to the needs of educators and students, (d) forms a multi-level resource network, which can be utilized in a most appropriate manner by district, school, teacher, student, and public audiences, (e) results in programs that contain substantive feedback and input from the K-12 and college-level educational communities, (f) creates an information flow which occurs in a smooth manner from research center to museums and intermediate astronomy education groups to the K-12 and college level community, and (g) results in programs which are more likely to reach diverse audiences.

3. Theoretical Framework

The introduction of technical innovations into the classroom has the potential to produce a dramatic redefinition of traditional roles for teachers and students blurring the boundaries between the processes of teaching and learning (Park & Hannafin (1993)). Major emphasis has been given to more relevant curriculum and models of learning that more closely reflect scientific workplace learning (Becker 1972; Lave & Wenger 1991; Collins, Brown, & Newman 1989). Park & Hannafin (1993) suggest that students and teachers benefit by learning how to use technological tools found in scientific, academic, and business workplaces. A specific scientific discipline has specialized and characteristic forms of discourse, including not only jargon, but ways in which research is carried out via the formulation of hypotheses and conjectures, the interplay between theory and observations, and the way in which new results are related to accepted knowledge (Schoenfeld 1985, 1987). Thus, the creation of an effective scientific curriculum that reflects the practices of the workplace can benefit from the integral participation of subject-matter experts.

Use of technology in the classroom can foster an environment that more closely reflects the processes scientists use in doing research (Linn, diSessa, Pea & Songer 1994). For instance, scientists rely on technological tools to model, analyze, and ultimately store data. Linn, diSessa, Pea & Songer (1994) suggest that technological tools be introduced to students from the earliest years because of their ability to facilitate scientific modeling, scientific collaborations, and electronic communications in the classroom. Our investigation is motivated by the need for effective participation of scientists in K-12 education using a methodology that prevents duplication of effort.

4. Brief Project Descriptions

We briefly describe three projects that have served as the foundation for the UC Berkeley Center for EUV Astrophysics Education Program.

4.1. *Satellite Mission Class for Teachers*

This semester-long class is offered to in-service teachers through UC Berkeley Extended Education where teachers obtain university credit they can apply toward their professional development credential requirements. During this class, funded through NASA's IDEA program, teachers collaborate with scientists, technical personnel and undergraduate students from UC Berkeley/CEA, as well as with graduate students from the UC

Berkeley Graduate School of Education. The teams work on development of lesson plans for the Internet/WWW based on the NASA EUVE and other satellite astronomy missions. Teachers learn about NASA missions and their data, the Internet, and the World Wide Web in the contextually-relevant process of lesson plan development, an important aspect of their professional growth. The sustained interactions with other professionals during the semester-long class results in a high level of training that the teachers take back to their schools and share with their peers.

4.2. *Science On-Line (SOL) Project*

The Center for Extreme Ultraviolet Astrophysics (CEA) at UC Berkeley received funding from NASA's Office of Space Science to support the Science On-Line – Earth and Space Science for the Classroom Project (SOL). Initiated in September 1994, SOL brings together formal and informal science centers, linking their unique and complementary assets. SOL joined the efforts of UC Berkeley's Center for Extreme Ultraviolet Astrophysics (CEA), the Lawrence Hall of Science and UC Museum of Paleontology, with the efforts of San Francisco's Exploratorium, and Chicago's Adler Planetarium for the purpose of creating on-line Internet-based resources for 6th-12th grade teachers and students. The Science On-Line (SOL) Project serves as the foundation for its follow-up project, the Science Information Infrastructure, which is described below. SOL was designed to achieve the following goals:

(*a*) Build on the experiences gained from the Satellite Mission Class for Teachers to design a robust methodology for supporting teachers in the development of Internet-based science lesson plans.

(*b*) Create a robust learning process that allows teachers emerging from the project and its successor programs to assume mentoring roles to other members of the K-12 community.

(*c*) Make certain that as schools increasingly gain access to the Internet/WWW, K-12 educators and students will find appropriate resources that are constructed according to their particular needs and that can be integrated with the science curriculum.

4.3. *Science Information Infrastructure (SII) Project*

The SII is a consortium of science centers (museums) linked together to bring cutting edge science and the technology associated with science research to the public. The purpose of the project is to create a network of resources created by teams of teachers, science museum personnel, scientists and technical staff and hosted by science museums. The resources being created are intended for use by anyone, but especially in the classroom setting as well as for exploratory learning. In the SII model, the small, local focus teams of teachers and individuals from the science museums and partner research institutions create educational resources for science and math classes based on NASA remote-sensing data and related information and networked science research services. The life cycle of the materials and on-line tools involves testing in the classroom, evaluation, modification and revision and further testing in different environments. These resources are made available through the host museum, and coordinated nationwide during the course of the program. The SII is intended to build a sustainable framework for this process and accommodate new and experimental activities that use networked technology in innovative ways. The SII national partnership, funded by NASA High Performance Computing and Communications Division, includes the following members:

• Research Institutions: UC Berkeley Center for EUV Astrophysics, Smithsonian Astrophysical Observatory, Space Telescope Science Institute, National Air and Space Museum Center for Earth and Planetary Studies

• Science Museums: Boston Museum of Science, Exploratorium, Lawrence Hall of Science, National Air and Space Museum, New York Hall of Science, Science Museum of Virginia

• Industry Partners: EOSAT, PacBell CalREN, Digital Equipment Corporation, Dun & Bradstreet

5. Key Elements of Our Partnership Model

Essential elements of our model include:

(1) *Engaging end-users (teachers and their students) in sustained interactions with scientists.* This strategy enables teacher participants to assume peer-mentoring roles at school sites, and develops a critical mass of end-users.

(2) *Allowing every member of the team (teacher, scientist, other educators) to contribute their unique expertise to the partnership.* This strategy facilitates team interactions to occur on a level playing field, and prevents duplication of effort. Through this strategy, the developed resources benefit from both scientific and pedagogical expertise.

(3) *Facilitating participation by scientists from other institutions within an established framework.* Through this strategy, scientists can contribute their unique assets (data, research results, mentoring) with potentially large impact and multiplier effect, taking advantage of an existing end-user support structure. This strategy also provides scientists with meaningful opportunities for interacting with educators from formal and informal institutions.

Our model is leveraged as follows:

(a) Having science museums adopt our methodology, having museum personnel support teachers in the long-term development of Internet based materials, and featuring SII resources as part of museum-led teacher workshops.

(b) Through a tiered resource presentation structure that encourages independent teachers to use our materials to develop their own lesson plans.

6. World Wide Web Information on the Science Education Program at CEA

For more information and resources, access the UC Berkeley Center for EUV Astrophysics World Wide Web Education Server at:

http://www.cea.berkeley.edu/Education

7. Results and Conclusions

One of this investigation's main objectives is to bring the K-12 community into closer working relationships with the scientific community. The relationships established through our programs bring teachers into contact with research institution workplaces where they have access to scientists and gain exposure to the technology and thinking of scientific communities. Our projects are resulting in Internet-based lessons based on science data and information, and our investigation suggests that many of the science resources posted on the Internet need to be adapted for effective use in the K-12 classroom. Such adaptation of scientific Internet-based materials for the classroom appears to be best accomplished by a team approach that involves experts in science content, lesson development, and pedagogy. Project teams that work closely with scientists produce more conceptually robust lessons. Conversely, several Internet-based tools offered by various scientific institutions for use in a broad, non-research oriented context require at times significant adaptation by teacher developers for use in the classroom. The findings of this investigation are enabling us to develop a coherent team approach for facilitating the effective access and use of scientific data and research practices by the K-12 community. This methodology provides teachers with a science curriculum that engages their students in research-oriented classroom activities that take advantage of a technologically-rich environment. Our goal is to establish a robust methodology that provides the scientific community with a viable mechanism for sharing research results with a broader audience in educationally-relevant ways.

We acknowledge funding for our programs by the following sources:
- NASA High Performance Computing and Communications
- NASA Office of Space Science
- NASA IDEA

REFERENCES

BECKER, H.S., 1972, *American Behavioral Scientist,* **16**, 85–105.

COLLINS, A., BROWN, J.S. & NEWMAN, S., 1989, Cognitive apprenticeship: teaching the crafts of reading, writing, and mathematics. *Knowing, Learning, and Instruction* (ed. Resnick, L.B.). 453-494, Erlbaum, Hillsdale, N.J.

LAVE, J. & WENGER, E., 1991, Situated Learning: Legitimate Peripheral Participation. Cambridge University Press, Cambridge.

LINN, M., DISESSA, A., PEA, R. & SONGER, N., 1994, Can Research on Science Learning and Instruction Inform Standards for Science Education? *Journal of Science Education and Technology.*

PARK, I. & HANNAFIN, M., 1993, Empirically-Based Guidelines for Design of Interactive Multimedia. *Educational Technology Research and Development,* **41(3)**, 63–85.

SCHOENFELD, A. H., 1985, Mathematical Problem Solving. Academic Press, Orlando, Fl.

SCHOENFELD, A. H., 1987, What's all the fuss about metacognition? In *Cognitive Science and Mathematics Education* ed. Schoenfeld, A.H., 189–216. Lawrence Erlbaum Associates, Hillsdale, NJ.

Astronomy Teaching in the Astronautics Club

By E.Yu. Aleshkina

Institute of Applied Astronomy of Russian Academy of Sciences, St.Petersburg, RUSSIA

1. Introduction

There is no necessity to argue about a vital need of an extension of astronomy and space knowledge equivalent to a modern state of the natural sciences. Astronomy teaching both professionally and for amateurs in the form of general courses is particularly needed nowadays because of the spread of various forms of mysticism in Russia.

The main goal of astronomy teaching is to help students to become aware of the place of humanity in the Universe. In this connection it is necessary to study not only astronomy but also other relevant courses simultaneously. Such complex astronomy study plays a significant role. It is necessary to show a close interaction of astronomy with other sciences such as traditionally mathematics, physics, chemistry and also biology and psychology, which just begin to be integrated in the field of space sciences. One cannot disregard other aspects of science development - the philosophy of science and the morality of any scientific research. These notions must be discussed with future scientists from the first steps in their education. Thus the association of astronomy and other subjects is rewarding. This purpose has been realized successfully in the Titov's Astronautics Club.

The Astronautics Club at the Sankt-Petersburg Palace of Youth Creativity is a supplementary education form for middle and high school students. This Club was founded after the space flight of the second Russian astronaut German Titov in 1961 and will celebrate in October its 35-year anniversary. Students attend the Club classes after school hours. The Club unites students who are interested in the study of space exploration and research. Such a specific audience permits us to combine two directions in astronomy teaching into the unified system. On the one hand, through astronomy we bring the scientific picture of our world to children. On the other hand, we start from early ages to teach the students who will specialize in astronomy. The connecting links between the general and special teaching are the history of astronomy and computer-assisted instruction.

2. The curriculum

Training in the Club is based on the three year system. In the first year, we usually have about 60-80 students in 4 groups. During the first year in the Club, the students acquire knowledge of the general astronomical notions, such as the celestial sphere, coordinate systems, apparent and real motion of the celestial objects, the grounds of the celestial mechanics such as Kepler's laws. A large part of the time is devoted to the history of astronomy. Students receive the minimum which all people should know about astronomy. This general astronomy course is taught for all first-year students in the Club along with the similar general aviation and astronautics courses.

After the first year, students make a choice of specialty according to their interests. There are two main directions: astrophysics-astronautics and aviation groups. In each group the students acquire deeper knowledge in these disciplines. The most important astrophysical topics are covered, such as

- Sun and the solar system;
- structure and evolution of stars;
- nebular physics;
- basics of cosmology.

The astrophysical course is taught along with the same astronautics subjects, such as rocket technics, theory of space flight, psychology.

At the end of the year students visit the Main Astronomical Observatory of the Russian Academy of Sciences, Pulkovo. Evidently some simple observations could help to understand better the topics studied theoretically. Unfortunately, at the present time the Club has no possibilities for this due to several reasons. Moreover, the Club is located in the historical center of one of the largest cities in Russia, so the conditions for observations are very poor.

3. The main purpose of the theoretical course.

There are several goals which the curriculum reaches. They are the following:
- creation of the terminological basis for further education;
- development of the skills required for work with astronomical literature (how to find an answer to your question in a book);
- development of creative thinking (suggest your own ideas, propose new explanations and hypotheses);
- getting experience in following a lecturer and writing notes on the subject of the lecture;
- development of a student's ability to give a short summary of a lecture with the emphasis on the main ideas. The practice shows that students under 13-14 years old cannot yet do this.

4. Different forms of knowledge checking

Our students pass tests twice a year. The forms of these tests are different.

The primary test is held in the form of individual oral answers to the preliminary given list of questions followed by a general discussion. Since there are 4 groups on the first year, we hold the second part of the annual test in a form of a common competitive game after the primary tests in each group separately.

Two computer systems are developed for use in a training process. One of them, "ASTREN", is practically complete. This system is designed for checking of the general theoretical knowledge. ASTREN is developed as a training system based on the widespread principle "check-and-point". The system contains various questions on the topics learned, and a student has to choose the correct answer from the five possibilities. "ASTREN" can be filled by questions on any subject wanted. The second system called "STARM" is planned to be more creative. It will include a choice of a space flight to real cosmic objects and several subsequent calculations tasks from the fields of celestial mechanics and astrophysics. The first system can be used after the first year of study. The second one requires deeper theoretical knowledge.

After three years of learning, the students write a paper on a chosen topic or give a talk at the traditional annual scientific conference "Human Being and Space" which the Club holds for school children from Saint Petersburg.

5. Conclusion

At the end of each school year, the students fill out a form. The suggested questions have much broader impact than just astronomy. In particular, I used to take an interest in science fiction which children read. It seems to me that the best of the world science fiction written by Bradbury, Lem, Asimov, Clarke, brothers Strugatzkie, Verne and Wells has a merit to be known and loved by future scientists. One of the main goals of the primary astronomy education is not only scientific training but widening the view on the world in general, on the human being and its place in the Universe. Astronomy is a very attractive science. I think, we have to prepare not only well-educated scholars but also people who are really devoted to the romantic spirit of astronomy. Otherwise, our science will be in trouble.

The TRUMP Astrophysics Project: resources for physics teaching

By E. Swinbank

Science Education Group, University of York, Heslington, York, UK

1. Introduction

Comets and quasars, black holes and the big bang, pulsars and planets all feature in the media and excite people to find out more – astronomy might be described as the popular face of modern science. In the UK, recent changes in Advanced Level (A-level) physics courses mean that many students have the option of studying astrophysics to a depth beyond the merely descriptive. This option is proving popular with teachers and students, but presents particular challenges shared by few other areas of A-level physics courses.

1.1. *Astrophysics within A-level physics*

A-level courses are taken by students who choose to stay in education beyond the age of sixteen. Students typically study three subjects at A-level over the course of two years. A-level is approximately equivalent to 12th grade and the first year of a bachelors degree in the USA. Students are awarded grades for their A-level work which depend on their performance in external examinations and on evidence of experimental skills collected by their teachers. The examinations are set, and the grades awarded, by independent examination boards which specify the content on which students are to be examined and the skills for which teachers are required to provide evidence. For many students, A-levels are a preparation for more advanced study at university.

Fifty percent of the content of all A-level physics syllabuses is now defined nationally (School Curriculum and Assessment Authority, 1994), whereas previously the examinations boards had a greater degree of autonomy. Current syllabuses have been discussed and summarized by Avison, 1994; most consist of a compulsory element, with a menu of optional topics of which students must study (and be examined on) a specified number. (The numbers of options vary between boards, but typically students study three from a menu of six.) The range of optional topics includes medical physics, particle physics, electronics, materials, environmental physics – and astrophysics.

Table 1. Areas of astrophysics in A-level physics syllabus

Making observations	Observational properties	Stars	Galaxies and Cosmology
Lenses and mirrors	Electromagnetic	Formation and	Models of the
Telescopes	radiation	evolution	universe
Fibre optics	Line spectra	Gravitation	Structure and scale
Diffraction and	Doppler shift	Nuclear reactions	of solar system,
Atmospheric effects	Black body radiation	Energy transfer	universe
Satellites	Luminosity, flux	Binary objects	Hubble expansion
Radar	and magnitude		Big Bang
Detectors	Stellar classification		Open and closed
	HR diagrams		universe

Before embarking on A-level work, students in England and Wales will have studied a range of subjects up to the age of 16 as required by the National Curriculum, which includes physics, chemistry and biology and some aspects of astronomy mainly of the what goes around what variety. The content of astrophysical options at A-level builds on this and is relatively sophisticated in terms of astronomical knowledge and of relevant physical principles. Table 1 lists some of the items in the syllabuses. The approach is generally quantitative; students are expected to use algebra, graphs and trigonometry, for example, but not calculus or much in the way of statistical techniques.

1.2. *The need for resource materials*

Students and teachers are very enthusiastic about including astrophysics in A- level work, but teachers, in particular, have some concerns. First, while most teachers of A-level physics are physics graduates, few have studied astrophysics and so lack the specialist knowledge needed to teach the topic with confidence. Second, there is a concern about what students actually do in class, since astrophysics does not seem readily to lend itself to the laboratory practical activities that are an integral part of much A-level physics work: few teachers wish to resort to lecturing their students, preferring to engage them in more active learning. Furthermore, some examination boards require that experimental skills be developed and assessed within each optional topic.

Teachers also have a concern about the financial and time demands of teaching astrophysics within A-level. Some schools and colleges are very well equipped (with telescopes and computers for example) and there are teachers with exceptional enthusiasm and expertise who involve their students in astronomical observations and visits outside normal school hours. But such institutions and teachers are unusual; the majority of schools and colleges operate on restricted budgets, and few teachers are able to spend significant extra time on an area that represents a relatively small fraction of their overall teaching commitment.

Despite the recent proliferation of astronomical material on CD-ROM, the Internet, and so on, as well as in books, there is little that meets the particular needs of A-level physics teachers and students: there is little between professional texts (which are too advanced) and popular materials which, while attractive and stimulating, do not address the underlying physical principles required for A-level study. The TRUMP Astrophysics project set out to address two distinct needs. First, the need of teachers to become familiar with aspects of modern astrophysics, and second, the need for activities in which, during normal class time and with easily- obtainable equipment, students are actively involved in learning astrophysics.

1.3. *The TRUMP Astrophysics project*

The Teaching Resources Unit for Modern Physics (TRUMP) is a collaboration between school teachers and academic experts whose aim is to support the teaching of modern physics in A-level courses. The forerunner to TRUMP was the Particle Physics Project which was instrumental in introducing particle physics into A-level courses (Swinbank, 1992) and which developed a resource package to support teachers and students in this exciting but unfamiliar area (Particle Physics Project, 1992).

Following the introduction of astrophysical options into A-level courses, the project team identified a need for a similar resource package devoted to astrophysics which would set out to address the particular needs of A-level physics teachers and students noted above.

A grant from the UK Particle Physics and Astronomy Research Council (PPARC) under the Public Understanding of Science and Technology (PUST) scheme has enabled us to develop one section of our planned resource package as a pilot. The pilot materials (which relate to observational properties of astronomical objects) are being evaluated by teachers during 1996 September and October. Additional funding from the UK Institute of Physics, PPARC, and industrial and charitable sponsorship, will enable us to develop materials relating to telescopes and instruments, stellar evolution, planets, galaxies and cosmology, and hence to complete the package.

One of the main strengths of the TRUMP project, which we would commend to others seeking to develop curriculum resources, is the range of experience within the team. The combination of practising school teachers and academic experts ensures that the materials are both scientifically rigorous and appropriate for use with students. Most of the TRUMP Astrophysics team work as consultants to the project in addition to their full-time jobs, so they necessarily bring a high level of enthusiasm and commitment which is another of the projects great strengths.

2. Designing an effective resource package

To meet the needs of the intended users discussed above, the TRUMP team believe that an A-level astrophysics resource package should aim to (a) help teachers to become familiar with the areas of astrophysics included in their syllabuses (b) support appropriate student activities and (c) be readily accessible to all teachers and institutions. A resource package was therefore designed with these aims in mind.

2.1. *The structure of the TRUMP Astrophysics package*

The TRUMP package has three main components: Study Notes and Teaching Notes (bound together in a Teachers Guide booklet) and a bank of loose-leaf Student Sheets. The Study Notes relate to the first aim. They are similar in style to Open University texts (Jones 1996) and are written so that teachers may study areas of astrophysics relevant to their syllabus prior to teaching it. Teachers can use the notes at times most convenient to them, and are free to skip parts that may already be familiar and to spend longer on aspects which are new. Teachers therefore have the flexibility to organise their study to suit their own particular circumstances, which many find preferable to attending in-service courses demanding full-time commitment at times that are not always convenient.

The second aim is addressed by the Teaching Notes and Student Sheets. The Teaching Notes are designed to help teachers plan a teaching programme appropriate to their particular students and examination syllabus. They suggest approaches to teaching, referring to the Study Notes and syllabus documents published by the examination boards and drawing attention to areas that may need particularly careful handling, and give details of a variety of student activities. The Teaching Notes also refer to some other resources that are appropriate for A-level physics. Many of the student activities are supported by Student Sheets - photocopy masters which may be reproduced for class use within the purchasing institution. All the materials are paper-based, so the package can be bought at relatively low cost and used without recourse to specialist equipment, thus addressing the third aim.

2.2. *Student activities in the TRUMP resource package*

While there are many observational and classroom resources in astronomy that have been described in this symposium and elsewhere, few meet the particular needs of A-level physics students in the majority of schools and colleges. Some exercises involve students only in the fairly passive receipt of images and information, many require telescopes and/or computing equipment, some can only be carried out at night and may require observations over an extended period, and (most important of all) the majority involve no thinking about physics. Activities appropriate for A-level physics students are therefore an important component of the TRUMP Astrophysics package.

In their A-level physics courses, students are required to develop experimental skills which include handling and interpretation of experimental data as well as using apparatus to make observations or measurements. While there may be few opportunities for students to carry out hands-on laboratory work in astrophysics, students can develop relevant skills using data obtained elsewhere and if the data are authentic and recent, students can gain an insight into current research in astrophysics. The TRUMP package therefore includes several data-related activities, using examples chosen carefully so that they relate to physics studied at A-level and can be tackled using appropriate mathematics.

In addition to experimental skills, the TRUMP activities help students to develop more general skills which include study skills, working with others and communication. This is not only a valuable part of students general education, but can also help them to transfer relevant skills (mathematical skills, in particular) from one context to another and so make their learning of astrophysics more effective. Following a recent review of qualifications for 16-19-year-olds in England and Wales (Dearing 1996) development of such 'key skills'will soon be a central feature of all A-level work.

Related to the development of study skills, many teachers encourage A-level students to take some responsibility for their own learning. The Student Sheets in the TRUMP package are therefore written so that they can be used by students working fairly independently: they contain guidance on ways of working as well as posing questions and problems for students to tackle. Where Student Sheets pose direct questions, answers and notes are provided - these are printed separately, though, as some teachers prefer to withhold them.

Finally, for reasons noted previously, the activities in the TRUMP package have been devised so as not to require specialised or expensive equipment, or extensive work away from school or outside normal class time. Table 2 lists the student activities developed for the pilot section of the resource package. Three examples are discussed further below.

2.2.1. *Broadband spectra*

In this activity a discussion of images and graphs over a wide range of wavelengths first gives students an opportunity to review their knowledge of the electromagnetic spectrum. Teachers then introduce the physics of black-body radiation, including Stefan's and Wien's Laws - the Study Notes and Teaching Notes discuss some conceptual difficulties and suggest some approaches to teaching. Students then return to the broad-band spectra provided on the Student Sheet, and are asked to decide which objects appear black-body-like over part or all of their spectra: some of the spectra provided resemble that of a black body (the microwave background spectrum from COBE is included), while others (for example the Crab Nebula) do not. They then estimate the temperatures of the black bodies' using Wien's Law with wavelengths or frequencies read from the graphs. Students thus have an opportunity to apply an important piece of basic physics in an astronomical context, at the same time learning that while many astronomical objects have black-body-like spectra, many do not, and that astronomical black-body temperatures range from 2.7 to many millions of kelvin. The activity also gives students practice in interpreting graphical information.

Table 2. Student activities in the TRUMP package

Activity	In this activity students will ...
Developing a summary	... devise summaries of key terms
Observing the night sky	... observe the colour and brightness of objects in the night sky
Comparing lab. light sources	... observe and discuss factors affecting the apparent brightness of lamps
Questions on luminosity	... use the inverse-square law in calculations relating flux, luminosity and distance
Questions on magnitudes	... use the magnitude scale in calculations involving distance and magnitude
3D model	... make and discuss a scale model of Cassiopeia
Cepheids(1)	... use Hubble data to estimate the distance to M100
Broadband spectra(1)	... discuss and interpret broadband spectra
Line spectra	... read a contemporary account of spectral classification and identify some line spectra
Oil-spot photometer(2)	... use simple apparatus to estimate the Sun's luminosity
HR diagrams(1)	... plot and interpret Hertzsprung-Russell diagrams

(1) Discussed below
(2) Reproduced from an Open University course (Jones, 1996)

2.2.2. *Cepheids*

This is a fairly open-ended problem-solving activity, intended to introduce students to the idea of a standard candle and to reinforce their understanding of brightness and distance. After reading some introductory information about standard candle techniques and about Cepheid variable stars, students encounter a variety of data including some Hubble measurements of Cepheids in the galaxy M100 and a graph of the general period-luminosity relation for Cepheids. They are invited to work in small groups, first to decide how best to use the data, then to estimate the distance to M100 and finally to present their result in a brief report. There are several ways in which the data can be used, so students have to think and understand rather than merely carry out a learned procedure. They have to apply graphical and numerical skills (including the use of logs), and to communicate their findings to others using skills of communication and presentation.

2.2.3. *HR diagrams*

Students are provided with luminosities, temperatures, magnitudes and colours of about a hundred bright and nearby stars and produce a Hertzsprung-Russell diagram using the data (the Teaching Notes suggest various ways in which this might be done). Students then work in small groups to tackle a variety of questions and exercises relating to the HR diagram: these include discussions of the quantities plotted and conventions for labelling the axes; identification of the main sequence, red giant and white dwarf regions; calculations using Stefan's Law to relate the luminosities of stars to their size and temperature; using a cluster HR diagram to estimate distance. From the start of this extended series of activities, students are invited to devise their own written account of the HR diagram for their own future use – they are asked to consider what information they should record and how this might be done most effectively. These linked activities therefore enable students to use skills of communication and presentation, to plot and interpret graphs, and to carry out calculations, while at the same time reviewing key areas of physics (Stefan's Law; brightness and distance) – all based around the HR diagram which is central to stellar astronomy and astrophysics.

3. Summary: key features of the TRUMP project

The TRUMP Astrophysics project provides one model for the development of effective curriculum resources. Key features, which could be transferred to other projects, include:

- active involvement of practising high-school teachers and academic experts
- low-cost materials that require no expensive or specialised equipment
- independent-learning materials for teachers' own professional development
- student activities that use suitable data and images to enable the learning of physical principles and the development of skills.

REFERENCES

AVISON, J. H., 1994, *A review of the new GCE A-level Physics Syllabuses for the 1996 examination in England and Wales*, Physics Education Vol. 29, 333- 346.

DEARING, R., 1996, *Review of Qualifications for 16-19 Year Olds*, SCAA Publications.

JONES, B. W., 1996 *Distance Education: At-a-Distance and on Campus*, IAU Colloquium no. 162.

PARTICLE PHYSICS PROJECT, 1992, *Particle Physics: a new course for schools and colleges*, Institute of Physics.

School Curriculum and Assessment Authority, 1994, *GCE Advanced and Advanced Supplementary Examinations: Subject core for Physics*, SCAA Publications.

SWINBANK, E., 1992, *Particle Physics: a new course for schools and colleges*, Physics Education Vol. 27, 87-91.

The Life in the Universe Series

By J. Billingham[1], E. DeVore[1], D. Milne[2], K. O'Sullivan[3],
C. Stoneburner[4] & J. Tarter[1]

[1] SETI Institute, Mountain View, CA

[2] Evergreen State College, Olympia, WA

[3] San Francisco State University, San Francisco, CA

[4] University of California at Santa Cruz, CA

Students, young and old, find the existence of extraterrestrial life one of the most intriguing of all science topics. The theme of searching for life in the universe lends itself naturally to the integration of many scientific disciplines for thematic science education. Based upon the search for extraterrestrial intelligence (SETI), the Life in the Universe (LITU) curriculum project at the SETI Institute developed a series of six teachers guides, with ancillary materials, for use in elementary and middle school classrooms, grades 3 through 9. Lessons address topics such as the formation of planetary systems, the origin and nature of life, the rise of intelligence and culture, spectroscopy, scales of distance and size, communication and the search for extraterrestrial intelligence. Each guide is structured to present a challenge as the students work through the lessons. The six LITU teachers guides may be used individually or as a multi-grade curriculum for a school.

Integral to the development process was the collection of evaluation data on draft materials from field test teachers, students, and scientists. These data led to revisions and further field tests. Responses indicate that the objectives for the materials were achieved, and that the materials were well received. The LITU project was conducted by the SETI Institute in Mountain View, CA; the project was funded by the National Science Foundation (NSF) and the National Aeronautics and Space Administration (NASA). The LITU Series is being published by Teachers Ideas Press, a division of Libraries Unlimited, Englewood, Colorado, USA.

1. Introduction

Dr. Jill Tarter walked into the classroom and 27 ten and eleven year-old students turned to look at her expectantly. They had been told by their teacher that Dr. Tarter was a nationally recognized radio astronomer specializing in SETI, the search for extraterrestrial intelligence. Many thoughts probably went through their lively minds: "Stephen Spielberg's movie ET about an extraterrestrial!" "Is it possible to visit an extraterrestrial civilization?" "What do scientists do when they search for extraterrestrial intelligence?" Most of the students had very naive conceptions about SETI but all of them had ideas and questions about it. After her presentation, Dr. Tarter responded to dozens of questions. As she worded careful answers it dawned on her that, no matter to whom she talked about SETI, there was always a great deal of interest in the subject.

Other SETI scientists concurred. During their talks, briefings and lectures they had also discovered that the topic of extraterrestrial intelligence stimulated more interest and excitement than most. It was obvious that an excellent way to capture the interest of students of all ages was to approach the teaching of science through the topic of SETI.

2. The Project

A plan to design a curriculum for pre-college students was proposed. Participation was solicited from the local community. The team organized to explore the possibilities consisted of scientists, curriculum developers, professors of preservice science teachers, and educators from elementary and middle schools. All agreed that exploring SETI in age-appropriate activities and lessons would motivate diverse students to study science. The planning team also saw the opportunity for thematic and integrated instruction as the topic of SETI uniquely combines many scientific disciplines and communication skills. Astronomy, biology, anthropology, physics, chemistry and paleontology are all part of SETI research. A curriculum based on SETI would help to fill the needs of teachers being pressed by state and national guidelines[Kober, N. (1994)] to improve the quality of science teaching and to increase student science literacy. These guidelines direct that science be taught, not as a collection of facts to be memorized, but as a lively, interactive subject appealing to all students regardless of race or gender. As stated in the guidelines, "Students (should be able to) connect science concepts with the natural world and explore how science and technology affect their lives and their society"[American Association for the Advancement of Science (1993)].

Table 1 Drake's Equation: $N = R_* \cdot f_p \cdot n_e \cdot f_l \cdot f_i \cdot f_c \cdot L$

Factor	Subjects relevant to study of this factor
$R_* =$ number of new stars formed in our galaxy each year	Astronomy, Chemistry, Mathematics, Physics
$f_p =$ fraction of stars that have planetary systems	Astronomy, Mathematics, Physics, Planetary Science
$n_e =$ average number of planets in each system that can support life	Astronomy, Chemistry, Physics, Planetary Science
$f_l =$ the fraction of such planets on which life actually originates	Astronomy, Biology, Chemistry, Ecology, Geology, Meteorology, Exobiology
$f_i =$ the fraction of life sustaining planets on which intelligent life evolves	Anthropology, Evolutionary Biology, Geology, Meteorology, Paleontology, Ecology, Atmospheric Sciences, Neurophysiology
$f_c =$ the fraction of intelligent life bearing planets on which the intelligent beings develop the means and the will to communicate over interstellar distances	Language Arts, Mathematics, Physics, Social Sciences, Behavioral Sciences, History
$L =$ the average lifetime of such technological civilizations	Astronomy, History, Mathematics, Paleontology, Social Sciences, Behavioral Sciences

The planning team requested joint funding from the National Science Foundation

(NSF) and the National Aeronautics and Space Administration (NASA) to develop a series of six teachers guides for use by teachers of 8 -16 year-old students. As proposed, each guide would focus on an aspect of the search for extraterrestrial intelligence in a way guaranteed to intrigue. Each guide would present a variety of scientific disciplines in the integrated manned exemplified by the Drake Equation which is the thread which binds the LITU series together. The Drake Equation provides a method of estimating the number of intelligent civilizations in our galaxy with whom we might make contact.

Both NASA and the NSF agreed that the SETI Institutes proposal was worthy, and funded the project, "Life in the Universe - An Exciting Vehicle for Teaching Integrated Science."

3. Curriculum Design

The first step was an intensive summer design workshop with a team of teachers, scientists, curriculum developers and project staff. The first few weeks of the workshop were spent brainstorming, listening to content lectures, testing experiments and creating lessons. From this collaboration arose four draft teachers guides: three for the ten to twelve year-olds, and one for the thirteen to fifteen year-olds. As a pilot test, the teachers who participated in the summer design workshop returned to their classrooms and implemented these lessons. They made careful notes on student response, availability of materials, and preparation time involved for each lesson. This process was repeated the following summer. Two additional draft guides were produced: one for eight to ten year-olds, one for fourteen to sixteen year-olds.

4. Testing and Evaluation

The original plan was to have the developer teachers spend a week of their mid-school-year vacation revising the draft materials. This proved to be impractical. One week was not enough time to revise materials and make them suitable for national testing. Additionally, most teachers had other commitments. The solution was to hire a former teacher, and later two additional teacher/writers, to collect all the pilot test information and use it to revise the teachers guides and produce consistent looking material. These revised drafts were sent to teachers across the United States for a second round of testing.

One of the most valuable investments made by the LITU project staff was attendance at state and national science teacher conferences. At most conferences, staff members presented LITU materials at a NASA booth in an exhibition area. Providing sign up sheets at these locations netted hundreds of names and addresses of interested teachers; 150 became national test teachers.

Accompanying these draft guides was a battery of carefully designed evaluation materials. First, the teachers completed a written assessment of individual lessons, detailing student reactions, effectiveness, materials and equipment demands, and so forth. A second form of evaluation was obtained from the students themselves. After they completed the unit, they wrote a letter to the project evaluator describing what they liked and didn't like, what they learned, and what changes they would recommend. Finally, the teachers' overall reactions were obtained by an independent consultant who contacted each teacher to conduct a telephone interview. These three sources of evaluation were very useful in producing revised materials.

5. Science Review

Lessons were reviewed by scientists who specialized in the subject areas addressed in the lessons including scientists involved in developing the guides and other scientists affiliated with NASA and the SETI Institute. Scientists from other institutions volunteered to review lessons as well. Using a checklist, reviewers evaluated the lessons for accuracy, up-to-date content, and age-level appropriateness of content. They also pointed out misconceptions and suggested additions and revisions.

For some scientists, it was difficult to account for age-level appropriateness in their suggestions due to their lack of teaching experience with younger children. However, the scientists comments on content were invaluable and the recommended changes were incorporated by staff writers.

6. Curriculum Product: The Life in the Universe Series

After three years of development and testing, six teachers guides were produced for grades 3-9. They are described briefly below.

In *The Science Detectives* (ages eight to ten), students trace the travels of Amelia Spacehart, an astronaut and radio astronomer, who is searching the solar system for the source of a mysterious radio signal. From her futuristic NASA spacecraft (by way of a videotape), Amelia provides clues that lead the students to explore features of the solar system, states of matter, lenses and magnification, and large scale measurements. Their challenge is to anticipate her destinations and track her journey for NASA.

The SETI Academy Planet Project trilogy is targeted for students aged ten to twelve. The three *Planet Project* guides represent an extended, interdisciplinary curriculum when used in series. (They can also be used independently.) In each, students are invited to participate in a "SETI Academy"education and research program in which they act as scientists who are exploring Earths history for clues to the origin and evolution of life to inform investigations about the possible existence of life beyond our solar system. In *The Evolution of Planetary Systems*, students learn about the evolution of stars and planets. In *How Might Life Evolve on Other Worlds?*, students study the evolution of plant and animal life. In *The Rise of Intelligence and Culture*, students explore intelligence, culture, technology and communication. In all three guides, the students learn what scientists have discovered about Earth and the evolution of life, intelligence, and culture. Throughout, they apply their knowledge about the Earth and its life and intelligent cultures to construct a fictitious planet with life forms and culture.

In *Life: Here? There? Elsewhere?–The Search For Life On Venus and Mars* (ages thirteen to fifteen), students investigate the characteristics of life through this introduction to the multidisciplinary sciences of comparative planetology and exobiology. Students simulate conditions on Venus and Mars and explore various means of detecting life in the atmosphere and soils of Earth. They apply this knowledge to propose a spacecraft design for detection of life on Mars or Venus. Students also review the results from the Viking missions to do their own analyses of whether there is life on Mars.

In *Project Haystack: The Search For Life In The Galaxy* (ages fourteen to sixteen), students investigate the questions: Are there intelligent civilizations out there and where might they be? If they are out there, how would they communicate with us and what would they say? How should we respond? Students study the scale and structure of the Milky Way Galaxy as they explore the cosmic "haystack". They construct a simple radio receiver, and learn about signal-to-noise problems in making observations. They study SETI science by using simple astronomical tools to solve some of the challenges of sending and receiving messages beyond our solar system.

As well as being enticing to students, each guide is carefully organized to provide teachers with the information they need to implement the program. The introductory section includes suggestions for preparation, teaching strategies, classroom management, and student assessment. This is followed by ten to fifteen hands-on, activity-based lessons. Every lesson begins with an introductory overview and is followed by a materials list, suggestions for materials preparation, detailed directions for every step of the activity, student worksheets, teacher answer keys, and suggestions for further explorations of the concept or concepts covered. The appendices contain comprehensive materials lists of the consumable and reusable supplies, ordering information and suggestions, annotated bibliographies, glossaries, and teacher information detailing the main scientific concepts underlying each lesson.

Additionally, each guide is provided with ancillary materials:
• all guides have an interactive poster, designed for the LITU Series by artist Jon Lomberg,
• *The Evolution of Planetary Systems, How Might Life Evolve on Other Worlds?, Life: Here? There? Elsewhere?* and *Project Haystack* are supported by video tapes of images of stars, galaxies, life of the past, and more that illustrate the lessons,
• *The Science Detectives* is based on an adventure story presented on video tape,
• *The Rise of Intelligence and Culture*, includes color transparencies of eight ancient human civilizations,
• *Life: Here? There? Elsewhere?* includes a set of orders-of-magnitude activity cards with views of Venus, Earth, and Mars designed by Jon Lomberg.

These ancillary materials were an unexpected but marvelous outgrowth of the curriculum development process. The design team members were exceptional in their capacity to be creative; they couldn't help imagining and inventing exciting accessory materials that enrich the lessons by reaching out to students who prefer a visual learning mode.

7. Publication

Publishers were first contacted at the National Science Teachers Association conferences in the exhibition hall. Most science education publishers exhibit at these meetings, and the personal contacts made during the conference were invaluable. As the guides neared completion, the LITU project manager contacted the publishers who had expressed interest in the curriculum materials. Along with an informative letter, samples of the materials were sent for their examination. Acquisition editors from three publishing houses responded. After many months of negotiations, Teacher Ideas Press was selected to publish and market the LITU Series and ancillary materials. The LITU Series is available directly from Teacher Ideas Press, a division of Libraries Unlimited, Dept. 9503, P. O. Box 6633, Englewood, Colorado, 80155-6633, USA.

8. Conclusion

The introduction of SETI-based science lessons to the elementary and middle school classes should help to persuade all students, including those in groups underrepresented in scientific fields, that science is interesting, challenging, and an attractive career possibility. The LITU series offers an introduction to the type of logical, disciplined thinking, both creative and analytical, that is the cornerstone of science. Further, by actively involving students-as- scientists throughout the LITU Series lessons, students learn about the nature of the universe and how scientists go about exploring it. These experiences and knowledge also allow students to discard common misconceptions regarding the search for extraterrestrial life. Perhaps equally important is that by developing a perspective of life on other worlds, students may gain a sense of the unity of life on Earth, particularly among human beings. One of the optimistic views of the SETI scientists themselves is that the discovery of an extraterrestrial civilization will underscore the unity of humanity.

Acknowledgements

Funding for this project was provided by NASA cooperative agreement NCC-2-336 and NSF grant # MDR-9150120. Additional acknowledgment goes to the SETI Institute for supporting the project. Special thanks to Dr. Cary Sneider and the staff of the Lawrence Hall of Science, University of California at Berkeley for their many contributions to the LITU series, and to all of the teachers and students who contributed ideas, tested lessons, provided critiques and learned about the search for life in the universe.

REFERENCES

AMERICAN ASSOCIATION FOR THE ADVANCEMENT OF SCIENCE *Benchmarks for science Literacy*, 1993, Oxford University Press, New York.

CALIFORNIA STATE BOARD OF EDUCATION. *Science Framework for California public schools kindergarten through grade twelve*, Sacramento,1990, California Department of Education.

RUTHERFORD, R. & AHLGREN, A. *Science for all Americans*, 1990, Oxford University Press, New York.

KOBER, N. *What we know about science teaching and learning*, 1994, Council for Educational Development and Research, Washington DC.

The Astronomy Village: Investigating the Universe

By S. M. POMPEA[1] & C. BLURTON[2]

[1]Pompea and Associates, 1321 East Tenth Street, Tucson, Arizona, 85719-5808 USA and Adjunct Faculty, Steward Observatory, University of Arizona

[2]NASA Classroom of the Future, Wheeling Jesuit University, Wheeling, West Virginia, 26003 USA

1. Introduction

The Astronomy Village multimedia program is designed to emphasize the process of science as much as its content (Pompea and Blurton, 1995; Pompea, 1996). It was designed for 14 year-old students, but has been used at slightly younger age levels and for older students, including university students. The investigations are flexible enough to be used at this wide variety of levels.

In this CD-ROM-based multimedia program, student teams can pursue one of ten research investigations. In each investigation they are guided by a mentor, receive e-mail, hear a lecture in the Village auditorium, and make observations using ground or space-based telescopes in a virtual observatory. The students also process data using the *NIH Image* image processing program. They keep a detailed logbook of their research activities and can run simulations on stellar evolution as well as manipulate 3-D astronomy visualization tools. At the end, they present their research results to their classmates and answer questions about their results at a press conference. The Astronomy Village builds upon previous work in the use of image processing for education (Pompea, 1994a), teaching techniques in astronomy (Pompea, 1994b) current research in astronomy (Pompea, 1995), and developments in optics education (Pompea and Nofziger, 1995; Pompea and Stepp, 1995; Pompea, 1996).

2. The Astronomy Village Process Model

The process model for the Astronomy Village program is that students become members of one of ten potential research teams that are pursuing front-line observational astronomy research. In the Astronomy Village, the research process is broken down into five stages.

(1) In the "Background Research" phase, students read about their topic in the Library, listen to an introductory lecture in the Auditorium, are asked probing questions about the fundamental thinking behind the research, pursue hands-on experiments in the Hands-On Laboratory, and look at images in the Image Browser at the Observatory.

(2) In the next phase, "Data Collection", students plan an observation and do more thought and hands-on experiments related to their project.

(3) In the "Data Analysis" phase, students use an image processing program to extract more information from their data.

(4) In the "Data Interpretation" phase, students are looking critically at their data and draw broader results from it.

(5) In the "Presentation" phase of the research, the students consolidate their knowledge, critically examine their research and report on their observations and conclusions to other students. They also answer questions on their project at their press conference.

The Village Auditorium (left) and the Village Computer Lab (below).

One of the Village Conference Rooms (above) and the Village Library (right).

FIGURE 1.

3. E-Mail and Visualization Tools

During these phases of the research process, the students receive electronic mail to welcome them to the research facility and as encouragement or guidance at key points in the research process. The e-mail is used to reinforce some key points about how research is done. It also gives a more realistic view of how new information can be injected into the research process.

During the research students have access to a variety of visualization tools. They use these tools to understand key astronomy concepts by manipulating objects in three dimensions. For example, through this manipulation they come to see constellations as random arrangements of stars in a spatial volume and the motion of these stars in three dimensions. They also have access to a stellar evolution simulator to follow the evolution of stars of different masses. In an orbital or gravity simulator, they are given a large variety of situations (solar system, binary encounter, asteroid swarm meets planet, etc.) programmed in for convenience. In this way students can better understand the dynamical evolution of a wide variety of planetary and stellar systems by letting the system go forward or backward in time under the influence of gravity.

We encourage students to write in much the same ways that scientists write. At the end of each research phase or even each activity in a phase, students are writing in the research notebooks, recording impressions, drawing, or pasting images. The student team uses their Log Books not only as a record of their progress, but also as a presentation tool when they have finished their research.

4. The Role of the Mentor and the Team

Mentors are an important part of the research process in the Astronomy Village. The mentor first introduces them to the investigation area. The mentors are muticultural and multiracial in appearance and name and provide a visual sign of diversity that students can identify with. There is one mentor for each investigation and this person appears in short movies to give the students guidance at each stage of the research process. The mentor also plays a valuable role, especially for younger students, in giving guidance and limiting the number of possibilities.

Like most scientists, the students work in cooperative research teams. The student teams have three members, with each student specializing in certain areas, such as image processing or using the telescope. The three roles are "observer", "researcher" and "data analyst". In each activity one student can take the lead. In this way the time on the keyboard controlling the computer can be shared by the students. The use of cooperative learning has proven to be an effective learning tool.

5. Assumptions About the Scientific Process

Students generally need some training in the general scientific process and in particular skills useful in astronomy. The general training on how to use the program is introduced in a 15 minute videotape and covered in more depth in skills training sessions in the Orientation Center. The skills training is given before the students begin their investigation. The Astronomy Village has certain fundamental pedagogical constructs about the scientific process which are also expressed in the training process. These include:

(a) Science is a contact sport. Students learn about science by actively constructing and discussing inferences, relations, comparisons, questions, and mental models. Knowledge is actively constructed by students, not passively received from the teacher or textbook.

(b) Science is a social activity. It is not an isolated activity, but occurs most often within groups and teams. The Village relies on a similar cooperative group process.

(c) Students should learn science the way that scientists do science. This is not to train them to become scientists, but to give them a more in-depth view of how science is done and the strengths and limitations of the scientific process. The Village five step model of the scientific process provides a sensible model of how science operates.

(d) Students need real-world problem solving skills. If we expect students to solve problems, they must be given practice and training in problem solving. The students working in the Astronomy Village are working with realistic astronomical data and problems and have access to a wide assortment of astronomical images and tools.

(e) Higher order thinking and creativity should be encouraged. Our experience is that few students at this age level have a solid understanding of the scientific process. We try to provide a sensible amount of structure but have tried to preserve the turns and twists inherent in the scientific process. We have done this by building in some blind alleys, some complexity, and some options and decisions for students. The realistic problems in the Village are designed to encourage higher order thinking skills.

6. The Research Investigations

The ten individual investigations in the Astronomy Village are related to current astronomy research topics:

(1) **Search for a Supernova** The supernova search team used data from neutrino detectors to locate a supernova. The team takes new images which are compared with archival images to find a supernova. If one is found, a light curve will be constructed to verify that it is a supernova and not another kind of variable star.

(2) **Looking at a Stellar Nursery** The team takes images of the Omega Nebula at several different wavelengths to identify which wavelengths are most useful for looking at different classes of young objects.

(3) **Variable Stars** This research team explores the properties of variable stars and uses Cepheid variables as "standard candles". The team searches for a Cepheid in another galaxy to in order to find the distance to that galaxy.

(4) **Search for Nearby Stars** This team is searching for nearby stars using parallax. Archival and current CCD images are used to identify several nearby stars.

(5) **Extragalactic Zoo** The research team explores different types of galaxies and clusters of galaxies to gain an understanding of galaxy environments and the scales of clusters of galaxies.

(6) **Wedges of the Universe** The team examines two sections or wedges of sky to see differences as the views go deeper and deeper into space. Students use these views to estimate the number of galaxies visible.

(7) **Search for a "Wobbler"** The team is looking for a star that is wobbling in its motion across the sky. The team uses image processing to see if a "wobbler" has a faint stellar companion or an orbiting planet.

(8) **Search for Planetary Building Blocks** This team examines the Orion Nebula in a search for protoplanetary disks. The teams constructs multicolor images of this star forming region to find these objects.

(9) **Search for Earth Crossing Objects** Will the Earth be hit by an earth-crossing asteroid? The team processes CCD data to find asteroids and try to determine if they will hit the Earth.

(10) **Observatory Site Selection** The team must make some decisions on site selection for a future observatory by evaluating five potential sites for light pollution, atmospheric stability, and accessibility.

The Astronomy Village can also be used as a database of astronomy articles and images. The Library has an on-line retrieval system that permits searching over 100 documents by author, subject, or title. The Observatory has over 300 images which can be searched by general topic. The teacher can also use the Astronomy Village as a presentation tool for an entire class or as a guide for student independent research projects.

7. Components of the Astronomy Village

The village metaphor is used to frame the research activities in each path. The interface allows students to interact with the program, navigate to different "facilities" within the village, and make use of various tools and resources. Here is a brief outline of the Astronomy Village components.

The Orientation Center The Orientation Center contains a Village tour that introduces the Village and its tools and resources to students. The tour is presented as a combination of computer graphics, audio, and video clips. The Orientation Center contains "skills building" activities that guide students through introductory activities to show them how to use the various facilities, tools, and resources within the Village.

The Conference Room The Conference Room contains the ten research questions from which teachers may assign projects or from which students may select projects.

The Observatory While in the Observatory, students are able to use a simulated telescope control panel to "capture" images of various objects at a variety of wavelengths. Both local observing and remote observing at a variety of sites are simulated. Students can also browse through an archive of over 300 images.

The Library The library contains an electronic catalog and search tools, the "COTF Times" newspaper (which contains articles illustrated with images), and a collection of NASA publications that focus on various aspects of astronomy (e.g. "Exploring the Universe with the Hubble Space Telescope"). There are over 150 articles in the Library.

The Computer Lab Within the Computer Lab students are able to access and use four tools: *NIH Image* (an image processing program), simulation programs, including a "Star Toolkit" which allows them to create stars of different masses and observe their life cycle, simulated e-mail, and a telecommunications program that allows them to connect to the NASA Classroom of the Future's server and to NASA SpaceLink.

The Hands-On Lab The Hands-On Labs contains directions for conducting experiments on and off the computer that teach, illustrate, or reinforce the astronomical concepts. Students will be able to print the directions to these experiments. There are over 25 Hands-On Labs.

The Auditorium The Auditorium provides students with twelve different 5-10 minute talks by astronomers related to the ten investigations. Each lecture is presented by means of audio recordings and illustrative computer graphics, video clips, and images.

The Cafeteria The Cafeteria is used by students to "overhear" astronomers informally talking about why they chose astronomy as a career, what being an astronomer is like, what it means to be engaged in scientific inquiry, and their specific research projects. There is also humor in the conversations, the menus, and the wall decorations.

The Press Conference At the end of each investigation, students get a chance to answer questions from the press. Sometimes the simplest questions are the hardest to answer!

Meta-tools There are four "meta-tools" students may access from any location within the Village. These include the Student Log Book, Calculator, Village Map, and Help. A fifth meta-tool is the "path" diagram, which outlines the types of activities possible when a more structured approach is desirable.

8. Running and Using the Software

The minimum system requirements are a Macintosh LC III running System 7 with 8 MB RAM, or a Power Macintosh with 16 MB of RAM; CD-ROM Drive; 20 MB on the hard drive; 13″ RGB color monitor (640 X 480 pixels, 256 colors); QuickTime™ 1.5; HyperCard™ 2.2 (not the HyperCard Player™). The Astronomy Software Village is available through NASA CORE and through NASA Teacher Resource Centers. The cost is nominal. For more information on obtaining the software, see the web pages on the Astronomy Village at the NASA Classroom of the Future: `http://www.cotf.edu`. Or, contact the Astronomy Village Project, NASA Classroom of the Future, Wheeling Jesuit University, Wheeling WV 26003. Phone (304) 243-2388; Fax (304) 243-2497.

The authors would like to acknowledge the funding of the project through NASA, to thank the entire Astronomy Village development team at the NASA Classroom of the Future, and to note the generous assistance of the many astronomers who aided us in the project.

REFERENCES

POMPEA, S. M., 1994a, Image Processing Exercises for Astronomy, West Publishing.

POMPEA, S. M., ed. 1994b, Great Ideas for Teaching Astronomy, 2nd edition, West Publishing.

POMPEA, S. M., 1996, The Astronomy Village. In Astronomy Education: Current Developments, Future Coordination, ed. Percy, J.R. ASP Conference Series, vol. 89, pp. 259-261.

POMPEA. S. M. AND BLURTON, C., 1995, A Walk Through the Astronomy Village. Mercury, Jan–Feb.

POMPEA, S. M., ed., 1995, Current Perspectives in Physics and Astronomy, West Publishing.

POMPEA, S. M. AND NOFZIGER, M. J., 1995, Resources on optics in middle school education. In Proceedings SPIE: 1995 International Conference on Education in Optics ed. by Soileau, M.J., vol. 2525.

POMPEA, S. M. AND STEPP, L., 1995, Great Ideas for Teaching Optics. In Proceedings SPIE: 1995 International Conference on Education in Optics, ed. by Soileau, M.J., vol. 2525.

POMPEA, S. M., 1996, Arizona Optics Industry Association (AOIA) Focus Group Activities on Education. In Proceedings of the SPIE, Global Networking of Regional Optics Clusters, ed. by Breault, R.P., vol. 1550.

Posters

Poster Review

By C. Iwaniszewska

Institute of Astronomy, Nicolaus Copernicus University, Chopina 12/18, 87-100 Torun, Poland

The reader of this proceedings volume might ask why was it thought interesting to publish a few pages about the posters presented at IAU Colloquium 162? It had been decided that indeed the history of the meeting would not have been complete without some words about the poster presentations. The final success of the entire Colloquium depended on all presentations, either oral or poster.

The posters themselves have been different but it has been interesting to note that sometimes similar projects and ideas have been elaborated at very distant places in the world.

The basis of our teaching should be related to our roots; we ought to mention old traditions in our own country, such as the cosmological ideas of old Guarani Indians or the story of the first South American 18th century. Observatory of F. Buenaventura Suarez which have been shown by *A. E. Troche Boggino* of the University of Asuncion, Paraguay. However, I found most interesting the history of evolution of the human mind as depicted by the diagram of A. E. Troche Boggino showing chronological sequences of contemporary scientists, philosophers, writers, painters, sculptors and composers, from the times of Copernicus to the present day. It is easy to make a perpendicular cut or cross-section of the diagram at a given epoch, for instance that of Copernicus, and get to know what other famous persons have been living during his life-time.

In another part of the world, at the University of Glamorgan, UK, *Mark Brake* had been also in favour of a historical/cultural approach when telling his audience scientific facts, both when dealing with university students and with the general public.

Our teaching should begin at primary and secondary schools. *Jean-Luc Fouquet* of Saint-Martin de Ré, France, wrote about the fun of introducing astronomy while playing with very young (5–12 yrs) children, while *T. Lacey* of Nottingham, UK, has displayed a set of astronomical work cards produced for young (7–11yrs) children to be posted to schools. *Vladimir Stefl* of Brno University, Czech Republic, urged teachers to introduce more astronomical examples when teaching physical laws in secondary schools. I must say that when I have expressed my disbelief whether a certain equation is used in secondary education, I was quickly convinced by looking at the content of textbooks from no less than 7 countries which V. Stefl had been comparing. *Michael McCabe* of Portsmouth University, UK, had written about his work: he had prepared 234 screens of interactive multimedia for showing fundamental astronomical notions. Two Spanish teachers from Alicante, *Bernat Martinez* and *Gullem Bernabeu* have told about their experiences when introducing two types of teaching, on practical and theoretical levels, as well as on the evaluation of students' alternative conceptions on astronomical observations. Practical worked at an Observatory was one of the prizes given for winners of the Astronomical Olympiads in Russia, where this type of contest has its 50th anniversary as announced by *A. V. Zasov* of Moscow University.

I have been happy to learn that many Observatories take an active part in the teaching process as described in four posters. *Francisco Diego* of University College London, UK, had a chance to enlist some of the staff members of the Physics and Astronomy Department of his University to go to schools for special lectures, *Margaret Metaxa* of Athens, Greece, has shown plans for a summer school for 16-18 yrs students at the Observatory of Athens, while *Marie-France Duval* of Marseille University, France, displayed special didactic programmes prepared by each of the nine French Astronomical Observatories. And finally, didactic activities of Abastumani Observatory in Georgia were mentioned in a poster by *E.K. Kharadze*.

Our teaching goes still farther, at University level, and here I have noted five posters. The introduction of new university programmes and practical work in astrophysics has been discussed by *Gabor Szecsenyi-Nagy* of *Eotvos* University, Hungary, while *Mary Kontizas* of Athens University, Greece, told about laboratory astronomical work for all physics students at her University. In order to get a better understanding of the physical processes in stars *Harinder Singh* and his collaborators at Delhi University, India, prepared for their students a visualisation of a flow in a box, meant to represent flows in stellar interiors. *John Wilkinson* of Central Queensland University, Australia, reported about astronomical courses for external students in physics, while *Valentin G. Karetnikov* from Odessa University, Ukraine, wrote on astronomy curriculum at his University.

A wide variety of approaches has been shown in relation to the general public. *E. Malamud* and his collaborators from the Open University, Milton Keynes, UK, proposed to raise the public interest in astronomy using the solar telescopes of an observatory, while *Gabor Szecsenyi-Nagy* of Budapest reminded us of the Total Solar Eclipse in 1999, whose path of totality runs so conveniently across Hungary, his country. *Julieta Fierro* of UNAM, Mexico, has been of the opinion that her students can develop many interesting ideas for the general public, so she let them prepare their own programmes for shows, etc, while broadcasting astronomical news on television was the topic of a poster by Alexandra Levell from Leicester University, UK. A whole complex of different astronomical interrelations for those who would go for a visit to the Laboratory - "Astronomical Village" has been proposed by Stephen Pompea of Tucson, USA, while *Nestor Camino* of the University de la Patagonia, Argentina, told of his experience to get a good contact with the public by means of an astronomical publication, published monthly in a local newspaper.

More sophisticated teaching projects have been shown by *Mark Jones* from the Open University, Milton Keynes, UK, who told of the possibility of conference contacts between students and their tutors by means of computers. Another poster by *Hans J. Fogh Olsen* of Copenhagen Observatory, Denmark, advertised the production of CD-ROMS with starmaps for this year. It has been a good idea to unite the efforts of more scientific institutions on giving information for the general public through a whole network which has been shown by *Carol Christian* and her colleagues from scientific institutions in the Eastern part of the US.

Two posters were meant to encourage simple observations. The first by *Roland Szostak* of Munster University, Germany, related simple experiments – solar measurements conducted by your (10 yrs) children where they could obtain a correct value of the length of the solar year. The second poster by *John R. Percy* of Toronto University, Canada, told about the project of the AAVSO to help secondary school students in observing variable stars with data obtained from AAVSO archives. For persons working in science it seems interesting to know more about themselves. Such a survey of motivation of scientists has been undertaken by *Ian Elliot* of Dunsink Observatory, Ireland, who has shown, *inter alia*, that a maximum interest in science is displayed at the age of 10-12 yrs, while the decision of taking a scientific career was taken by the majority towards 15-17 yrs. *Artur Chernin* of Moscow University, Russia, has written some reminiscences of George Gamov as a teacher.

At the end of the posters presentation I would like to mention those related to our artificial sky, to the Planetariums. *Tallas Saygac* and his colleagues of Istanbul University, Turkey, had listed the most important conditions for the future work of a large planetarium which they hope to be able to have one day in their city, while *Francisco Diego* of University College London, UK, recalls his past experience at the Luis E. Erro Planetarium in Mexico City, which he proposes to introduce in future in the Greenwich Planetarium in London.

A review of posters does not give the whole picture of many colourful panels, with diagrams and photographs, that have been displayed for five days in the south Wing of University College London. But I hope that all participants will remember the words of Julieta Fierro: "When we shall speak to our various audiences let us try to let them *feel the excitement of understanding!*"

SECTION EIGHT
Final Address

The Role of Astronomy in Education and 'Public Understanding'

By Martin J Rees

Institute of Astronomy, Madingley Road, Cambridge, CB3 0HA, UK

Although it comes at the end of the programme, this contribution is in no sense a 'summary' of the meeting. It addresses some issues that were covered by earlier speakers, but is written from the individual perspective as a research astronomer working in the UK.

1. Astronomy and Young People

A few comments first on education in schools – this is a special worry here in the UK, where our international rankings are disappointing. An appreciation of science is vital not just for tomorrow's scientist and engineers, but for everyone who will live and work in a world even more underpinned by technology – and even more vulnerable to its failures and misapplications – than the present one. Even more important, the option of higher education in science and technology should not be foreclosed to them. There is widespread concern particularly about the 16-18 age group. Many of us put strong emphasis on broadening the curriculum for this group, which currently enforces unduly early specialisation here in England. Young people opting for humanities should not drop all science when they are 16. (I have carefully said 'England' rather than 'the UK' because the curriculum is already broader in Scotland. Scottish education has its admirers here, but few in Scotland advocate a switch to the English system!)

It is crucial that enough of the brightest young people go on to acquire some professional expertise in science and technology. They will not do so unless , when making the key decisions at age 16 or 18, they perceive a range of appealing opportunities. They will be discouraged if the courses do not inspire them. They will be discouraged if scientists seem valued less than accountants. And they need to feel that science is humanly relevant – that it meets their ethical concerns. (A separate issue is the depressingly low proportion of girls among those who opt for physical sciences – the proportion of women in science and technology will always remain low unless the trends and choices made by 16 year olds can be changed.)

Astronomy has a specially valuable role to play. It attracts wide public interest. It has a positive and non-threatening public image. In this latter respect it has the edge over other high-profile sciences such as genetics, and nuclear physics. It is also inspiring to bright students. An interesting survey was recently carried out carried out among those in their first term at UK universities who had chosen to study physics. They were asked what had influenced their choice. Astronomy and space ranked high. (It was also clear that many had been enthused by particular teachers; text books ranked low, but 'popular' books and magazines were major influences.)

It is right that astronomy should be part of the formal school curriculum. But young people are a receptive and important target for informal initiatives, of the kind addressed at this conference. There are many innovative schemes for bringing individual research scientists in contact with schools. There is growing scope here: telecommunications allow remote access to large facilities, so that individuals – amateurs at home, as well as young people at school – can participate in scientific discovery.

Virtual reality offers new opportunities for science centres, etc. This, however, raises

344

the important issue of whether 'virtual reality' is a supplement and enhancement to what we have already, or whether it may be a counter-educational substitute for traditional 'real' (hands on) reality. Scientists already spend much of their time in cyberspace – almost instant contact with colleagues and collaborators around the world. This affects communication and journal publication (though it has not made conferences obsolete). But it also affects how we do science, what we value. At its worst, the information highway could just smother us in shoddy work. At its best, it can offer marvellous opportunities, especially (if they can surmount the threshold of resources needed for access) to developing countries.

2. Public Perceptions of Science Generally

I will come back to say a bit about college education, but I would like to say something about popular science and the media generally. I have been influenced by recent experience as President of the British Association for the Advancement of Science: this is an organisation, dating from 1830, whose mission is to promote understanding of (and debate about) science, engineering and technology.

In the British Association's Victorian heyday, the national scientific enterprise was minuscule by today's standards. But the commitment to public understanding was not. The marvellous national and civic museums – cathedrals of discovery and invention – consumed large resources by the standards of that time. Our forebears believed that science, engineering and technology deserve wider appreciation, that science is part of our culture, and that how it is applied should concern us all.

Science and engineering had a high profile. Most people have heard of the great 19th century engineers – Brunel, Telford and so forth. It is actually harder to name living engineers – even though their marvels surpass those of earlier centuries.

And it was not just the practical men – the 'wealth creators '– who earned public acclaim. Think of Darwin: his insights had no practical payoff, but he was a revered figure because he changed the way humans see their place in nature. There was also wide interest in exploration of remote parts of the Earth.

Astronomy and cosmology maybe play the same role in contemporary culture as Darwinism and terrestrial exploration did a hundred years ago. We can now probe our cosmic environment and origins; our explorations of the cosmos with telescopes and spaceprobes can be vicariously shared by a wide public. Just as Darwin attracted interest so now do the discoveries of astronomers – setting our entire Earth origin in a cosmic context.

Manned space-flight is, of course, the highest profile and most expensive aspect of our field. The Apollo moon-landing programme was a spin-of, and indeed an inspiring one, from the superpower rivalry of the cold war era. But NASA's current space station seems neither inspiring nor a step towards any worthwhile longer-range goal. There is one striking feature of people's perception of the Apollo programme. Along with everyone else who has now reached middle age, I grew up thinking of 'men on the moon' as a futuristic concept. It became reality in 1969. Even the last lunar landing was 1972. Nobody much under 35 can remember it. To all young people today – to my present students – it is a remote historical episode. They know the Americans landed men on the Moon, just as they know the Egyptians built the pyramids. But the national motivations seem almost as bizarre in the one case as in the other. And the recent film of Apollo 13 – the mission that nearly met disaster – seems to them as dated, in technology and in values, as a traditional 'Western'.

What is the general state of public attitudes to science, engineering and technology? We have all seen quizzes that check what science people know; these are sometimes inflicted, to their embarrassment, on politicians and other dignitaries It is sad if such people's astronomical views are pre-Copernican – or if they cannot tell a proton from a protein. But they can (partially) excuse themselves by claiming that the facts in themselves are not the essence. What matters for everyone is having a rough 'intellectual map' : so that we can appreciate our natural environment; so that the artifacts that surround us do not seem mysterious; and so that we can participate in shaping how technologies are developed and applied.

Everyone needs a basis for assessing when scientific claims are credible and when they are not. Noisy controversy does not always signify evenly-balanced arguments; but most issues that rightly concern us involve genuine scientific uncertainties, and major tradeoffs. The ethical and social implications of (for instance) environmental degradation can and should be widely appreciated and discussed, even by people who do not understand (and may not be specially interested in) the science *per se*. (The same is true even more of biomedical issues.)

3. The Media

Here in Britain there is a long tradition of science popularisation. In astronomy it goes back to Eddington and Jeans, and continues, through Fred Hoyle and others, to Stephen Hawking and Patrick Moore. There are also equally impressive figures in other fields.

There is also a strong tradition of science journalism. But there is an impediment – these dedicated journalists are up against the problem that few in editorial positions have any real background in science. The editors of even the 'highbrow' press feel they cannot assume that their readers possess the level of scientific knowledge that we might hope for in a fourth-former, whereas the same journals would not 'talk down' to its readers on an economic topic or on the arts pages: economic articles on the money supply are quite arcane; the music critic would be thought to be insulting his readers if he defined a concerto or a modulation. About half of the readers of the quality press have some scientific education, or are engaged in work with a technical dimension It is those who control the media (and those in politics) who are overwhelmingly lacking in such basic knowledge.

(There is perhaps an interesting lesson to be learnt from the ''computer pages' of many newspapers, whose success may have caught editors by surprise, just as the enthusiasm for home computing has indicated the enthusiasm and talent of young people untapped by formal education).

There has in this country been growing 'official' encouragement for Public Understanding of Science – from government bodies, scientific and professional societies, etc. Promotion of 'public understanding' is in the formal mission statements of the research councils. There are even small amounts of money to encourage initiatives of this kind. The Particle Physics and Astronomy Research Council, the body that funds astronomy and particle physics, takes this issue particularly seriously, and encourages all astronomers and particle physicists to use every opportunity to disseminate their work broadly. This research council has a special obligation, because its research, though expensive, has less short-term 'spin off' and relevance than some other branches of science; but it also has a special opportunity, because of the public interest in astronomy and space. (We try to cover particle physics, but that offers a good deal more of a challenge than astronomy.)

Public Understanding of Science, as a phrase, is slightly unfortunate or at least suboptimal. Not only does it have an ugly acronym, but it falsely implies a demarcation between science and public – between a priesthood and an unwashed populace.

The adult 'public' is very heterogeneous. All of us here are part of it. Professional scientists are depressingly 'lay' outside their specialisms – we all depend on 'popular' presentations for biomedical topics. Likewise, many of the 'consumers' of popular astronomy have some scientific and technical expertise in other areas.

Broadcasts or newspaper articles about astronomy deepen my respect for journalists who successfully cover all the sciences, working to tight deadlines. I know from experience – and probably most people here know – how hard it is to explain, non-technically, even something in one's specialist field.

Science generally only earns a newspaper headline, or a place on the TV bulletins, as background rather than as a story in its own right. Indeed, coverage restricted to 'newsworthy' items – newly-announced results that carry a crisp and easily summarisable message – cannot avoid distorting how science develops. Scientists cannot reasonably complain about this any more than novelists or composers would complain that their new works do not make the news bulletins. The place of science is in features, documentaries, etc, rather than news. (News coverage of astronomy and other sciences is of course further distorted because some institutions – NASA, for instance – are specially effective in relating to the press. Unfortunately the scientists themselves sometimes 'hype up' their own contributions – science reporters now have to be as sceptical of some scientific claims as they routinely are of politicians'.)

A 'Daily Telegraph' poll last year asked people on what topics they'd like to see more newspaper coverage. Top choice was 'medicine'; 'science and invention' tied with 'crime' for second place.

We are often told that science has to be made relevant to everyday life. That is true, but only up to a point. It is often the utterly 'irrelevant' subjects that fascinate people most. Dinosaurs have been high in the popularity charts ever since Richard Owen discovered them in 1841. Cosmology and astronomy rank high too, of course, so does human origins. All utterly fascinating – all seemingly quite unrelated to practical issues.

As I've emphasised, I feel great admiration for 'professional communicators'. But many of us who are professional astronomers (or indeed working scientists of any kind) do spend some time as 'amateur communicators', presenting our work to general audiences. I would personally derive far less satisfaction from my work if it only interested a few other specialists. It is a challenge – just as teaching is harder at the elementary level than at the more advanced level.

Whatever the audience reaction, the experience is certainly salutary for us as speakers. It helps us to see our work in perspective. Researchers – in astronomy or in any field – do not usually shoot directly for a grand goal. Unless they are geniuses (or unless they are cranks) they focus on bite-sized problem that seem timely and tractable. That is the methodology that pays off. But it carries an occupational risk – we may forget we are wearing blinkers and that our piecemeal efforts are only worthwhile insofar as they're steps towards some fundamental question. Arno Penzias, co-discoverer with Wilson of the microwave background, plainly made a really great discovery. But he said that he did not himself appreciate its full significance until he read a 'popular' description of it in the New York Times. (We need to oversimplify, but should not be too dogmatic. Niels Bohr said that you should speak as clearly as you think, but no more so. That is a good maxim – though Bohr himself took caution to excess by mumbling inaudibly and incomprehensibly!)

One often gets asked very 'fundamental' questions: Is there life in space? Will the universe go on for ever? Why didn't the big bang happen sooner? This reminds us of our ignorance. Also, when even the specialists are at sea, there's less of a gap with general audiences.

Claims to understand anything about the early universe might seem presumptuous.

But cosmology is actually one of the more tractable sciences. Inside a star (and in the early universe) conditions are so extreme that everything is broken down into its atomic constituents, and governed by simple laws. It is complexity that makes things hard to understand, not size. Understanding a frog is a far more daunting intellectual challenge than a star or the early universe. The atoms that made the young Earth are stardust – to have understood this is a triumph of 20th century science. But elucidating how those atoms combined, via Darwinian selection, into progressively more intricate forms, and eventually into creatures that could ponder their origin, is an unending quest that has barely begun. This perspective should caution us against scientific triumphalism – against exaggerating how completely we will ever understand anything really complex.

Incidentally, I think it is crucial that expositions of cosmology, and indeed of any 'frontier' science, should avoid conflating things that are fairly well understood (like, for instance, the broad evolution of stars, and the Hubble expansion) with those that are not (like the physics of the ultra-early universe). Otherwise, credulous readers will accept flakey ideas too readily; those who are more sceptical may, on the other hand, fail to appreciate that at least some parts of astronomy now rest on quite secure foundations.

As well as cosmology, there is also interest in more 'accessible' astronomical topics. There will, for instance, be many 'outreach' events this Autumn linked to the 150th anniversary of the discovery of Neptune. Britain's celebrations will be somewhat ambivalent. The story is well known of how John Couch Adams, then a Cambridge student, predicted that a new planet should exist, but parallel calculations were done by Le Vernier, and the planet was discovered in Berlin. Adams failed to activate the interest of either the then Astronomer Royal, Airy, or the then professor at Cambridge, Challis. We still have in Cambridge a 12 inch telescope, which I describe to visitors as the telescope that failed to discover Neptune.

4. Astronomy at College Level

This conference has covered many aspects of college-level education in astronomy. Introductory courses of the kind normally known in the US as 'astronomy 100' have a great value not only for their intrinsic content, but because they convey the flavour of frontier research at an elementary level. Here we have an advantage over particle physics, chemistry, or molecular biology. These courses have spawned many excellent textbooks. In the UK, there is not the same scope for these broad elementary courses because university degrees are more specialised. Those specialising in the humanities take no science courses – indeed, as I've mentioned already, they may have had none since the age of 16.

There has, however, been a growth, in the UK, of astronomy teaching at undergraduate level in conjunction with physics. Many universities (here at UCL was among the first) have joint physics/astronomy honours degrees. Astronomy offers scope to 'enrich' the physics curriculum, and its inclusion has benefited enrolments in physics departments.

5. A Sociological Note

We also, I believe, have a mission towards our academic colleagues in other fields (particularly in social sciences) – to convey the way we perceive the nature of the scientific enterprise.

The way we approach science, what problems strike us as interesting, what styles of explanation are culturally appealing, and (more mundanely) what fields attract funding, plainly depend on a range of political, sociological and psychological factors. Some projects, especially big international ones, are a byproduct of activities driven by other imperatives. Space science is a byproduct of the superpower rivalry and rides along on a large application-led programme. Supercomputers have transformed much of our subject, etc.

It is important, as well as enlightening, to appreciate how pervasive these social and political factors are. Scientists are a fascinating topic for anthropological study. But for us 'in the zoo' science nonetheless moves towards a culture-independent outcome; it is, albeit fitfully, advancing.

We have not altogether succeeded – as Gerald Holton in the US has reminded us in his eloquent writings – in asserting and clarifying the role of science among other intellectuals.

In his book 'Dreams of a Final Theory', Steven Weinberg gives an apt metaphor: "A party of mountain climbers may argue over the best path to the peak, and these arguments may be conditioned by the history and social structure of the expedition, but in the end either they find a good path to the summit or they do not, and when they get there they know it."

Perhaps I might venture another analogy. It is fascinating to study how the development of music — for instance, the emphasis on operatic versus liturgical music; the increase in the scale of orchestral compositions that stemmed from the transition from private patronage to public concerts, etc – was moulded by social and economic factors. But this is in a sense peripheral to the essence of the music itself.

6. Conclusion

In conclusion, astronomy is a fundamental science; it is also the grandest of the environmental sciences. It has – especially during its current phase of unprecedented scope and progress – a key role to play in education at all levels, and in public understanding. We will surely all leave this meeting fully mindful of (in the previous speaker's words) the 'excitement of understanding'; and with renewed enthusiasm to spread this understanding still more widely.

Authors

A	Abati, L.E.	261
	Aleshkina, E.Yu	317
	Andrews, F.	267
	Andronov, I.L.	235
	Ashton, D.	185
	Atkin, K.	185
B	Baruch, J.E.F.	128
	Battle, R.	312
	Baxter, J.H.	139
	Bennett, M.A.	249
	Berthomieu, F.	89
	Billingham, J.	326
	Blurton, C.	332
	Brecher, K.	37
	Bretones, A.L.K.	191
	Bretones, P.S.	191
	Broughton, M.P.V.	111
	Brück, M.T.	20
	Buckland, R.	237
C	Camino, N.	277
	Carter, B.	214
	Christian, C.A.	74
	Comins, N.F.	118
	Cunow, B.H.L.	54
D	Dodd, R.	214
	Dow, K.L.	230
F	Fadeeva, A.A.	302
	Fierro, J.	180
	Fogh Olsen, H.J.	106
	Fouquet, J-L.	283
	Fraknoi, A.	249
G	Gerbaldi, M.	60, 256
	Gougenheim, L.	1, 256
	Greenstein, G.	16
	Gulyaev, S.A.	26
	Gutsch, W.	153
H	Hall, R.	214, 267
	Harrison, B.	185
	Hawkins, I.	88, 312
	Hoff, D.B.	273
	Hufnagel, B.	124